GUIDELINES FOR

# Fire Protection in Chemical, Petrochemical, and Hydrocarbon Processing Facilities

GUIDELINES FOR

# Fire Protection in Chemical, Petrochemical, and Hydrocarbon Processing Facilities

**Center for Chemical Process Safety**
of the
American Institute of Chemical Engineers
*3 Park Avenue, New York, NY 10016-5991*

**CCPS Publication Number G-83**

**Library of Congress Cataloging-in-Publication Data**

Guidelines for fire protection in chemical, petrochemical, and
hydrocarbon processing facilities.
p. cm.
Includes bibliographical references and index.
ISBN 0-8169-0898-2 (Hardcover)
1. Chemical plants—Fires and fire prevention. 2. Chemicals—Fires
and fire prevention. I. American Institute of Chemical Engineers.
Center for Chemical Process Safety. II. Title.
TH9445.C47G85 2003
60'.2804—dc22
2003017934



PRINTED IN THE UNITED STATES OF AMERICA

16

# CONTENTS

# 3
## Fire Protection Strategy



# 4
## Overview of Fire Prevention Elements

# 6
## Fire Risk Assessment



# 7
## Fire Protection Fundamentals

## 10
## Inspection, Testing, and Maintenance

## 11
## Fire Emergency Response



## A
## Case Histories

# B
## Understanding Fires



# C
## Computer Tools for Design

# PREFACE

The American Institute of Chemical Engineers (AIChE) has helped chemical plants, petrochemical plants, and refineries address the issues of process safety and loss control for over 30 years. Through its ties with process designers, plant constructors, facility operators, safety professionals, and academia, the AIChE has enhanced communication and fostered improvement in the high safety standards of the industry. AIChE's publications and symposia have become an information resource for the chemical engineering profession on the causes of incidents and means of prevention.

The Center for Chemical Process Safety (CCPS), a directorate of AIChE, was established in 1985 to develop and disseminate technical information for use in the prevention of major chemical accidents. The CCPS is supported by a diverse group of industrial sponsors in the chemical process industry and related industries who provide the necessary funding and professional guidance for its projects. The CCPS Technical Steering Committee and the technical subcommittees oversee individual projects selected by the CCPS. Professional representatives from sponsoring companies staff the subcommittees and a member of the CCPS staff coordinates their activities.

Since its founding, the CCPS has published many volumes in its "Guidelines" series and in smaller "Concept" texts. Although most CCPS books are written for engineers in plant design and operations and address scientific techniques and engineering practices, several guidelines cover subjects related to chemical process safety management. A successful process safety program relies upon committed managers at all levels of a company who view process safety as an integral part of overall business management and act accordingly.

A team of fire protection experts from the chemical industry drafted the chapters for this guideline and provided real world examples to illustrate some of the tools and methods used in their profession. The subcommittee members reviewed the content extensively and industry peers evaluated this book to help ensure it represents a factual accounting of industry best practices.

# ACKNOWLEDGMENTS

The American Institute of Chemical Engineers wishes to thank the Center for Chemical Process Safety (CCPS) and those involved in its operation, including its many sponsors whose funding made this project possible; the members of its Technical Steering Committee who conceived of and supported this Guidelines project, and the members of its Fire Protection Subcommittee.

*If this Guideline prevents one chemical, petrochemical, or hydrocarbon processing facility fire, the efforts of all those involved in preparing this work will be deeply recognized and rewarded.*

The members of the CCPS Fire Protection Subcommittee were:

Robert M. Rosen, Chair, *BASF Corporation*
Siegfried Fiedler, *BASF Corporation*
Gene Hortz, *Rohm & Haas Company*
Duncan L. Hutcheon, *ExxonMobil*
Joel Krueger, *BP Amoco*
John Sepahpur, *ChevronTexaco Energy Research & Technology Company*
John Sharland, *FM Global*
William A. Thornberg, *Industrial Risk Insurers*
Della Wong, *Aon Reed Stenhouse*
Jeffrey Yuill, *Starr Technical Risks Agency, Inc.*

John Davenport was the CCPS staff liaison and was responsible for overall administration of the project. Additional contributors to the subcommittee were Charles E. Fryman, *FMC,* and Dave Moore, *Acutech.*

Risk, Reliability and Safety Engineering (RRS), of League City, Texas (www.rrseng.com) was contracted to write this guideline. The principal RRS authors of this guideline were:

John Alderman, PE, CSP
Bill Effron, CSP
Christy Franklyn
Tim McNamara

Additional RRS staff that supported this project includes Donna Hamilton, Marlon Harding, Ted Low, and Tom Lawrence.

Daniel T. Gottuk, PhD and Joseph Scheffey, PE of Hughes Associates were the primary authors of Chapter 5.

CCPS would like to thank Bud Slye, PE, Loss Control Associates, who provided technical quality review.

CCPS also gratefully acknowledges the comments and suggestions received from the following peer reviewers; their insights, comments, and suggestions helped ensure a balanced perspective to this Guideline:

Dr. Ezikpe Akuma, *New Jersey Department of Environmental Protection*
Reginald Baldini, *New Jersey Department of Environmental Protection*
Michael P. Broadribb, *BP America, Inc.*
Keith L. Farmer, *DuPont Engineering Technologies*
Les Fowler, *BASF Corporation*
Eric Lenoir, *AIU-Energy*
Darren Martin, *Shell Chemical Company*
Lisa M. Morrison, *NOVA Chemicals, Inc.*
Dave Owen, *Exxon-Mobil*
Asit Ray, *New Jersey Department of Environmental Protection*
Thomas Scherpa, *DuPont Engineering Technologies*
Milt Wooldridge, *MRW & Associates, Inc.*

The members of the CCPS Fire Protection Subcommittee and the peer reviewers wish to thank their employers for allowing them to participate in this project.

# ACRONYMS

| | |
|---|---|
| ALARP | As low as reasonably practical |
| AIChE | American Institute of Chemical Engineers |
| AISC | American Institute of Steel Construction |
| AHJ | Authority Having Jurisdiction |
| ANSI | American National Standards Institute |
| API | American Petroleum Institute |
| BI | Business Interruption |
| BLEVE | Boiling Liquid Expanding Vapor Explosion |
| CCPS | Center for Chemical Process Safety |
| CFD | Computational Fluid Dynamics |
| CFR | Code of Federal Registry |
| CMPT | Center for Marine and Petroleum Technology |
| DCS | Distributed Control System |
| DOT | Department of Transportation |
| EANS | Emergency Alarm Notification System |
| EHS | Environmental, Health, and Safety |
| EOC | Emergency Operations Center |
| EPA | Environmental Protection Agency |
| ERP | Emergency Response Plan |
| ERT | Emergency Response Team |
| FCC | Fluid Catalytic Cracking (Unit) |
| FHA | Fire Hazard Analysis |
| FMEA | Failure Mode and Effects Analysis |
| FM | Factory Mutual |
| FPS | Fire Protection Strategy |
| FRP | Fiberglass Reinforced Plastic |
| GRP | Glass Reinforced Plastic |

| | |
|---|---|
| HVAC | Heating, Ventilating, and Air Conditioning |
| HAZID | Hazard Identification |
| HAZOP | Hazard and Operability Study |
| HSSD | High Sensitivity Smoke Detection |
| HAZMAT | Hazardous Material |
| ICS | Incident Command System |
| IEEE | Institute of Electrical and Electronic Engineers |
| I/O | Input/Ouput |
| IR | Industrial Risk |
| LEPC | Local Emergency Planning Committee |
| LFL | Lower Flammability Limit |
| LOPA | Layer of Protection Analysis |
| LPG | Liquefied Petroleum Gas |
| MERITT | Maximizing EHS Returns by Integrating Tools and Talents |
| MFL | Maximum Foreseeable Loss |
| MOC | Management of Change |
| MSDS | Material Safety Data Sheet |
| NICET | National Institute for Certification in Engineering Technologies |
| NFPA | National Fire Protection Association |
| NLE | Normal Loss Estimate |
| NOAA | National Oceanic and Atmosphere Administration |
| OSHA | Occupational Safety Hazard Association |
| P&ID | Piping and Instrumentation Drawing |
| PC | Personal Computer |
| PDA | Personal Digital Assistant |
| PE | Professional Engineer |
| PHA | Process Hazard Analysis |
| PML | Probable Maximum Loss |
| PFD | Process Flow Diagrams |
| PPE | Personal Protective Equipment |
| PSM | Process Safety Management |
| PVC | Polyvinyl Chloride |
| RMS | Risk Management System |
| RP | Recommended Practice |
| RVP | Reid Vapor Pressure |
| SFPE | Society of Fire Protection Engineers |
| SI | Standard Instrumentation |
| SIS | Safety Instrumented System |
| UFL | Upper Flammability Limit |
| UL | Underwriters Laboratories |
| UK | United Kingdom |
| VCE | Vapor Cloud Explosion |

# 1

# INTRODUCTION

This Guideline provides tools to develop, implement, and integrate a fire protection program into a company's or facility's Risk Management System. Figure 1-1 highlights the guidance provided in this Guideline.

For the thirty-year period of 1970 through 1999, 116 fires resulted in large-scale property damage (greater than $10MM) in the hydrocarbon and petrochemical onshore industries, and totaled over 4.5 billion dollars adjusted for year 2000 dollars (Marsh Risk Consulting, 2001). This is an average of approximately 39 million US dollars per occurrence, and includes losses in refineries, petrochemical plants, gas plants, marine terminals, and offshore oil and gas operations. Consequential business losses are two to three times property damage losses.

During the five-year period of 1995 to 2000, 50 large-scale fire losses have resulted in losses totaling approximately 2 billion dollars, or an average of 40 million dollars per occurrence. These numbers indicate that although the average dollar loss per occurrence is about the same for both time frames, the number of large losses is increasing. These incidents reinforce the importance of utilizing a systematic approach for addressing fire hazards in the hydrocarbon and petrochemical industries.

A Risk Management System (RMS)[1] is vital for effective loss prevention. Fire protection is an essential part of an RMS. Appropriately designed, installed, and maintained fire protection systems are paramount to mitigating the direct consequences, and preventing the escalation, of fires in processing facilities.

[1] Some companies use the term *Hazard Management System* or *HSE Management System*.

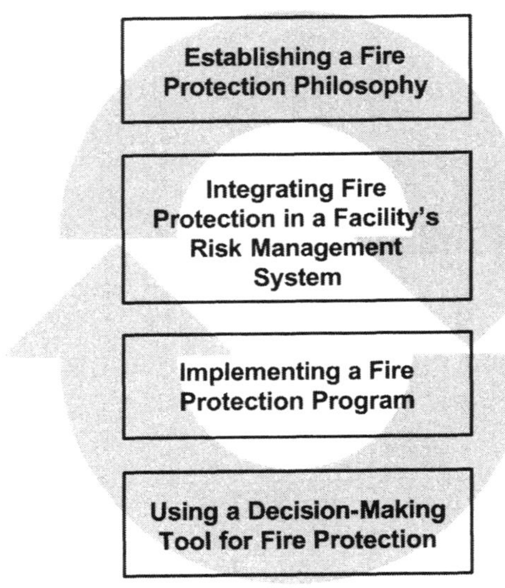


**Figure 1-1.** Fire Protection Guidance in This Guideline

## 1.1. Scope

Information on fire protection codes and standards are available from several sources, including the National Fire Protection Association (NFPA), the Society of Fire Protection Engineers (SFPE), the Fire Suppression Systems Association (FSSA), and the American Petroleum Institute (API). Jurisdictions that provide requirements for fire protection include federal, state, and local agencies. This Guideline bridges the regulatory requirements and industry standards with practical application and provides:

- A useful tool for making fire protection decisions
- Specific examples of fire protection criteria

While life safety issues are not a primary focus of this Guideline, they are an integral part of good fire protection design.[2]

There is a very close relationship between fires and explosions. In many instances, an explosion is the initial event, followed by a significant fire. Sometimes the fire can be the trigger that causes the explosion, such as a Boiling Liquid Expanding Vapor Explosion (BLEVE). This Guideline does not address the prevention of explosions, methods to quantify the severity of explosion, or explosion suppression techniques. Explosions are specifically addressed in

[2] For additional information on life safety issues, refer to NFPA 101.

*Guidelines for Evaluating the Characteristics of Vapor Cloud Explosions, Flash Fires, and BLEVEs* (CCPS, 1994) and *Understanding Explosions* (CCPS, 2003a).

## 1.2. Who Will Benefit from This Guideline?

Because fire protection is an important aspect of risk management and loss prevention, this Guideline will benefit many different people within an organization.

- *Corporate Leadership*—Senior executives define the basis for the development of fire protection philosophies. Their commitment and recognition of the value of fire protection is vital to integration into an RMS and implementation of fire protection strategies.
- *Site Managers*—Site Managers are responsible for developing and maintaining the facility's fire protection philosophy and strategies.
- *Line Management*—Line Managers are responsible for maintaining fire protection systems and for assuring personnel are trained on their use. Line Managers are the champions of a facility's entire RMS. They ensure that policies and procedures, including fire protection, are integrated and implemented. They also ensure that fire protection systems are tested and maintained.
- *Project Managers*—Project Managers are responsible for executing projects, usually from design through startup and commissioning. A Project Manager is responsible for determining the basic fire protection design concepts to apply in the execution of a project. The Project Manager is responsible for implementing the decisions and abiding by the project procedures associated with amending and adding to the fire protection system.
- *Engineers*—Engineers are responsible for specifying and designing fire protection systems that meet their company's fire protection requirements. This still leaves room for making decisions when designing fire protection systems and knowledge of performance vs. prescriptive methods is beneficial.
- *HSE Professionals*—Health, Safety, and Environmental (HSE) Professionals provide technical guidance to engineers and typically are in an assurance role for fire protection systems.

***All fire protection decision makers will benefit from this Guideline.***

Figure 1-2 provides an overview of the contents of this Guideline and also provides examples of how each Chapter can assist in establishing fire protection programs, fire protection decision making, design, installation, etc.

| GUIDELINE CHAPTER | QUESTIONS THIS CHAPTER WILL ANSWER |
|---|---|
| Chapter 1<br>**Introduction** | *What is fire protection?*<br>*How can this book help me?*<br>*Why is fire protection important?*<br>*What other resources are available?* |
| Chapter 2<br>**Management Overview** | *Why is management commitment essential?*<br>*What are the impacts of over- versus under-protection?*<br>*What are the cost benefits?*<br>*How is fire protection integrated into other Management Systems?* |
| Chapter 3<br>**Fire Protection Strategy** | *Why is establishing a company philosophy important for a Hazard Management Strategy (HMS)?*<br>*How does an HMS integrate with other Management Systems and into the lifecycle of a facility?* |
| Chapter 4<br>**Overview of Fire Prevention Elements** | *What are the elements of a Fire Prevention Program?*<br>*How do the elements of a Fire Prevention Program integrate with other Management Systems?* |
| Chapter 5<br>**Fire Hazard Analysis** | *What is a Fire Hazard Analysis (FHA) and how are they used?*<br>*How do you determine the potential impact of fires?*<br>*What are examples of some typical fire hazard problems and how do you solve them?* |
| Chapter 6<br>**Fire Risk Assessment** | *What is a risk assessment and how is it used in decision-making?*<br>*What are the advantages/disadvantages of consequence-only decision-making?*<br>*How can a Layer of Protection Analysis (LOPA) be used?* |
| Chapter 7<br>**Fire Protection Fundamentals** | *What are some design concerns and control mechanisms?*<br>*What are active and passive fire protection systems?*<br>*What are the specific requirements when designing active and passive fire protection systems?* |
| Chapter 8<br>**Specific Design Guidance** | *What are the specific criteria for installing fire protection systems for process areas, storage, buildings, distribution, and utilities? (Examples are included)* |
| Chapter 9<br>**Installation of Fire Protection Systems** | *What is the process for notification/sequencing and timing?*<br>*What are the qualifications of the installers?*<br>*What are some lessons-learned during installation?* |
| Chapter 10<br>**Inspection, Testing and Maintenance** | *Who has ownership of fire protection systems?*<br>*How are they inspected, tested, and maintained?*<br>*What are the impairment procedures?* |
| Chapter 11<br>**Fire Emergency Response** | *How does fire protection impact emergency response?*<br>*How are response activities coordinated?*<br>*What are fire plans and how do you use them?* |

**Figure 1-2.** Guideline Overview and Contents

### 1.2.1. What Is Fire Protection?

This Guideline focuses on fire protection. For the purpose of this Guideline, fire protection and fire prevention are defined as:

- *Fire Protection*—The science of reducing loss of life and property from fire by control and extinguishment. Fire protection includes fire prevention, detection of a fire, providing systems to control or mitigate the fire, and providing manual firefighting capabilities.
- *Fire Prevention*—Activities whose purpose is to prevent fires from starting. Fire protection and fire prevention go hand-in-hand. All fire protection programs include a fire prevention program. For example, control of ignition sources is very important in minimizing the risk of fire, but does not meet the definition of fire protection in this Guideline.

Much of process safety deals with the prevention of catastrophic events, such as fires and explosions. This is accomplished by containing hazardous materials within the process system. The Center for Chemical Process Safety (CCPS) has developed many Guidelines that assist companies in this effort (see Section 1.3 and References).

### 1.2.2. Examples

Fire protection is often driven by the likelihood of potential consequences. Examples of incidents resulting in fire are provided in Table 1-1.

## 1.3. Relation to Other CCPS Guidelines and Resources

Other CCPS Guidelines provide additional resources for topics discussed in this Guideline. Some of these include:

*Guidelines for Engineering Design for Process Safety*

*Guidelines for Evaluating the Characteristics of Vapor Cloud Explosions, Flash Fires, and BLEVEs*

*Guidelines for Facility Siting and Layout*

*Guidelines for Technical Planning for Onsite Emergencies*

*Guidelines for Integrating Process Safety Management, Environment, Safety, Health and Quality*

*Guidelines for Technical Management of Process Safety*

*Guidelines for Safe Warehousing of Chemicals*

**Table 1-1**
*Examples of Major Fire Incidents*

| Year / Location | Incident Description |
|---|---|
| 1984<br>Mexico City, Mexico | *LPG Terminal*—A major fire and series of catastrophic Boiling Liquid Expanding Vapor Explosions (BLEVEs) killed 500 people and destroyed the LPG terminal. |
| 1998<br>New Brunswick, Canada | *Process Facility*—A fire originated in the feed heater of a hydrocracker and resulted in one fatality and significant damage to a Hydrocracking Unit. |
| 1998<br>Ras Gharib, Egypt | *Terminal*—16 tanks, containing approximately 30,000 barrels of crude oil each, caught fire after being struck by lightning. |
| 1999<br>California, USA | *Process Facility*—A fire in a process unit resulted in three fatalities, significant downtime, and public scrutiny of refinery operations. |
| 2000<br>Ohio, USA | *Warehouse*—A pharmaceutical warehouse fire resulted in damage to adjacent warehouses and a total property loss of 100 million dollars. |

Additional resources include the National Fire Protection Association (NFPA), the Society for Fire Protection Engineers (SFPE), Fire Suppression Systems Association (FSSA), and the American Petroleum Institute (API). Refer to the References section of this Guideline for specific resources.

# 2

# MANAGEMENT OVERVIEW

Fire protection is a science that stretches as far back as the Roman Empire. The aqueducts and the Corp of Vigilantes gave the Romans what they needed for fire protection and control. Through the years, the practice of fire protection has evolved from a problem-solving approach to a mature, systematic discipline.

Most processing facilities, due to the materials being handled, have a high potential for loss due to fire. Management teams (like individuals) tend to believe that major incidents, such as fires, are unlikely to occur at their facility. This perception is not accurate. The statistics related to the number of fires do not vary widely year-to-year and losses continue to occur. To effectively implement a fire protection program, it is important to understand that a significant loss is possible. Top management personnel should view fire protection as a benefit - *an integral part of the recovery of operations after an incident* - and not just as a cost.

This book will assist organizations in making informed, risk-based decisions to determine the appropriate level of fire protection.

## 2.1. Management Commitment

All responsible organizations will have a fire protection program to protect their assets. For some, it may be fire extinguishers in the warehouse; while for others it is a department of professionals supplementing numerous automatic fire detection and suppression systems. Due to the nature of the program, it is often necessary to involve several individuals, each having a specific, assigned responsibility.

Management commitment to support the fire protection program is necessary if fire protection is to be available when needed. The commitment includes ensuring adequate staffing, resources, and technical support is provided. While fire protection is included with new capital projects, sufficient resources to maintain these systems must be included in the facility's budget for maintenance.

Management has a responsibility to fully define the roles and responsibilities of each individual. These duties should not be assigned as an add-on or left to chance as this creates the impression that fire protection is not a priority issue. No matter the size of the organization nor the complexity of the program, the need for an effective fire protection program is always present.

## 2.2. Integration with Other Management Systems

While the implementation of risk management systems may vary from company to company, they are a fundamental activity in the chemical, petrochemical, and hydrocarbon processing industries. A company's approach to risk management reflects its beliefs and values.

An organization needs to develop a strategy for fire protection. This allows for cost-effective and efficient implementation and continuous improvement in fire protection systems. This strategy must be reviewed and updated periodically because of the many changes that take place within processing industries.

A strategy for fire protection is only one part of an overall framework of guidance to allow consistent, methodical evaluation and management of hazards and risk. There are many ways to approach risk management; however a strategy and procedures for fire protection must be established and followed.

## 2.3. Balancing Protection

Three factors contribute to the extent of any fire loss. The first involves an act, omission, or system failure allowing an ignition source and fuel to combine. The second involves the potential for continued fire growth and escalation. The third factor is extinguishment.

Providing the right level of protection can be a delicate balancing act. Overprotection results in unnecessary capital expenditure and higher ongoing costs. The larger the system and the more complex its components, the more capital will need to be invested and the greater the requirements for training on fire protection system operations, testing, and maintenance. Overprotection may result in an over confidence in the ability of the system to address all situations and a subsequent deterioration in readiness. There are minimum requirements for maintaining each system and the more systems or more complex the system, the more maintenance that is required.

While less protection may initially reduce the capital investment and the ongoing maintenance costs, the additional risk to company assets, employees, the environment, and the public could be substantial. The potential for escalation increases due to the lack of fire protection systems. Should a company choose less protection, potential adverse affects such as damage to reputation, increased insurance costs, loss of business and customers, as well as possible charges of criminal negligence could become a factor in the event of an incident.

## 2.4. Cost-Benefit

The use of cost-benefit analysis plays an important role in the decision-making process for fire protection systems. A cost-benefit analysis sums the expected benefits and is divided by the sum of the expected costs. A challenge often lies in determining what "expected" means and estimating the value of money over the time period the fire protection is in use. In fire protection, the expected benefits can be defined as the difference between the cost of a loss without protection and the cost of a loss with protection. The expected costs include the initial costs of the fire protection as well as any annual testing and maintenance costs. The likelihood of an incident is factored in to obtain residual risk. This residual risk is compared to the benefit to determine what benefit is available each year versus the annualized cost.

# 3

# FIRE PROTECTION STRATEGY

Management of risk is a fundamental activity of companies in the chemical, petrochemical, and hydrocarbon processing industries. Most companies have developed a formal strategy of how risks will be managed. These strategies reflect relevant corporate beliefs and values.

The way these strategies are implemented may vary from company to company, but come under the classification of a Risk Management System (RMS), which is the nomenclature used in this guideline. An RMS provides a framework of guidance to allow consistent and methodical evaluation and management of hazards and risk. A fire protection strategy is considered an integral part of an RMS. There are many ways to approach the development of an RMS, however key decisions must be made and strategies established. Figure 3-1 illustrates an example of an RMS and how a fire protection strategy fits within an RMS.

The objectives of this Chapter are to clarify the considerations involved in developing a fire protection strategy and provide guidance on how that strategy can be integrated into other management systems.

Section 3.1 discusses key factors a company may consider in the development of their fire protection strategy. Section 3.2 discusses how to develop a fire protection strategy. Section 3.3 discusses the need for integration with other facility management systems and Section 3.4 outlines the need for fire protection through the lifecycle of the facility.

## 3.1. Key Factors in a Fire Protection Strategy

Key factors that should be reviewed by a company in determining their fire protection strategy are discussed in this section. These factors will assist company

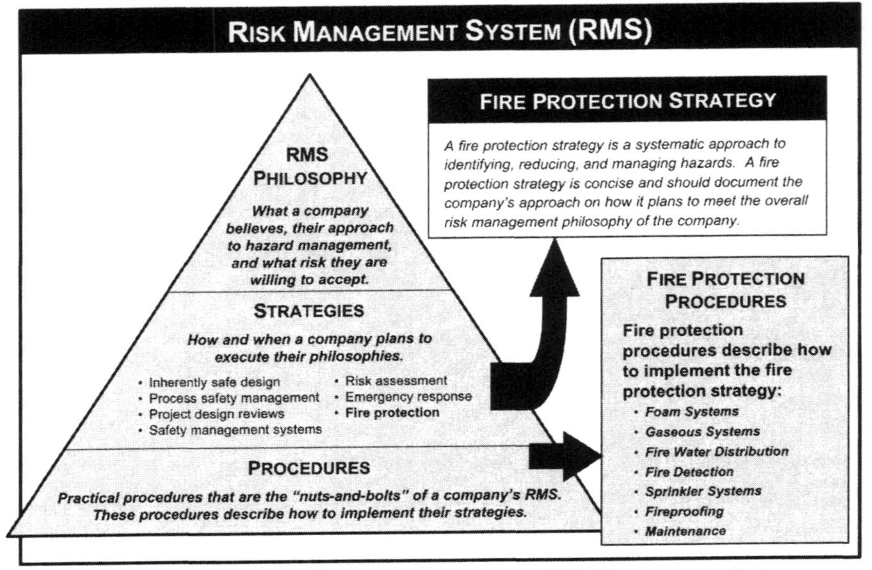


**Figure 3.1.** Integration of Fire Protection into a Risk Management System

management in determining their approach to fire protection. Determining an appropriate fire protection strategy involves considering and balancing a number of technical and economical factors, including:

- Protection of personnel
- Environmental impact
- Impact on national economy
- Value of a business
- Nature and cost of major incidents that could potentially occur
- Potential business interruption
- Loss of market share
- Loss of company reputation
- Amount of loss acceptable to company
- Insurance coverage limits (risk transfer)
- Type of fixed fire protection, automatic or manual, passive or active
- Emergency response capabilities
- Maintenance and testing capabilities

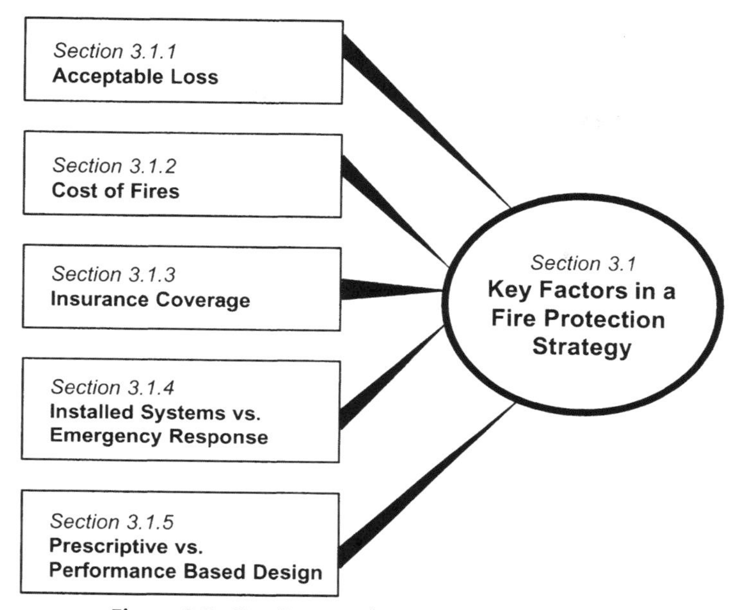


**Figure 3.2.** Key Factors of a Fire Protection Strategy

These factors can be highly interrelated and should not be considered individually. Figure 3-2 illustrates key factors of a fire protection strategy that will be discussed in this guideline.

### *3.1.1. Acceptable Loss*

One approach in beginning the development of a fire protection strategy is to define the level of risk that the company is able or willing to accept. Acceptable loss is defined as the cost of a loss event (repair/replacement, including demolition and debris removal, plus consequential business loss) that is within the capability of the company, business unit, or division to absorb financially and culturally. This loss can be retained within the company or partially transferred to others through insurance.

For example, a company, based on financial requirements, may choose a low deductible ($1 million/incident); another company, with different financial requirements, may elect a higher deductible ($5 - $10 million). For business interruption (BI), market conditions may influence the need for insurance coverage. For example, when capacity is high, a company may elect to have BI coverage, however, if market conditions are weak, BI coverage may not be warranted.

### *3.1.2. Cost of Fires*

The costs associated with a fire can be measured in a number of ways. These vary depending on the corporate culture and can include the following components (Lees, 1996):

- Impact to personnel
- Damage to plant assets
- Delay in plant startup
- Plant downtime
- Business interruption
- Loss of markets
- Loss of public reputation
- Fines
- Legal actions

It is important that a company maintains a consistent philosophy for estimating potential fire loss in their facilities to establish fire protection strategies. There are different approaches for estimating fire loss, but most are a combination of insurance and industry approaches.

These estimates are prepared for different reasons. The insurance estimates are intended to guide insurers in establishing the amount of liability they are willing to accept and the premium they will charge for that coverage. The industry estimates are intended to inform management of the potential fire loss and liability. The results of these two approaches should not necessarily be compared. However, both are useful to management in determining a fire protection strategy.

#### *3.1.2.1. Insurance Loss Estimate Approach*

The insurance industry looks at various levels of potential losses:

- A fire loss that occurs with all fire protection systems in service, often described as the Normal Loss Estimate (NLE).
- A fire loss that occurs with one active fire protection system out of service, often described as the Probable Maximum Loss (PML).
- A fire loss that occurs with all active fire protection systems out of service, often described as the Maximum Foreseeable Loss (MFL).
- A fire loss that occurs from a worst-credible incident.

The steps in estimating the insured fire loss are:

- Determine the value of the company's process facilities that produce its major product(s). This is usually the replacement cost or, if reasonably recent, the construction costs of the selected process unit and its equipment.

- Identify the fire scenario for the level of potential loss.
- Estimate the cost of repair or re-building of the facility after a fire and the amount of lost production from the incident until the process is re-started.
- Estimate the lost income from the affected product sales due to business interruption (BI) (see 3.1.2.3). The typical way to do this is to assume standard production rate, sales price, and raw material costs, but not take credit for utilities, maintenance, and similar costs that may not be incurred during the downtime. The BI may be estimated as:

$$BI = [(\text{Product sales income per month}) - (\text{Nonincurred monthly expenses})] \times (\text{Downtime months})$$

- An estimated fire loss estimate can then be calculated as:

$$\text{Fire loss estimate} = (\text{Estimated cost to rebuild}) + (BI) + (\text{Extra expense})$$

### 3.1.2.2. *Industry Loss Estimate Approach*

Industry approaches to estimating fire loss generally fall into two key areas:

- Calculations to determine loss, such as fire hazard analysis, consequence modeling, etc.
- Company design standards based upon industry and company experience.

While some companies will use one or the other approach, many adopt a combination of both.

These approaches essentially identify fire scenarios for all units and the consequences of those fires. If escalation is deemed possible, then additional damage is determined. Once the total damage is determined, a cost for replacement can be calculated.

The estimated fire loss estimate establishes an upper limit of cost which can be tested against company, division, or business unit management criteria to determine whether additional fire protection features are necessary.

### 3.1.2.3. *Business Interruption (BI)*

The cost to repair or rebuild a unit after a fire is often significantly less than the cost of business interruption. There are a number of ways to calculate business interruption. Some factors to consider when determining the cost of business interruption include the following:

- *Daily interruption cost*—the cost associated with producing the final product. This could include raw material contract penalties and fixed costs (salary, maintenance, taxes) for the plant.
- *Seasonal production*—the cost of not being able to produce when demand for the product is very high, such as low Reid Vapor Pressure (RVP) gasoline in Northern states during the summer months.
- *Lost profit during periods of no production.*
- *Customer losses*—the loss associated with customers who locate alternative suppliers and will not return to buy your product when you start production again or the customer who negotiates a better price in order for you to keep the contract.
- *Extra expense incurred to replace the lost product.*
- *Impact on upstream or downstream facilities*—a significant impact on cost is when the unit that had the fire is either a supply to a downstream unit or the receiver of an intermediate product. In either case, several units may be impacted and the cost for total loss of production needs to be considered.

### 3.1.3. Insurance Coverage

While complex in detail, a basic understanding of insurance coverage is necessary in determining a company's fire protection strategy. The insurer is, in effect, wagering that a company will not have a loss or that it will be smaller than the maximum estimated. The company, on the other hand, buys insurance (peace-of-mind) to protect investments in the event something happens.

The basic principal of insurance is to spread the risk of losses. This is generally accomplished by a combination of deductibles, direct insurance, and reinsurance. The insured chooses the amount of loss they are willing to retain by setting the deductible and the cost of coverage. When one insurer is not able to provide full coverage, the direct insurance is often shared by a number of direct insurers. In reinsurance, the direct insurers spread their risks to other insurance companies that specialize in reinsurance.

Most companies have some level of insurance coverage. The amount of insurance coverage varies year to year depending on the cost of the desired coverage. Usually, the level of fire protection provided by the company has little effect upon the cost of insurance, which is much more market driven. Many companies choose not to pay for full insurance coverage and carry some of the risk themselves. This can be either self-insurance through a captive insurance company or reserves, or by the use of a deductible amount of coverage. Individuals are faced with the same decision when buying automobile insurance. They can choose whether to purchase collision insurance and the amount of deductible desired. The decision process is the same for determining the amount of self-insurance a company should purchase.

Underwriters conduct inspections of the properties being insured to determine the estimated fire loss for each location. The risk of insuring a facility is determined by estimating the fire loss and combining it with the loss experience of a company, the particular industry, and other related industries. The underwriters then determine how much of the risk they take or layoff. The premium a company pays for insurance is determined by the insurer(s). Figure 3.3, originally developed in 1986 and still relevant today, shows the typical breakdown of insurance premiums (Norstrom, 1986).

The extent to which insurers should give credit for preventive and protection measures is a long-standing question. Companies want to hold premiums down *and* receive credit for preventive and protection measures at their specific facilities. On the other hand, the insurance industry considers all premiums on the basis of the total loss to which they are exposed, rather than any specific facilities. Additionally, premiums go through periodic cycles. Historically, the price of insurance will vary with respect to market conditions rather than being directly related to the risks. Accordingly, comparison of premiums for similar risks set at different times is unlikely to present a meaningful trend. Furthermore, where like facilities have their respective premiums set at approximately the same time, the premium for the best-protected may well be only 10% lower than that set for the least-protected. However, in hard markets, poorly protected risks may not find insurance or may only find partial coverage.

Insurance does not cover all costs associated with the loss. Losses from personnel time in firefighting operations, environmental impact, loss of company reputation, loss of customers, and finally the cost of potential legal actions may not be covered.

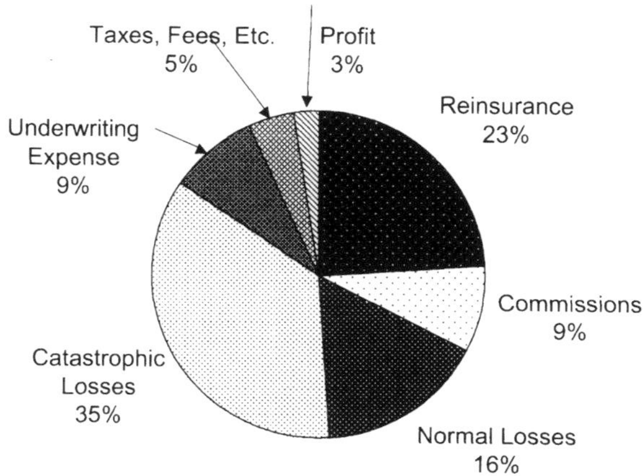


**Figure 3.3.** How Premium Dollars Are Spent by Insurance Companies

### 3.1.4. Installed Systems versus Emergency Response

There are three main strategies in providing adequate fire protection of a facility. These are:

- Installed active fire protection.
- Installed passive fire protection.
- Install isolation systems to minimize fuel.
- Emergency response by organized firefighting teams composed of both onsite and offsite fire brigades using mobile units to extinguish fires at the facility.
- Incipient firefighting, and if the fire becomes too large, protect exposures and walk away.

#### 3.1.4.1. *Active versus Passive*

An active fire protection system requires some action to occur before it functions per its design intent. This action may be taken by either a person or control system. Examples of active fire protection systems are monitors, water spray systems, foam systems, emergency isolation valves, and ESD systems.

A passive fire protection system requires no action to occur for it to function per its design intent. Examples of passive fire protection methods are fireproofing, spill containment, and physical separation of units and buildings.

Additional information on active and passive systems is contained in Chapter 7.

#### 3.1.4.2. *Manual versus Automatic Activation*

Active fire protection systems can be installed to provide the desired mitigation by either manual or automatic activation.

Manual activation requires a person to activate the system by pushing a button or opening a valve. The advantage to this type of system is the reduction in nuisance activations and maintenance and testing costs. The disadvantage is the potential for delay in activation of the system that allows fire growth. The decision to use manual over automatic activation may require additional emergency response capability because the fire may be larger and escalation of the fire is more likely. For manual systems, the delay in activation may result in larger and longer fires, hence more property damage (cost).

Automatic systems have a detection system that should automatically activate fire protection systems, significantly reducing the chance of a larger fire. The disadvantage is the potential for nuisance activations that can damage the equipment being protected, shutdown production, and require the fire protection systems to be recharged.

When the decision is made to install an automatic detection system, the fire/gas detection technology should be carefully selected to match the expected hazard and the environment in which it will be installed. The design and installation of a reliable fire/gas detection system in an industrial facility should only be done by experienced personnel. Otherwise, the facility owner may have an inappropriate system and may be plagued by nuisance alarms and high maintenance costs.

When evaluating a facility, a review of the existing fire/gas detection system is recommended. Due to recent technology developments, the performance of fire/gas detection systems has significantly improved over the last few years.

The use of automatically actuated systems involves an increased cost of initial installation (e.g., because of the control unit), as well as additional maintenance costs throughout the life of the automatic systems.

#### *3.1.4.3. Emergency Response*

Emergency response has a significant influence on a company's fire protection strategy. The question is "to fight, or not to fight" fires at a facility. Emergency response is one of the last layers of protection available to a facility (CCPS, 2001a).

No matter how a company may provide emergency response, they must comply with federal, state, and local regulations. These regulations cover evacuation, fire brigades, hazardous materials response teams, confined space rescuers, and medical responders. These regulations generally apply to training, equipment, and other administrative matters affecting personnel and how they operate (CCPS, 1995a).

Emergency response can be internal resources, external resources, or a combination of both. The degree of internal response can vary from the basic ability to properly respond with fire extinguishers and sound the alarm to a fully fledged industrial fire brigade, with internal structural firefighting capability. External response may include local fire departments and mutual aid organizations. Before relying on external response to perform firefighting actions within the facility, a relationship should be built with the responding organization. This will better allow the facility opportunities to understand the capabilities of the external responders and the external responders to better understand the hazards they will face. It will also allow responders to jointly agree on a practical protocol for command and decision making during a fire.

A decision analysis technique for assisting companies in deciding options for emergency response is contained in *Guidelines for Technical Planning for On-Site Emergencies* (CCPS, 1995). Factors that should be considered are divided into three broad categories:

- Response effectiveness
- Management issues
- Cost considerations

Additional information on the decision process for emergency response organizations is contained in Chapter 11.

### *3.1.5. Prescriptive versus Performance-Based Design*

Prescriptive fire protection is standardized guidance or requirements without recognition of site-specific factors. For example, providing two 2,000 gpm fire pumps per facility is one company's approach to fire water systems. The size of the facility, hazards posed by each unit, or specific water demand is not considered. Prescriptive approaches to fire protection generally are a result of compliance with regulations, insurance requirements, industry practices, or company procedures. Performance-based design adopts an objective-based approach to provide a desired level of fire protection. The performance-based approach presents a more specific prediction of potential fire hazards for a given system or process. This approach provides solutions based on performance measured against established goals rather than on prescriptive requirements with implied goals. Solutions are supported by a Fire Hazard Analysis (FHA) or, in some cases, a fire risk assessment. A fire risk assessment takes account of more than just the consequences, and includes the likelihood of the fire scenario occurring. A performance-based approach determines the need for fire protection on a holistic basis, reviewing the fire protection needs of the complete facility and not just each unit.

Performance objectives and metrics allow the designer of fire protection systems more flexibility in meeting requirements and can result in significant cost-savings as compared to the prescriptive approach. Conversely, for small projects, the performance-based design may not be cost-effective. However, a fire protection system design should consider potential future changes and hazard modifications. For example, there is only a small, extra cost to install a higher density sprinkler system, but it will allow much more flexibility in the future use of a warehouse.

An example of a performance-based design for fire protection is a stated objective that:

- A process unit should not exceed $100,000 in damage for any fire in the unit.
- A fire should not escalate to additional units.

The designer has the option to meet this objective in a variety of ways. For example, it can be accomplished through inherently safer design, providing fire protection mitigation systems, emergency response, or a combination of all three.

## 3.2. Developing a Fire Protection Strategy

A fire protection strategy is a systematic approach to identifying, reducing, and managing fire hazards. The objective of a fire protection strategy is to ensure that:

- Protection of personnel is considered in the design of fire protection systems.
- Credible hazards have been identified, assessed, understood, and documented.
- Every opportunity to minimize the hazards has been identified, considered, and, where practical, implemented.
- The capital investment and estimated operating expense is optimized with a view to minimizing hazards.
- Corporate goals can be met for the facility's lifetime.
- Potential adverse effects on neighbors, community, and the environment are controlled.

Each facility should develop a fire protection strategy and maintain it evergreen throughout the lifecycle of the facility. If multiple facilities are similar in design and operation, then a fire protection strategy may be applicable to more than one. The fire protection strategy should be given to and used by a project team for expansions at the facility. A fire protection strategy should be developed for new projects, if one does not already exist.

A fire protection strategy should be concise and generally ranges in length from two to five pages. A fire protection strategy should demonstrate how it is integrated into the company's overall risk management philosophy.

A common aspect of any fire protection strategy is defining how hazards are managed and describing the order of priority for managing those fire hazards. Table 3-1 identifies the priority of how hazards should be managed.

A fire protection strategy serves as a bridge between the company's perceptions of fire-related risks and the details of how to manage specific risks. The fire protection strategy should be considered as the tool that defines *when* certain protection levels are required for a facility. The fire protection strategy should attempt to define general performance requirements or controls for specific situations. Below are examples of the types of statements that could be used in the fire protection strategy.

- Perform fire hazard analysis (FHA) for all facilities.
- Perform testing and maintenance in accordance with recommended and generally accepted good engineering practices.
- Limit loss from fire to one process unit or one storage tank.
- Provide fixed protection vs. manual response.

**Table 3-1**
*Prioritizing Hazard Management Approaches in a Fire Protection Strategy*

| Method | Definition |
|---|---|
| *Elimination* | The removal or minimization of a hazard through inherently safer design. |
| *Prevention* | The minimization of the likelihood of an event through the installation, integrity, and application of prevention systems to address each hazard through design practices. |
| *Detection and Control* | The limitation of the severity of incidents by early detection of the incident and the operation of safety systems, including fire protection systems through fire and gas detection and process isolation. |
| *Mitigation* | The protection of people and physical assets from the effects of any remaining credible hazards through fire protection systems. |
| *Emergency Response* | The protection of people from the effects of an escalating or catastrophic incident. |

- Use NFPA codes except as modified by company guidelines.
- Provide manual response for tank fires.
- Provide incipient fire response and call fire department.
- Process areas with a probable maximum loss of less than $*X* do not require automatic suppression systems.
- Facilities with a maximum foreseeable loss of greater than $*X* shall be protected by a looped fire water system equipped with redundant fire pumps.
- Processes susceptible to a runaway reaction during fire exposure shall be protected by water spray systems and fire monitors.
- Flammable gas detection is required in areas where a credible release exceeds *X* lbs/sec (kg/sec) and the material being released will form a vapor cloud.
- An onsite fire brigade shall be provided when the maximum foreseeable loss exceeds $*X*.

Note that the above statements define *when* a specific control should be implemented. The fire protection strategy should avoid using general statements about *how* something should be protected. As shown in Figure 3-1, the fire protection procedures are intended to describe how to implement the fire protection strategy. The following are some examples of the types of statements that should be avoided when developing a fire protection strategy:

- Facilities shall be protected by fire water systems that are designed in accordance with NFPA 20 and NFPA 24.
- Structural steel shall be protected with a 3-hour rated fire resistive system per UL 1709.

The first bullet indicates that the NFPA standards should be used to determine the fire water requirements for a facility. The NFPA standards do not define when a system is required. The first bullet does not achieve the result of defining when protection is necessary. In addition, the level of protection needed for a process facility is not covered by the cited NFPA standards. The NFPA standards simply describe *how* to design and install system components.

The second bullet implies that all structural steel requires a 3-hour fire resistive coating. Again, the statement does not define when the specific design feature must be used. These types of statements normally apply to design specifications for projects.

## 3.3. Integration with Other Management Systems

In almost every region, state, and country, regulations have been introduced, or are being introduced, that require formal management systems for managing risk. Within a company, it is not unusual for the management programs for process safety, environmental, health, and safety to have been developed separately. Thus, rather than create a whole new system, every effort should be made to integrate existing programs, including fire protection, into one Risk Management System.

*Guidelines for Integrating Process Safety Management, Environment, Safety, Health, and Quality* (CCPS, 1996a) may be of assistance in integration of management systems.

## 3.4. Integration with the Lifecycle of a Facility

A Risk Management System (RMS) is a lifecycle process and should be applied from a facility's conception to decommissioning. The lifecycle of a facility is shown in Figure 3.4. The RMS is an iterative and proactive process that requires the design team to take full ownership of the hazards and actively seek ways to assess and manage hazards with an ultimate goal of ensuring that the facility achieves corporate criteria and goals.

The integration of fire protection in the RMS needs to be considered during all phases of the lifecycle of the process or facility.

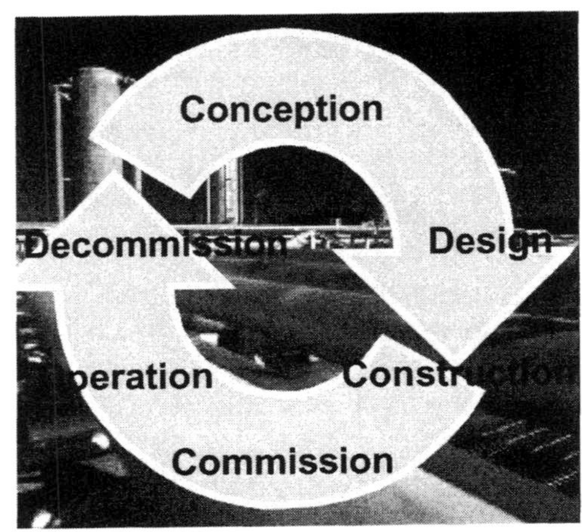


**Figure 3.4.** Lifecycle of a Facility

### *3.4.1. Design*

*Making EHS an Integral Part of Process Design* (CCPS, 2001b) discusses the concept that environmental, health, and safety (EHS) should be addressed early in the design process. Experience indicates that late changes to the design are costly and usually do not integrate well into the overall design.

*Making EHS an Integral Part of Process Design* introduces the concept of MERITT (Maximizing EHS Returns by Integrating Tools and Talents). The objective of MERITT is to enhance the process though more effective integration of EHS evaluations throughout the design process.

> ***Fire protection is much the same in that it needs to be part of the MERITT process and will have impacts on EHS issues throughout the lifecycle of the process unit or facility.***

#### *3.4.1.1. Project Management*

There are practical reasons why fire protection issues need to be addressed early in a project (at the front-end loading stage). These may not always be obvious, especially at facilities where design is not governed by a building code.

All codes that apply to the facility should be identified early in the design phase to ensure that all code requirements are included in the initial design, avoiding add-on costs of compliance. For example, if a sprinkler system is required, but not identified during front-end loading, the cost and potential project delay to add underground piping and install the sprinkler system increases drastically.

Engineering dollars spent early in the project cycle will payoff in reduced fire protection costs because hazards are designed out or managed as early as feasible. The hazard identification and management process should begin early, with understanding of the general facility hazards, and progress in detail as the facility design becomes more complete. This approach is most effective when all personnel involved in facility design take ownership in understanding the mechanisms that affect the hazards. Figure 3-5 depicts the cost of safety performance during facility design phase.

Project specifications and approvals should specifically address the need for fire protection. It is the responsibility of the Project Manager to ensure that the design team addresses the fire hazards, including consequences to the operation and potential for escalation. In turn, it is the responsibility of the design team to convincingly demonstrate to the Project Manager that the design includes mitigation to reduce the severity of the consequences, an inherently safer design has been accomplished, or the likelihood of the scenario occurring has been reduced to acceptable levels.

Hazard reviews are generally conducted progressively and in increasing detail during a project to ensure that hazards have been identified and that appropriate hazard mitigation has been applied. The outcome of hazard reviews should form the basis for the initial design and subsequent development of the fire protection system. It is not unusual that, as the overall project

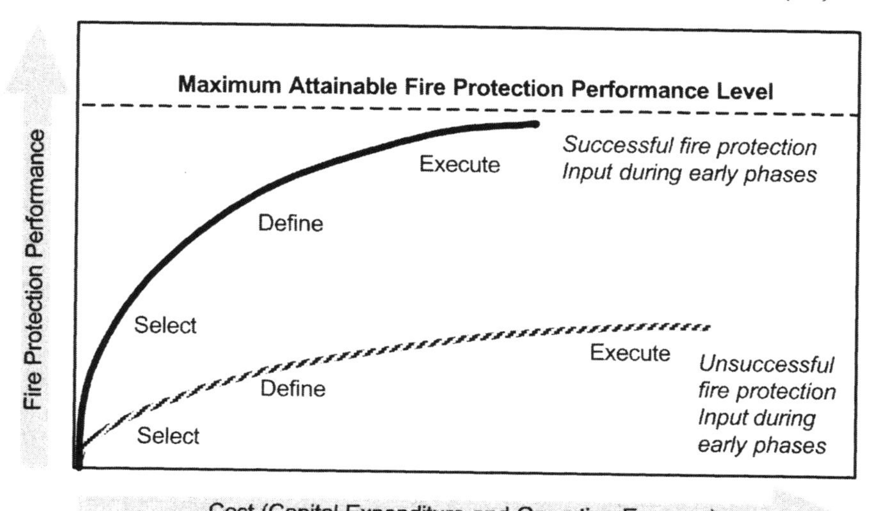


**Figure 3.5.** Cost of fire protection performance

design progresses, the performance requirements of the fire protection system will be affected, and the design modified. The impact on the project of such changes will be minimized by the early inclusion of the fire protection system in the design process. Further detail on the hazard review process is discussed in Chapters 5 and 6.

Organizations often have a phased design process and that will generally consist of four phases. The most effective management of the major hazards will be attained during the early design phases. The fire protection activities should be integrated with design activities in each phase as noted in Table 3-2.

***The reliability and availability performance requirements of safety systems and especially fire protection systems should be considered in the design. Fire protection systems are not used on a regular basis, but must work when required. It is also important that the design of the fire protection system facilitate a testing and maintenance program for achieving these goals.***

#### *3.4.1.2. Third-Party Requirements*

Quite often, insurance companies, the local Authority Having Jurisdiction (AHJ), corporate staffs, and others will have the responsibility to review and approve fire protection system designs to ensure compliance with their specific requirements. Designers should identify such requirements and ensure that these are adequately addressed in developing the design for the fire protection system.

#### *3.4.1.3. Operating Procedures*

Operating procedures for the fire protection system need to be developed and maintained. These operating procedures should provide sufficient guidance on the design, system operations, and maintenance requirements.

It is important that documentation of the fire protection system be developed so that future additions or questions can be answered. Chapter 9 identifies documentation that should be maintained for the life of the fire protection system.

### 3.4.2. Construction and Commissioning

It is essential that construction follows recognized and generally accepted good engineering practices. In most instances, construction will be monitored by the design engineer or HSE professional to ensure the installed equipment meets the design intent. It is not unusual for changes to occur during construction that

**Table 3-2**
*Timing of Fire Hazard Studies*

| Goals | Study | Design |
|---|---|---|
| **Phase 1—Concept** | | |
| Provide early fire protection engineering input<br>Identify major fire risks associated with each concept under consideration<br>Generate a comprehensive listing of major fire risk issues to be addressed in Phase 2 | Concept Risk Analysis (CRA)<br>Hazard Identification (HAZID)<br>PFD review | Spacing and layout requirements<br>Philosophy for fire protection (passive vs. active, inherently safer design, emergency response) |
| **Phase 2—Preliminary Design** | | |
| Define all key fire hazards for selected concept<br>Develop strategy to manage each fire hazard in design<br>Maximize opportunities to improve inherently safer design<br>Ensure engineering decisions are simple and robust<br>Optimize configuration and layout<br>Identify key fire protection decisions for project | Checklist Fire Hazards Analyses (FHA)<br>Layout reviews<br>Coarse fire risk analysis<br>Detailed fire risk analysis near the end of this phase<br>PFD/P&ID review | Layout and spacing review<br>Conceptual fire water design and computer model<br>Inherently safer design review<br>Develop specification for:<br>▪ Fire protection systems<br>▪ Fire pumps<br>▪ Detection<br>▪ Alarm systems |
| **Phase 3—Detailed Design** | | |
| Ensure final design fire hazards are identified and addressed<br>Design fire protection systems | Detailed HAZOP<br>Fire Risk Assessment | Layout and spacing finalized<br>Model review<br>Procure fire protection equipment<br>Drawing approval:<br>▪ Fire protection systems<br>▪ Fire pumps<br>▪ Detection<br>▪ Alarm systems<br>Electrical classification review |
| **Phase 4—Construction and Startup** | | |
| Confirm the facility is ready for start-up | Pre-startup safety review | Installation of fire water underground as early as possible<br>Performance acceptance test<br>Quality assurance inspections<br>Emergency plan/drill |

require modifications to be made to the design of a fire protection system. It is vital that such changes are made in such a manner that the overall design intention of the fire protection system is maintained or, if necessary, better matches the hazards resulting from these changes. Management of Change (MOC) procedures must exist or be developed in order to manage changes in fire protection during construction.

Commissioning of the system includes verifying the design drawings, providing training to plant personnel on operation of the fire protection system, and conducting acceptance tests. The acceptance test provides for testing the detection system and activation of the fire protection system to ensure the system meets the design criteria.

Additional information on installation, acceptance testing, and construction are contained in Chapter 9 of this Guideline.

### *3.4.3. Operations*

Process units and facilities now have a potential operating life of many years. During the course of operation, it can be expected that changes will be made that effect fire protection capabilities. These changes could result in increased hazards, resulting in changes to:

- Fire protection strategy
- Emergency response capabilities

Additionally, changes in insurance companies (hence changes in protection desired) or a change in ownership that requires different levels of protection can impact the facility's fire protection strategy.

Fire protection should be included as an item for review and consideration during the Management of Change process at the facility. Where an FHA exists, it should be updated as changes to the facility are made.

### *3.4.4. Decommissioning*

Fire protection systems should be maintained until materials that pose a hazard are removed. A fire water system should be the last item removed or deactivated. When necessary, decommissioning plans must be reviewed with the AHJ. Fires have occurred during removal of equipment due to cutting operations, material still in the equipment, and other hazards. A separate fire hazard analysis should be conducted to determine fire hazards that may be present during a decommissioning of a unit or plant.

# 4

# OVERVIEW OF FIRE PREVENTION ELEMENTS

A fire prevention program is essential for the continued safe, successful, and profitable operation of a processing facility. Fire prevention is also a critical part of a process facility's overall loss prevention and control program and is closely linked to process safety and overall Health, Safety, and Environmental (HSE) programs.

A successful loss prevention and control program should include the following fire prevention elements:

- Audit program
- Layout and spacing
- Control of ignition sources
- Employee training
- Housekeeping
- Incident investigation
- Inherently safer design (as it applies to fire safety and protection)
- Plant maintenance/inspection
- Management of change
- Material hazards
- Alarm and surveillance

Additional elements of a loss prevention and control program are covered in other Chapters:

- Process hazard analysis (*Chapter 6*)
- Risk analysis (*Chapter 6*)

- Fire protection equipment inspection (*Chapter 10*)
- Impairment handling (*Chapter 10*)
- Emergency response (*Chapter 11*)

Since fires are frequently identified as the dominant type of loss event in many process facilities, the fire prevention program should be considered an essential part of a Risk Management System. Fire prevention often includes procedures and features to support other types of emergency management and loss prevention efforts.

For example, the system that provides emergency notification to employees of a fire is generally the same as, or operated in conjunction with, the system for providing alarms in the event of natural peril, such as tornados.

On the other hand, there are a number of elements of fire prevention that overlap with elements of other programs. For example, Incident Investigation, Management of Change (MOC), Process Safety Information (PSI), and Process Hazard Analysis (PHA) input required for fire prevention can be derived from these same elements of the Process Safety Management (PSM) program.

## 4.1. Audit Program

An audit is a systematic, independent review to verify conformance with established guidelines or standards. An audit uses a well-defined review process to ensure consistency and allow the auditor to reach defensible conclusions. An audit evaluates the procedures, operations, and activities performed in the management and execution of a program in order to verify conformity to established criteria. Such evaluations are intended to provide feedback to management and those responsible for the status of the audited program.

### *4.1.1. The Audit Process*

The audit process is generally described in the following steps:

- Understand the management systems and procedures for control
- Assess the adequacy of these control systems
- Identify critical tasks and activities that are necessary to maintain the effectiveness of controls
- Gather and evaluate information to assure critical tasks and activities are being performed and documented
- Report findings and exceptions, or deviations, from the established criteria
- Confirm that changes have been implemented

An effective audit requires the development of a plan or protocol for the audit, establishing:

- The steps that need to be taken
- How each step is to be accomplished
- Who will take each step
- In what sequence the steps need to be performed

### 4.1.2. Qualifications and Staffing

Those leading audit teams should be experienced in the audit process. All auditors should have some training in the audit process, particularly in the gathering of data:

- Interviewing techniques
- Observation—physical examination or inspection
- Verification—data selection, sampling strategies, analysis

It is also important for the auditor to have some subject-matter expertise, in this case knowledge of fire protection.

The audit staffing requirements will vary based on the scope of the audit (facility size, complexity, and degree of risks).

Audits may be conducted by one person experienced in the audit process with general knowledge of the type of facility to be audited. More often, audits are conducted by a small team of people. The audit team should not be associated with the program or facility under audit in order to provide an unbiased independent perspective.

For companies having multiple plant sites, it can be useful to staff the audit team with competent people from other sites. In this way, the audit process can provide an outside perspective and a learning experience for the audit team members.

### 4.1.3. Frequency of Audits

In practice, the frequency of fire prevention audits ranges from annually to every third year. Longer audit cycles are possible; however the longer the audit cycle, the more intense and independent the audit process should be.

The audit frequency is determined by a variety of criteria, including:

- Age of facility
- Experience level of facility staff
- Maturity level of the facility's overall loss prevention, process safety, and other programs

- Rate of change or frequency of modifications to the facility, its operations, personnel, processes, and equipment
- Nature of hazards and degree of risks
- Results from previous audits
- Recent history of unplanned operation or process incidents, upsets, and near-misses, particularly those involving flammable release, fire, or explosion

### 4.1.4. *Application to Fire Protection*

The fire protection audit element can be conducted with the scope of the audit focused only on fire protection. However, many companies in the process industries audit fire prevention as an integral part of the overall loss prevention effort and combine it into audits with other programs.

Additional information on auditing can be found in *Guidelines for Auditing Process Safety Management Systems* (CCPS, 1993a).

## 4.2. Layout and Spacing

The overall design of process facilities, particularly those handling flammable liquids or gases and combustible dusts, must consider layout and spacing as a key means of preventing the spread of fires. Hazard analyses of planned changes, modifications, and new projects should specifically review the adequacy of layout and spacing of process and support equipment. Some basic layout and spacing considerations are:

- All site buildings and structures are constructed of noncombustible materials, particularly exteriors and structural support systems.
- Control rooms, operating offices, and their occupants are separated from potential hazardous processing areas.
- Storage of large volumes of flammable or combustible materials is separated from high value operating or processing areas and personnel occupancies.
- Fired process heaters and boilers, incinerators, flares, and other equipment with flame burners are located at an appropriate distance from high value operating or processing areas, large volume storage of flammable or combustible materials, control rooms, operating offices, and their occupants.

For additional information, refer to Chapter 7 and *Guidelines for Facility Siting and Layout* (CCPS, 2003b).

## 4.3. Control of Ignition Sources

A fundamental element of fire prevention is the control of ignition sources. The process should be designed, installed, and operated to minimize or prevent the release or spill of flammable gases, liquids, or combustible dusts, as well as eliminate or control ignition sources. The basic controls for these unwanted ignition sources are:

- Electrical area classification
- Control of personal ignition sources
- Control of hot work
- Control of static electricity and stray electrical currents

### *4.3.1. Electrical Area Classification*

Electrical equipment can be the source of ignition where flammable materials are handled. Wherever possible, electrical equipment and wiring are best located outside hazardous locations. Where these ignition sources cannot be located outside the hazardous area, then ignition sources are controlled by defining electrically classified areas and designing, procuring, and installing equipment accordingly. The National Electric Code (NFPA 70) divides hazardous locations into three classes according to the nature of the hazard:

- Class I Flammable Liquids and Gases
- Class II Combustible Dusts
- Class III Easily Ignitable Fibers and Flyings

Additional information on electrical classification can be found in Chapter 7.

### *4.3.2. Personal Ignition Sources*

Controls are required for ignition sources that may be carried into a hazardous area. These ignition sources include any material, object, or device that is potentially incendiary or capable of producing a spark.

Personal electronic or electrical devices that may require control are pagers, cellular phones, personal digital assistants (PDAs), and personal radios or music players. Few, if any, of these devices are evaluated to determine if they may be safely used in hazardous areas. Typically, such devices do not claim to be "intrinsically safe" or of "nonincendive circuit" design (NFPA 70, Articles 500 and 504).

Fires in the workplace continue to be caused by matches, lighters, and carelessly discarded cigarettes and other smoking materials that ignite near combustible materials. Control of these potential ignition sources is essential for an effective fire prevention program. Facility management should consider involving those working in the facility in the development of a policy on smoking and related issues in order to ensure support and compliance.

### 4.3.3. Hot Work

Operating, maintaining, repairing, and modifying a typical process facility frequently involves activities that produce sparks or use flame. The portability of spark and flame producing equipment and its inappropriate or careless use in areas not specifically designed for its safe use can increase the likelihood of a fire.

Temperatures sufficient to start fires or ignite explosive materials may come from a number of sources including:

- Open flame of a torch used for heating or thawing process lines
- Torch cutting
- Welding
- Improperly applied electric arc welding grounding clamps
- Molten slag or metal that flows from the work piece
- Improperly handled soldering iron or propane torch
- Grinding sparks that fly from the work
- Electric motor-powered hand tools
- Portable heaters
- Forklift trucks or other industrial powered vehicles not rated or classified for use in a potentially hazardous area
- Vacuum tank trucks removing spilled flammable/combustible material
- Roofing installation or repair using hot-mopped asphalt or using open-flame heating devices to seal roofing sheet membrane seams
- *There are many others*

Diesel engines are used extensively at most sites for welding, air compressors, etc. These engines need to be remotely sited or provided with flame arrestors, insulation on hot surfaces/exhausts, etc.

The principal hazard associated with these and other flame-, heat-, or spark-producing work is the introduction of unauthorized ignition sources into areas of the facility. Control of hazards related to portable equipment and hot work requires developing and maintaining a comprehensive hot work procedure.

To ensure that necessary hot work for maintenance, construction, or modifications is done safely to prevent ignition of fires and protect life and property, the hot work control procedure should include the following as a minimum:

- Assigned responsibility for the program
- A permit system requiring:
  - Job site to be inspected before work begins
  - Testing for the presence of flammable vapors and inspection for combustible materials
  - Personal protective equipment appropriate to the job
  - Additional temporary protections, e.g., a firewatch with fire extinguisher and emergency notification procedure that includes covering sewer openings, construction of fire boxes of flame resistant material to contain sparks, etc.
  - A time limit for the duration of the permit
  - Signed approval by a designated authorized person
  - Close-out of work permit
- Training of personnel
- Providing/maintaining necessary equipment, e.g., flammable vapor detectors
- Auditing and periodic review of program.

Refer to NFPA 51B for additional guidance on hot work.

### 4.3.4. Static Electricity

Static electricity discharges and unexpected electrical currents are frequently overlooked as potential sources of ignition that must be controlled. Some of the conditions that may result in sufficiently intense electrical discharges or arcing are:

- Flow of liquids in piping
- Pneumatic conveying of dusts, powders, or particulates
- Splash or free-fall filling of tanks, vessels, or containers
- Mixing and blending of powders
- Use of wet steam
- Moving nonconductive rubber belts, e.g., conveyors or drive belts
- Personnel wearing nonconductive shoes
- Static generated by clothing
- Atmospheric lighting strikes
- Stray electrical currents from faulty equipment, improperly applied electric welding leads, or other sources

A program to ensure that a well-designed and effective electrical earth-grounding system and equipment bonding system is in place is an essential first step in the control of these potential ignition sources. Such a system must be regularly checked to ensure ground straps are fully secured.

For more information, refer to *Guidelines for Engineering Design for Process Safety*, Chapter 11 (CCPS, 1993b); API Recommended Practice 2003, *Protection Against Ignitions Arising Out of Static, Lightning and Stray Currents* (API, 1997); *Avoiding Static Ignition Hazards in Chemical Operations* (CCPS, 1999a), and *Electrostatic Ignitions of Fires and Explosions* (Pratt, 1997).

## 4.4. Employee Training

Training to provide knowledge of process operations and job execution skills is an important aspect of incident and fire prevention. In addition to job skills training, employees must be trained to properly execute fire protection tasks. A training requirements matrix for some of the elements of fire protection and the personnel that require training is illustrated in Table 4-1.

**Table 4-1**
*Training Requirements Matrix*

| | Some Fire Protection Program Training Elements | | | | | | | | | |
|---|---|---|---|---|---|---|---|---|---|---|
| **Typical Groups To Be Trained** | **Test/Inspect FP Equipment** | **Maintain FP Equipment** | **FP Impairment Procedure** | **Testing and Operation of Fire/Emergency Alarm System** | **Recognition and Activation of Fire/Emergency Alarms** | **Testing and Operation of Vapor, Smoke, Heat Detection Systems** | **Operation and Activation of Fixed FP Systems** | **Use of Fire Extinguishers** | **Control of Ignition Sources** | **Emergency Plan** |
| Fire protection equipment technicians | X | X | X | X | X | X | X | X | X | X |
| Operating personnel | | | X | X | X | | X | X | X | X |
| Technical staff | | | X | X | X | X | X | X | X | X |
| Mechanical maintenance | | X | X | | X | | X | X | X | X |
| Instrument and electrical maintenance | X | X | X | X | X | X | | X | X | X |
| Security/guard personnel | | | X | X | X | | | X | X | X |
| Contractors | | | | | X | | | X | X | X |
| Other facility personnel | | | | | X | | | X | X | X |

An aspect of employee training involves hazard recognition. As personnel are trained in recognizing hazardous situations and how to respond effectively, the number and severity of incidents decreases. Training effectiveness is directly impacted by the resources dedicated to the training program. Allocating resources to training is part of a facility's overall Risk Management System.

## 4.5. Housekeeping

Industrial experience and insurance loss records indicate that poor housekeeping contributes to an increased frequency of loss and greater loss potential. The added quantity and distribution of fuel in the facility caused by poor housekeeping practices can result in the following issues:

- Greater continuity of combustibles that makes fire spread easier and increases the area of involvement.
- Impaired ingress and egress.
- Increased overall combustible loading that provides more fuel to feed a fire and can increase the severity of the fire.
- Increased potential for severe secondary dust explosions when dust accumulates.
- Increased probability of fire.
- Increased probability of spontaneous ignition in residue accumulations or thick dust layers.

Poor housekeeping may also be a symptom of other fundamental problems, such as careless operation, frequent temporary repairs, and generally inadequate maintenance. These conditions can result in process leaks, releases, and spills, missing or open covers on equipment and electrical panels, unpainted rusting metal, and nonfunctional gauges and instruments.

### *4.5.1. Housekeeping Program*

Proper housekeeping does not just happen. It requires the leadership and wholehearted support of facility management, staff, and operating and maintenance supervision and the cooperation of all employees.

To develop a good housekeeping program as an element of fire prevention, the following actions should be taken:

- Appoint specific personnel to be responsible for proper housekeeping; participation could be on a rotated basis with a designated chairperson. Inform all employees of their authority and responsibility.

- Establish acceptable levels of cleanliness and orderliness in conjunction with the housekeeping committee with particular regard to:
  - Process liquid spills and residue accumulations
  - Dust control
  - Storage and handling of combustibles and flammables
  - Program for handling empty containers
  - Blockage of fire protection systems and equipment
- Require specific personnel to conduct regular periodic inspections of the facility and record results by area.
- Report the housekeeping inspection results by area to recognize improvements and encourage competition.
- Actively demonstrate support of proper housekeeping practices through regular, positive reinforcement. In addition to verbal reinforcements, written commendations and award for individuals, areas, and departments can be effective positive reinforcement.

### *4.5.2. Process Area Housekeeping*

Inadequate housekeeping controls in laboratories, process, or operating areas can result in process waste, leakage, and spillage accumulations that can lead to increased fire losses. Such accumulations are typically from one of several causes:

- Doors left open
- Dust or other material released from normally closed containers or systems
- Improper or excessive storage of materials, i.e. materials, including flammables, stored under stairways/escape paths
- Improper or inadequate removal of accumulated process wastes or residues
- Leakage of process or lubricating fluids, steam, or condensate
- Trash or debris left because of carelessness
- Trash or debris resulting from an inadequate pickup schedule
- Unnecessary scaffolding still in place

Housekeeping practices that allow spilled or released combustible or flammable process materials to accumulate could provide fuel for a fire to start or allow more rapid or vigorous fire spread than protection systems can manage.

### 4.5.3. Dust Control

Areas where combustible powders are handled in bulk quantities or areas containing dust-producing equipment should be cleaned on a regularly scheduled basis. Horizontal overhead surfaces, such as tops of beams, and concealed or other out-of-sight spaces where dust can accumulate, should be identified and included in the cleaning schedule or changed to minimize the potential for dust collection.

Using compressed air to blow dust off surfaces is discouraged because dust suspended in the air during blow-down operations may produce a potentially explosive mixture. Vacuum cleaners should be used. These may be either portable or attached to a central system. When portable units are used, they should be appropriately approved for use in any hazardous location at the facility.

### 4.5.4. Inappropriate Storage and Handling

Poor housekeeping practices can allow materials, containers, debris, or unused equipment items to be stored, placed, or handled so that they impair fire protection systems. Examples of such inappropriate storage and handling are:

- Open drain lines
- Leaks on tank roofs
- Pallets jammed against fire doors that prevent them from closing when needed.
- Materials, particularly combustibles, or equipment placed inside storage tank dikes.
- Materials or equipment blocking access to sprinkler control valves.
- Materials or containers stacked or placed so that they block the effective discharge of sprinkler or deluge fire protection systems or fire monitor nozzles.
- Sample storage
- Trash bins not emptied or removed from process area

Although it might be tempting to use the "free space" in switchgear, boiler, compressor, and other equipment rooms for the storage of brooms, paint, drawings and manuals, spare parts, and various other utility and maintenance supplies, it is essential to resist such practices. Suitable broom closets, spare parts storerooms, and utility rooms should be constructed for such storage.

Small quantities [1 gal (4 l) or less] of flammable or combustible paints, solvents, or cleaning materials, including aerosol cans, should be stored in approved flammable liquid storage cabinets. Large quantities [more than 1 gal (4 l)] of these materials should be stored in separate, remote, or fire-rated rooms or areas that will not present a fire hazard to other areas of the facility (NFPA 30).

### 4.5.5. Housekeeping and Equipment

Poor housekeeping may increase the failure or breakdown of electrical and mechanical equipment. Even without considering the possibility of a resulting fire or explosion, electrical and mechanical breakdowns can result in damage or destruction of major pieces of equipment or injury to personnel. Poor housekeeping can lead to breakdowns in the following ways:

- Accumulation of dust and other debris can create a thermal blanket or block air flow resulting in inadequate cooling of electrical equipment and cause the equipment to fail or run hotter reducing efficiency and life expectancy.
- Oil, grease, and other contaminants can damage electrical insulation on cables and in motor windings, resulting in electrical short-circuits and failures.
- Rags and debris in tower skirts provide a fuel source.
- Dirt, soot, moisture, and other contaminants can provide paths for electrical flashover or short-circuiting in switchgear and other electrical equipment.
- Accumulation of water, certain vapors, and other materials can damage paint and promote corrosion. Undetected corrosion has led to building collapse, pressure vessel failure, mechanical linkage separation, and electrical breakdown.
- In machinery, debris of any sort can lead to accelerated wear or direct breakdown.
- In plugs and drains, improper isolation practices can lead to spills.

### 4.5.6. Cleaning Materials

Cleaning materials and their methods of use can present significant and, frequently, undetected fire hazards. Any cleaning chemical or material brought into a facility should be reviewed for potential hazards using Material Safety Data Sheet (MSDS) information as a part of the material's hazard identification element of the overall fire prevention program. Cleaning activities should not be allowed to add unreasonable hazards to a facility.

For additional information on housekeeping refer to IRInformation IM.1.14.0, *Proper Housekeeping* (Industrial Risk Insurers, 1998a).

## 4.6. Incident Investigation

Facilities typically investigate incidents to determine their causes and prevent their recurrence. An incident is an unusual or unexpected event, which either resulted in, or had the potential to result in:

- Serious injury to personnel
- Significant damage to property
- Adverse environmental impact
- A major interruption of process operations

This definition is meant to include both accidents and *near misses*. Thus, the definition covers all cases where there was, or could have been, injury, damage to property, or the release of hazardous or toxic material to the environment.

### *4.6.1. Incident Investigation Process*

A facility's incident investigation process should be based on a documented procedure defining the goals and requirements of incident investigations and providing detailed steps outlining how incident investigations will be performed and reported. The facility Incident Investigation Procedure should clearly establish the process, responsibilities, and accountability for incident investigations.

The process of incident investigation generally utilizes a team-based approach involving gathering and analyzing the evidence determining the causes of the incident, generating corrective or preventive actions, and documenting the findings. Implementation of the recommended corrective actions should then result in reduction of risk.

### *4.6.2. Application to Fire Prevention*

Most facilities already have an established Incident Investigation Program as part of the Process Safety Management program. These existing Incident Investigation Programs should incorporate fire prevention requirements.

One concern with any incident investigation process is the incident severity threshold at which the investigation process is triggered. Where spills, releases, or fires that do not result in injury, significant property loss, or require environmental reporting are not investigated, then causes that may lead to serious incidents are being ignored.

For a description of many of the available methods of incident investigation techniques and how they can be applied refer to *Guidelines for Investigating Chemical Process Incidents* (CCPS, 2003c), *Guide for Fire Incident Field Notes*, (NFPA 906), and *Guide for Fire and Explosion Investigations*, (NFPA 921).

## 4.7. Inherently Safer Design

A system or operation is considered an inherently safer design if it remains in a nonhazardous state after the occurrence of unacceptable deviations from normal operating conditions.

Risk is defined as a measure of economic loss, human harm, or environmental harm in terms of both the incident likelihood and the magnitude of the loss or injury. Thus, any effort to reduce the risk arising from the operation of a processing facility can be directed toward reducing the likelihood of incidents (incident frequency), reducing the magnitude of the loss or injury should an incident occur (incident consequences), or some combination of the two.

In general, the strategy for reducing risk, whether directed toward reducing frequency or consequence of potential accidents, falls into one of the following four categories:

- *Inherent* or *Intrinsic*—eliminating the hazard by using materials and process conditions that are nonhazardous (e.g., substituting water for a flammable solvent).
- *Passive*—eliminating or minimizing the hazard by process and equipment design features that do not eliminate the hazard, but do reduce either the frequency or consequence of the hazard without the need for any device to function actively (e.g., the use of higher pressure-rated equipment).
- *Active*—using controls, safety interlocks, and emergency shutdown systems to detect potentially hazardous process deviations and take corrective action. These are commonly referred to as engineering controls.
- *Procedural*—using operating procedures, administrative checks, emergency response, and other management approaches to prevent incidents or to minimize the effects of an incident. These are commonly referred to as administrative controls.

Theoretically, a facility is considered an inherently safer design when all hazards have been eliminated. In practice, a facility can only be made inherently safer with respect to specific hazards and not all possible hazards. For example, many processing operations use heat transfer fluids for either heat addition or removal. These systems have historically been operated at elevated temperatures above the flash point of the heat transfer fluid. In the event of a piping failure, the released material can form a flammable atmosphere and, if ignited, result in an explosion or fire. Substituting a material with a higher flash point (higher than its operating temperature) will reduce the potential for forming a flammable atmosphere and the probability of a fire or explosion.

The best time to consider inherently safer design is during the development of new operations and the preliminary design of facilities. Safety can be built into an operation by using, where feasible, less hazardous process materials or less severe operating conditions. An operation can also be made inherently safer by careful site selection, plant layout and equipment, and building design. For additional information refer to *Guidelines for Engineering Design for Process Safety* (CCPS, 1993b) and *Guidelines for Inherently Safer Chemical Processes, A Life Cycle Approach* (CCPS, 1996b).

## 4.8. Plant Maintenance

All facility buildings, equipment, and systems require maintenance. As with all activities, there is the potential to do either too much and not be cost-effective or too little and incur an unnecessary risk. For example, a maintenance program can expend too high a percentage of its resources on unneeded, noncritical, or less important work. On the other hand, "run-to-failure" assures unnecessarily frequent and severe failures, equipment outages, and loss exposures. Neglected standby and emergency systems and equipment, including fire protection and control systems, will not perform reliably.

The fact that plant maintenance is essential to loss prevention has been emphasized in recent years because:

- Process plant utilization rates and capacities have increased. These increases occur primarily through small and large debottlenecking projects.
- Plant staffing levels have decreased.
- Process facilities continue to age.

Equipment and machines typically deteriorate as they age with potentially increasing rates of failures, shutdowns, or process interruptions. Buildings also age, but more slowly. Effective maintenance manages this aging process. Managing the aging process helps prevent failures that occur during service. These in-service failures are the ones more likely to result in unwanted incidents, including fires.

Overall plant-wide maintenance is an element of fire prevention. It is shared with the business need to maintain the production process, as well as with Process Safety Management and other health, safety, and environmental programs.

### 4.8.1. *Poor Maintenance*

Some symptoms of inadequate maintenance are:

- Change in key indicators, i.e., mean time between failure, overdue inspections, reduced equipment availability
- Frequent or temporary repairs

- Process leaks, releases, and spills
- Missing covers on equipment
- Electrical panels left open
- Insulation left off after maintenance
- Unpainted rusting pipework and structural metal
- Nonfunctional gauges and instruments

### 4.8.2. Good Maintenance Program Elements

The facility should be maintained in a way that cost-effectively controls the probability of failure-induced incidents and their likely consequences. A cost-effective maintenance program allocates maintenance resources according to the level of risk.

An effective facility-wide maintenance program should include the following key components:

- High ratio of preventive to repair maintenance work
- A maintenance organization with leadership that can implement and support an effective maintenance program and appropriately trained personnel who will consistently *"do the job right the first time, on time."*
- An ongoing risk analysis and risk ranking system that focuses and supports maintenance program needs.
- Risk-based maintenance priorities that ensure sufficient resources are applied to items identified as high risk (critical equipment).
- Clear management support and commitment for critical equipment maintenance, testing, and inspection, since these activities often require production downtime in order to be performed.
- Written procedures to describe how critical equipment maintenance will be performed, quality-controlled, and safety-ensured, such as use of decontamination, hot work, line-breaking, and lockout/tagout procedures.
- An efficient work order system that provides adequate description of work to be performed, the parts required, and the procedures to be followed. This work order system should also document completed work information in equipment history files.
- Controls and sign-offs in the work order system that ensure Management of Change procedures are followed.
- Precautions and practices to ensure that equipment worked on has been restored to its normal conditions before it is returned to service.

- A maintenance information system that details equipment and component maintenance scope and frequency, documents work completed, and provides feedback on maintenance program effectiveness.
- Controls and surveillance procedures that ensure contractor-performed work also adheres to all facility health, safety, and environmental, and loss prevention programs.

For additional information refer to IRInformation IM.1.3.0, *Maintenance* (Industrial Risk Insurers, 1998). For additional information on maintenance of fire protection systems, see Chapter 10.

## 4.9. Management of Change

The fire prevention program can be adversely affected by the inevitable changes to a process or facility unless these changes are reviewed with consideration for fire prevention and protection. Management of Change (MOC) procedures ensure that changes and modifications to operations receive appropriate review and approval before implementation. Typically, Management of Change is part of a facility's Process Safety Management program. Some companies, however, have implemented Management of Change systems for all processes, not just those covered by Process Safety Management.

The Management of Change process should be utilized to ensure that proposed changes are analyzed for their possible impact on fire prevention.

This may require alterations to the Management of Change procedure to improve its effectiveness for identifying and examining changes that effect fire protection. Possible procedure alterations may include addition of the following:

- Specific questions relating to fire prevention and protection and emergency response added to the Management of Change procedure's process for review of potential hazards.
- Approval by a designated person or function with fire prevention and protection responsibility.

### *4.9.1. Personnel Changes*

Personnel changes are a change that should trigger an organization's MOC program. These personnel changes can impact a company's fire protection strategy, including the effectiveness of the hot work permitting process, testing and maintenance of fire protection systems, hazard identification, etc. The MOC process should identify the additional training and resources that may be necessary when personnel changes occur.

Emergency response programs may be adversely affected by personnel changes. Companies may implement productivity and efficiency improvements that result in reduced staffing. Cumulatively, these operating staff reductions can limit how aggressively a facility's emergency teams can respond to serious incidents. Reviewing the conceptual framework for emergency response is important following changes in personnel. For example, one particular company had an interior structural fire brigade in the 1970's, but experienced years of staff reductions and budget cutbacks that eliminated the training and equipment necessary to support the original mission. The company had to reevaluate its emergency response strategy and, subsequently, had to depend on a community volunteer fire company.

### *4.9.2. Process Changes*

Changes occur when modifications are made to the physical plant, operation, equipment, personnel, or procedures. The most obvious changes occur when a new plant, or a major addition to an existing one, is constructed. Other, more subtle changes can occur when new alarm settings, interlock logic, or suppliers are used, when procedures are modified, and when equipment is repaired or replaced. All such changes, if they are not carefully implemented, can increase the risk of a loss. Experience has demonstrated that inadvertent, unintended, erroneous, or poorly performed changes have resulted in many catastrophic fires, explosions, and other losses.

Incremental increases in the amount of combustibles can eventually render the existing fire protection systems inadequate. A more subtle change is the gradual replacement of metal parts in a warehouse with plastic ones of the same size and shape.

Seemingly inconsequential changes to a process, such as a slight increase in temperature, pressure, or flow rate, can lead to a major increase in hazards. Materials of construction also need to be considered. Many fires have been caused by the replacement of a material or part with what was assumed to be an "equivalent replacement."

### *4.9.3. Maintenance Turnarounds*

Incidents can and often do occur during maintenance turnarounds or construction work. One company estimated that 80% of their fires occurred during startup, shutdown, and maintenance phases of operation. Most companies establish emergency systems for normal operations, but these may be inadequate when plants undergo expansions or maintenance turnarounds. Personnel responsible for plant operations should review their response strategies and capabilities before temporary or long-term maintenance and construction projects begin.

For additional information refer to *Guidelines for Technical Planning for Onsite Emergencies* (CCPS, 1995a) and *Plant Guidelines for Technical Management of Chemical Process Safety* (CCPS, 1995b).

## 4.10. Material Hazards

A program is necessary for identifying all materials in the workplace, and making employees aware of the hazards of these materials and the necessary precautions to be taken to prevent or control personnel exposure. Materials Hazard Identification and information gathering is an essential element of fire prevention. The hazardous properties of all chemical substances used in the workplace should be known in order to develop the appropriate design, routine handling practices, and fire prevention plan.

Flammability properties of materials are clearly important for fire prevention; but there are other properties that are also significant. There have been a number of severe fire incidents initiated by a material's reactivity properties that were previously unrecognized or unknown to the user. The development of a Materials Hazard Identification program requires knowledge of a material's toxicity and reactivity, as well as flammability.

### *4.10.1. Materials Hazard Evaluation Program*

Some key elements of a Materials Hazard Evaluation program are:

- Assign responsibility for the program to determine the physical and chemical properties of each material handled at the facility.
- Collect available information, evaluate the hazardous properties, and identify the relative hazard levels of each substance and any necessary handling precautions.
- Identify those potentially hazardous materials for which important properties are unknown and conduct appropriate material hazards evaluation tests.
- Distribute material hazard information and handling precautions to employees, emergency response organization, local community response agencies, and others as appropriate.

### *4.10.2. Material Safety Data Sheets*

Facilities should obtain data about a substance from the chemical manufacturers' Material Safety Data Sheets (MSDS) or from other published sources. In order to identify, evaluate, or respond safely to incidents involving hazardous material brought onsite by contractors, it is also important to include outside contractors in the Materials Hazard Identification program.

A summary of evaluating hazardous materials can be found in the book *Guidelines for Chemical Reactivity Evaluation and Application to Process Design* (CCPS, 1995c) and *Essential Practices for Managing Chemical Reactivity Hazards* (CCPS, 2003d). For additional information, refer to IRInformation IM.1.8.0, *Hazardous Materials Evaluation* (Industrial Risk Insurers, 1998).

## 4.11. Alarm and Surveillance

Alarm and surveillance systems are an important element of fire prevention. These alarm and surveillance systems:

- Provide notification of emergency events
- Can be used manually by people observing the emergency
- Can automatically activate protection systems
- Notify those onsite of an emergency and communicate actions to take
- Provide surveillance of the facility for fire
- Notify offsite emergency response organizations

Failure to promptly report a fire could result in greater damage and, more importantly, could delay warning affected personnel. The alarm and surveillance element of fire prevention triggers emergency response and has a major impact on the control of property losses, safety of personnel, and community impact. Some key components of alarm and surveillance are:

- A continuously manned location for receiving and acting on reported incidents and emergencies.
- Automated detection and protection systems to signal at an offsite central alarm station service for continuous monitoring.
- A reporting system for personnel to report incidents and emergencies to the manned station. This could include an "alarm pull-box" system, plant telephones, or radios.
- An alarm system for notifying personnel of an emergency in progress and for communicating action required, such as information only, shelter-in-place, or evacuate. This could include bells, sirens, whistles, horns, or public address systems.
- A documented procedure for periodically and systematically testing the reporting and alarm systems to confirm their functionality.
- Assurance of an acceptable level of surveillance for the facility by appropriate resources, procedures, and facility design features.

Each type of emergency alarm or signal must clearly inform those onsite of the actions to be taken. This requires training and testing of the alarm so personnel can recognize the alarm and take appropriate action. Some of these alarms may be automatic. For example, detection of a fire may be signaled directly by the protection or detection system rather than by an individual. This alarm signal may alert not only personnel in the immediate area, but all facility personnel and the community fire department.

The alarm and surveillance procedure should also describe how to use the warning and alerting equipment, which may include telephones, alarms, buzzers, lights, horns, public address systems, radios, and pagers. A useful addition to this procedure is a simple flow diagram indicating how information is distributed, an emergency call recording form, and a regulatory reporting requirements form.

For additional information, refer to NFPA 72, NFPA 101, and IRInformation IM.1.11.0, *Fire Protection and Security Surveillance* (Industrial Risk Insurers, 1998).

### 4.11.1. Security

It is not the intent of this Guideline to deal in depth with facility security issues. However, effective fire prevention in a processing facility depends on people in addition to systems to detect developing fires and other incidents and to detect unauthorized intrusion into the facility. Intruder-caused vandalism, damage, spills, releases, or fires are not common, but are a credible threat. The potential fire prevention and protection requirements to manage the risk of security events from terrorism need to be considered in the overall fire protection system design.

The CCPS book, *Guidelines for Analyzing and Managing the Security Vulnerabilities of Fixed Chemical Sites* (CCPS, 2002), is a resource for determining the potential vulnerabilities of a processing facility to security events. Key concepts explained in the book include the following:

- *Layers of protection*—a concept whereby several different devices, systems, or actions are provided to reduce the likelihood and severity of an undesirable event.
- *Security layers of protection*—also known as concentric *Rings of Protection,* a concept of providing multiple independent and overlapping layers of protection in depth with prevention and mitigation to both increase the reliability of the safeguards as well as to lessen the likelihood of an event escalating to extreme consequence. For security purposes, this may include various layers of protection such as counter-surveillance, counterintelligence, physical security, and cyber security.
- *Delay*—a security strategy to provide various barriers to slow the progress of a perpetrator in penetrating a site to prevent an attack or theft, or

in leaving a restricted area to assist in apprehension and prevention of theft.

- *Detect*—a security strategy to identify a perpetrator attempting to commit a chemical security event or other criminal activity in order to provide real-time observation as well as postincident analysis of the activities and identity of the perpetrator.
- *Deter*—a security strategy to prevent or discourage the occurrence of a breach of security by means of fear or doubt. Physical security systems, such as warning signs, lights, uniformed guards, cameras, and bars are examples of systems that provide deterrence.

Some commonly used physical security measures for processing facilities include:

- Perimeter fences with anti-climbing features
- Adequate illumination of perimeter and key areas at night
- Locked gates at road and railroad entrances
- Surveillance video cameras at gates, perimeters, and strategic locations
- Guard/security personnel sufficient to staff a central station and provide routine checks at key points in the facility
- Motion detectors

During weekdays there are larger numbers of people in the facility and a good chance that any fire or other incident would be detected by someone at an early stage. On the weekends and at night there are less people in the plant, fewer people in each area, and no people in many support, storage, and perimeter areas. In effect, the level of fire detection and security may be lowered during these off-hours.

Facility management must assess its unique security needs and establish an appropriate level of security protection service.

Depending on the strategies employed for security, along with the hazards of the facility and consequences of security events and the attractiveness of the targets, it is necessary to provide fire protection that may exceed the ordinary fire protection for accidental events alone.

The protection of fire protection equipment from tampering to render it ineffective during a security event needs to be considered. Examples of this include locking water supply valves or installing fire detection cabinets in controlled areas.

# 5

# FIRE HAZARD ANALYSIS

Understanding fire hazards is essential to risk reduction and fire protection decision-making. A fire hazard analysis (FHA) is a tool used to understand fire hazards. The process of quantifying the fire hazard is typically motivated by the need to determine the overall hazard of a process or facility or to have a decision-making tool for fire protection systems (Chapter 6). An FHA is an important element of a risk assessment and can also be used as a stand-alone hazard evaluation tool.

Figure 5-1 shows how the FHA is integrated into an overall risk assessment. A process hazard analysis is required to identify likely fire scenarios that are carried forward to the FHA. An FHA provides the tools to characterize the hazards and evaluate consequences. The results are incorporated into an overall risk assessment. See Chapter 6 for more information on fire risk assessment.

This chapter provides:

- Guidance on how to develop different types of fire scenarios common to process facilities.
- Steps and tools for conducting an FHA.

An FHA is used to document the inventory of flammable or combustible material, calculate the potential magnitude of the fire, and determine the probable impact of the fire on personnel, equipment, the community, and the environment. An FHA can be performed on proposed or existing designs. Based on the impact, fire losses can then be estimated. The basic elements of FHA are illustrated in Figure 5-2.

The FHA accomplishes three objectives:

- Provides an understanding of the hazards
- Enables the specification of performance-based fire protection
- Forms part of an overall risk assessment

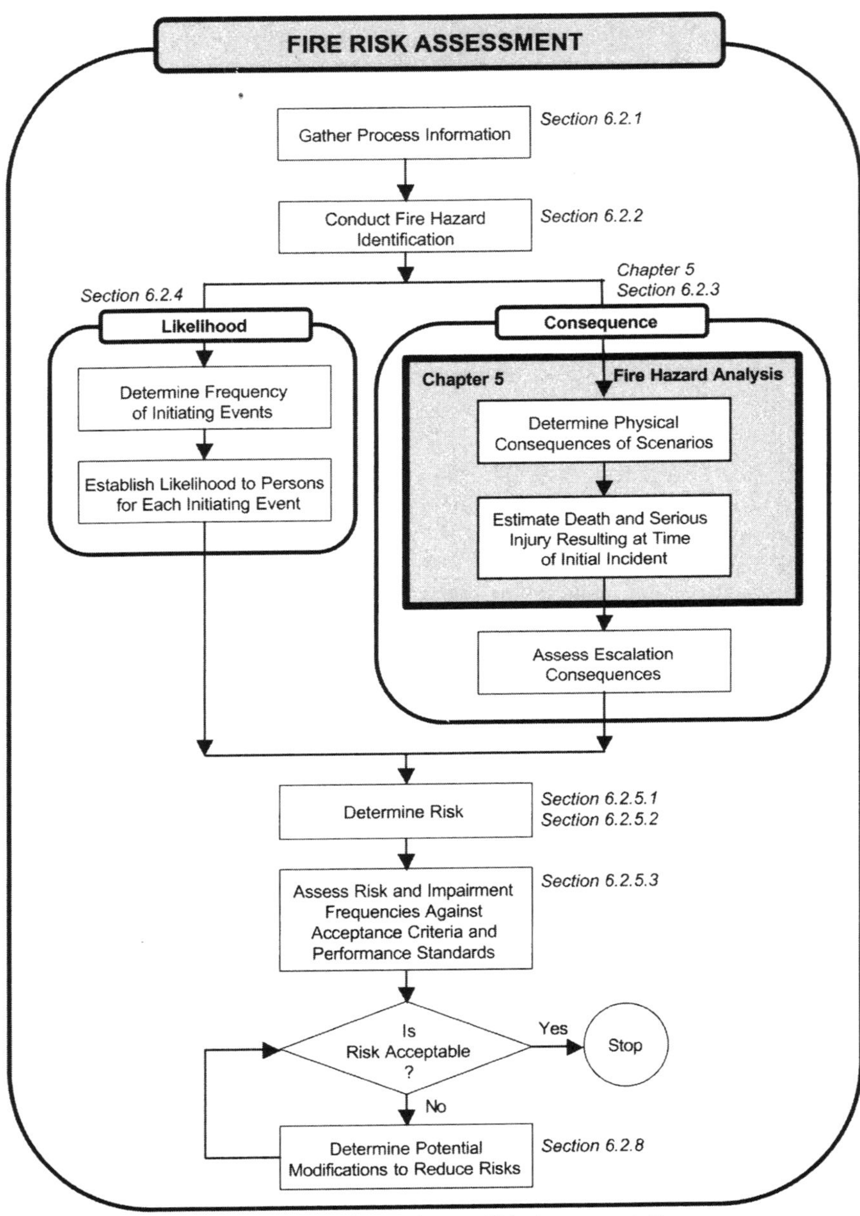


**Figure 5-1.** FHA Relationship to Risk Assessment Process

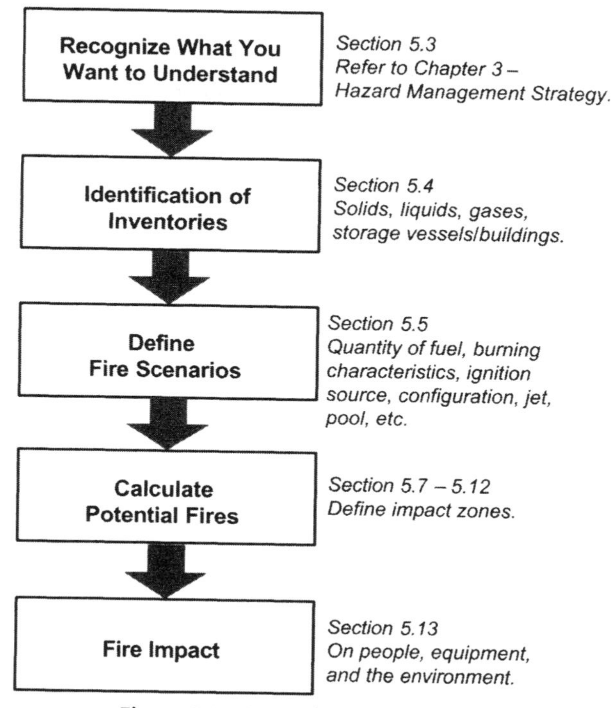


**Figure 5-2.** Basic Elements of an FHA

The benefits of conducting an FHA include:

- An inventory of fire hazards, including quantities.
- A comprehensive understanding of the fire hazard, including potential magnitude and duration.
- An estimate of the potential impact of a fire on personnel, equipment, the community, and the environment.
- Development of a list of appropriate mitigation options.

The equations presented in this chapter are intended to provide a simple tool for preliminary assessment of hazards. As such, most of the techniques are conservative. The units in this chapter are metric and have not been converted to English units because most equations used in fire modeling are based on metric units. The following sources provide more detailed information:

- *The SFPE Handbook of Fire Protection Engineering* (SFPE, 2002)
- Center for Chemical Process Safety (CCPS) *Guideline for Consequence Analysis of Chemical Releases* (CCPS, 1999)

- Center for Marine Technology (CMPT) *Guide to Quantitative Risk Assessment for Offshore Installations* (Spouge, 1999)
- *SINTEF Handbook for Fire Calculations* (SINTEF, 1997)

Before discussing the FHA process in detail, it is important to have an understanding of the fire hazards present at chemical, petrochemical, and hydrocarbon processing facilities.

Process fires are very similar whether they occur outside or in enclosed buildings. The major differences are that products of combustion (toxic fumes, smoke, CO, $CO_2$) build-up in an enclosure very quickly and can incapacitate personnel and hinder escape. Depending on the location and size of the fire, personnel will not have much time (less than one minute) to escape the building. It is important that life safety issues be handled by following the applicable building code and NFPA 101, Life Safety Code.

Another difference is that heat will build-up within an enclosure and the temperature on equipment and structures will increase quicker. The effects of both products of combustion and heat will be impacted by the venting that occurs in the building.

## 5.1. Hazardous Chemicals and Processes

Chemical, petrochemical, and hydrocarbon industry processes involve the handling of a vast number of flammable and combustible materials. Processing and storage operations involving these materials provide innumerable opportunities for their release and subsequent ignition. It is important to analyze all materials and processes for the potential for fire including production, manufacturing, storage, or treatment facilities.

The release of a flammable gas or liquid can lead to different types of fire scenarios. These are dependent on the material released, the mechanism of release, the temperature and pressure of the material, ambient conditions, and the point of ignition. Types of fires include:

- Jet fire
- Flash fire
- Pool fire
- Running liquid fire
- Boiling liquid expanding vapor explosion (BLEVE) or fireball
- Vapor cloud explosions

Figures 5-3 and 5-4 illustrate the potential outcomes of a gas and liquid release, respectively (CCPS, 2000). Unconfined vapor cloud explosions

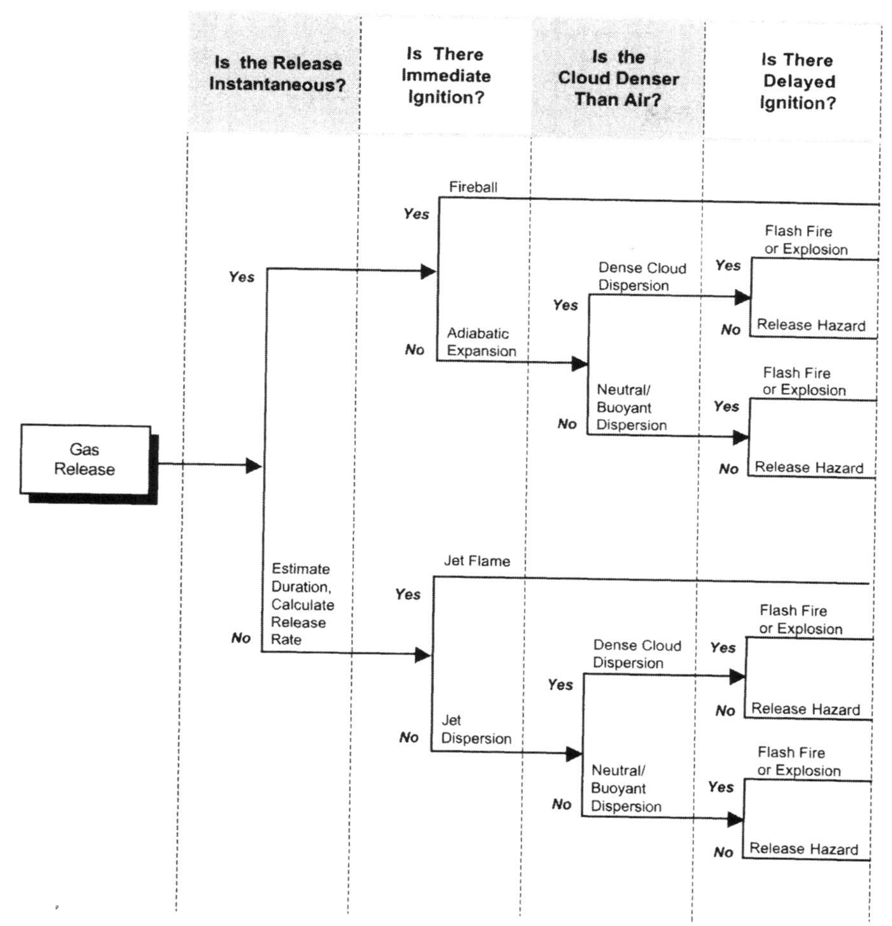


**Figure 5-3.** Event Tree of Gas Release

defined by Figure 5-3 as an outcome of gas or vapor release need the following conditions to occur:

- Sufficient quantity of material
- Adequate mixing of gas or vapor and air
- Delayed ignition
- Degree of confinement (by equipment or structures)
- Degree of turbulence

When ignition occurs without these conditions, flash fire results.

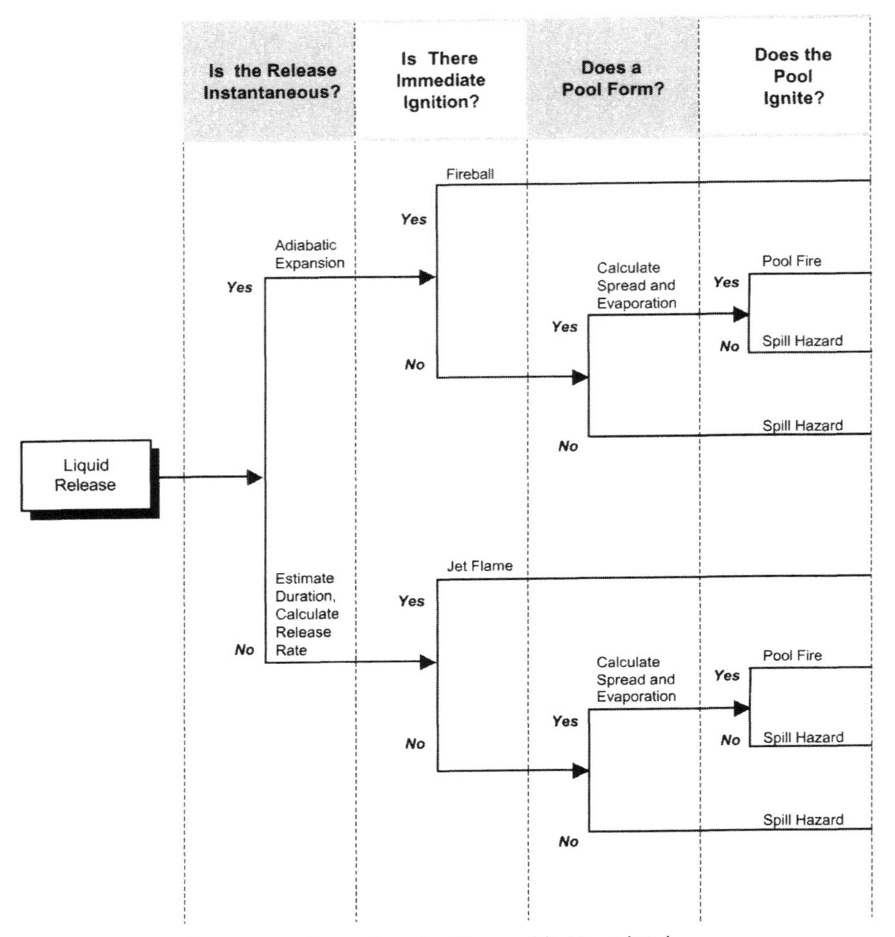


**Figure 5-4.** Event Tree for Flammable Liquid Release

Other fires that can occur in specific areas within a process plant include:

- Fires involving oxygen, e.g., systems for oxygen addition to a fluid catalytic cracking (FCC) unit
- Fires involving combustible metals, e.g., sodium used as a coolant or catalyst
- Fires involving pyrophoric materials, e.g., aluminum alkyls used as catalysts
- Fires involving hydrocarbons, e.g., hot oil pumps in crude units

- Solid fires, e.g., cellulose fires involving material, such as wood, paper, dust, etc.
- Material storage fires in warehouses or buildings
- Electrical equipment fires, e.g., transformer fires

## 5.2. Recognize What You Want to Understand

As is the case with any analysis, it is important to clearly state the purpose of the study. Several general reasons for conducting an FHA have been discussed earlier. More specifically, an FHA will provide information for:

- Understanding the fire hazards and determining how they should be controlled.
- Determining if personnel have time to escape the building or facility.
- Determining the potential damage to structure and equipment.
- Determining the level and extent of passive fire protection which may be required.
- Defining the active protection systems needed to limit/control fire spread.
- Determining fire water demand and duration of worst-case fire scenarios.

A clear statement of the objectives of an FHA should ensure those executing the study are properly focused and will conduct it in an efficient and cost-effective manner.

## 5.3. Identification of Inventories

In order to estimate the duration of potential fires, it is necessary to identify inventories that can be isolated. An inventory of flammable and combustible materials should be developed for each process unit and storage area within a facility. This list should contain the quantity, storage configuration, material characteristics, and location.

The inventory of contained liquid and gas can be calculated between isolation points to determine the amount of material that can be released for process-related scenarios. The time frame to close manual valves—depending on accessibility, operating conditions, and availability of personnel—is typically 10–30 minutes. For remotely operated valves actuated from the control room, 60–90 seconds is used for isolation. The release rate will decrease based on the pressure decay of the system once the system has been isolated.

It is important to document the assumptions used in determining the inventories for the FHA. The inventory list should be maintained throughout the lifecycle of the facility by the Management of Change (MOC) process.

## 5.4. Define Fire Scenarios

Fires range in size and consequence from those that are small, easily controlled, and result in minor damage to those that are large, difficult to control, and create a major loss.

Fires in process facilities usually follow a loss of containment. While the consequences of such fires are dependent upon a number of factors (weather, wind, leak orientation, etc.), the most significant are the:

- Rate at which the spill occurs
- Total amount spilled

In theory, there are an infinite number of leak sizes, ranging from a tiny pinhole to a full rupture of piping or equipment. It is clearly impractical to investigate them all. Thus, some practical guidance is necessary in selecting leak sizes that will allow a reasonable range of fire scenarios to be evaluated.

The process hazard analysis can be a starting point for the selection of fire scenarios. The process hazard analysis can be reviewed to develop a list of scenarios that result in fire as a consequence. Generic release sizes for small, medium, and large releases have been proposed as shown in Table 5-1 (Spouge, 1999). This saves time by eliminating the need to develop a detailed scenario. The analyst can use these release sizes to perform fire modeling calculations and determine the impact by moving the release point locations. The release criteria are considered to be representative of scenarios that could reasonably be expected to occur.

The worst-case scenario is typically taken to be full-bore rupture of gas or liquid lines. There are circumstances where full-bore rupture is *not* the worst-case, particularly when there is a restricted inventory to consider. For example,

**Table 5-1**
*Typical FHA Release Categories (Spouge, 1999)*

| Type | Release Size, inches (millimeters) |
|---|---|
| Small | 0.1–0.4 (3–10) |
| Medium | 0.4–2 (10–50) |
| Large | 2–6 (50–150) |
| Rupture | Full-bore (equipment diameter) |

while a 15-second jet fire resulting from a full-bore rupture can cause much damage, a small leak, long duration scenario [e.g., ¼-inch (6.4 mm) jet that lasts 20 minutes] may result in more damage, such as initiating a structural collapse.

Other CCPS guidelines provide discharge calculations for various releases, including gas, liquid, and two-phase flow. Dispersion calculations and pool spreading on different materials are also described.

The approach used in an FHA is to assume ignition of releases. In reality, not all releases result in a fire. The likelihood of ignition can be addressed in the quantitative risk assessment process. However, in an FHA it is important to identify if ignition sources are present for the fire scenarios to occur. In some instances, fire scenarios can be eliminated from analysis because of the lack of a credible ignition source.

## 5.5. Calculate Potential Fire Hazard

Fire hazard calculation techniques for combustible and flammable liquids and gases range from the basic rule-of-thumb to the sophisticated, including computer modeling techniques. A relatively simplistic approach is adopted for this FHA framework in recognition of the uncertainty of other inputs to the FHA (e.g., leak sizes, orientations, ignition delays, and total volume of discharge).

### *5.5.1. Ignition and Combustion*

The combustion process is the rapid reaction of a vapor with oxygen. Combustion of a material will only be initiated when a flammable mixture has been formed, that is, the concentration of vapor in air is within the flammable limits of that material. Most solids and many liquids that are initially at ambient temperature must be heated to produce sufficient vapor to form a flammable mixture.

Ignition is the process of initiating self-sustaining combustion. A sufficient amount of energy (minimum ignition energy) must be applied to the mixture to transition it from a stable state to the unstable, dynamic state known as combustion. When this energy is supplied by an external source, such as a flame or spark, it is referred to as "piloted-ignition." Ignition can also be initiated in a flammable mixture when its bulk temperature is progressively raised. The temperature at which this occurs is referred to as the autoignition temperature and is a unique property of each flammable gas and liquid.

The mass burning rate of a fuel is a key factor in the correlations that have been developed for calculating the energy released from a fire. Mass burning rates for some materials are provided in Appendix B.

### 5.5.2. Heat Transfer

The ignition, burning, and extinguishment of fires can be explained in terms of classical heat transfer principles. Heat is transferred by one or more of three mechanisms:

- *Conduction*—transfer through a solid
- *Convection*—heat transfer in a fluid
- *Radiation*—heat transfer by electromagnetic radiation. Radiation does not require a transfer medium

Heat transfer is always from hot to cold. The main mechanisms of heat transfer in a hydrocarbon are thermal radiation and direct flame contact (convection). Heat transfer to personnel can cause burns. Heat transfer to equipment and structures can lead to failure of hydrocarbon-containing equipment that can further feed the fire.

Detailed discussion of conduction, convection, and radiation is contained in Appendix B.

### 5.5.3. Fire Growth and Heat Release

The growth and steady-state burning of fuels is dependent on the mass-burning rate. For flames in air, heat release rate is a relatively straightforward calculation based directly on the rate of supply of gas. Fires involving liquids all exhibit rapid-fire growth in open air. Once the ignition is sustained, even high flash point pool fires exhibit relatively rapid fire growth. This is shown conceptually in Figure 5-5. The duration of a fire is a function of the quantity of fuel and the

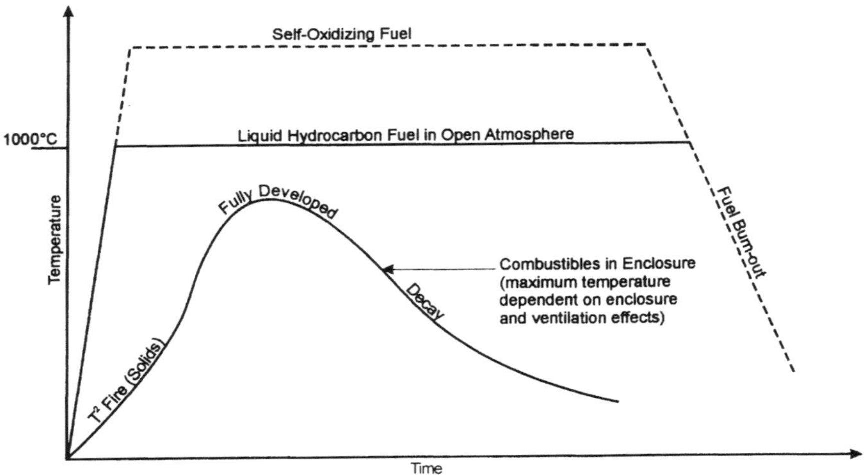


**Figure 5-5.** Generalized fire growth characteristics for solid, liquid, and self oxidizing

mass-burning rate. Except for unusual circumstances (e.g., restriction of air supply due to structural collapse of process equipment), an outdoor fire will burn at or near the peak heat release for the duration of the fuel supply.

### 5.5.4. Solid Materials

Solid materials include carbon containing polymers, electrical equipment, cable insulation, and cellulosic materials. It has been shown through experimentation that the rate of heat release from a burning material is strongly related to the heat of combustion and heat required for vaporization. These vary widely for different materials. The heat required for vaporization is considerably higher for solids than for liquids. Characterizing the burning of cellulosic materials, particularly wood, is further complicated by the nature of the material and by the formation of char when combustion occurs. Experimental results can be used for estimating the burning rates of solid combustibles in open air.

### 5.5.5. Enclosure Effects

Enclosed fires may exhibit fire growth characteristics as shown in Figure 5-5. Unlike gaseous or liquid fuels, there may be a considerable fire growth period in which temperatures and overall heat release is low and the fire is localized. As the fire becomes fully developed, the entire room volume can become engulfed in flames. Finally, as air is depleted or fuel is consumed, a decay period occurs. In many cases, an enclosure fire will be starved for air ("ventilation-limited"), and the available airflow becomes the limiting factor for the fuel-burning rate.

Limited by oxygen in air, complete combustion may not occur in the enclosure even though fuel is vaporized. This unburned fuel may ignite once air is available, for example at a door or window opening. As a result, large volumes of flames may discharge from enclosure openings.

Techniques are available to calculate conditions under which enclosed fires are ventilation- or fuel- controlled. Computer models are available to estimate compartment fire growth and temperature effects. In particular, the zone fire model C-FAST (Jones et al., 2000) is widely used. Additional information on models is contained in Appendix C.

## 5.6. Flash Fires

A flash fire occurs when a vapor cloud of flammable material burns. The cloud is typically ignited on the edge and burns toward the release point. The duration of a flash fire is very short (seconds), but it may continue as a jet fire if the release continues. The combustion of an unconfined vapor cloud will generate overpressures [0.1 to 0.3 psi (0.7–2.1 kPa)] that are not considered significant in terms of damage potential to persons, equipment, or structures. The major hazard from a flash fire is direct flame impingement. Typically, the burn zone is estimated by performing dispersion modeling where the burn zone is defined as the area the vapor cloud covers out to one-half of the LFL. The use of one-half of the LFL provides a conservative estimate, allowing for fluctuations in modeling. Other criteria can also be used. Even where the concentration may be above the UFL, turbulent induced combustion mixes the material with air and results in a flash fire.

In order to compute the thermal radiation effects produced by a flash fire, it is necessary to know the flame temperature, size, and dynamics during the propagation. Since flash fires are of very short duration, the effect of radiation is much less than from a jet or pool fire.

## 5.7. Fireballs

A fireball is an intense spherical fire resulting from a sudden release of pressurized liquid or gas that is immediately ignited. The best-known cause of a fireball is a boiling liquid expanding vapor explosion (BLEVE). Fireball durations are typically 5–20 seconds, thus the heat loads are unlikely to damage any equipment. However, the impact damage from fragments of the BLEVE can be substantial. No model of this effect is available; a simple assumption is made that the damage effects may cover the same area as the fireball. Personnel may be injured by short duration, high intensity fireballs as described in Section 5.11.1.

*Fireball diameter* is based on correlations with observed fireball size in experiments and actual accidents. The most recent correlation (Spouge, 1999) is:

$$D = 5.8M^{0.333} \tag{5-1}$$

where

$D$ is fireball diameter (m)
$M$ is mass released (kg)

*Fireball duration* is based on a mixture of small-scale experiments and dimensional analysis (Spouge, 1999):

$$\begin{aligned} t &= 0.45M^{0.333} \quad \text{for } M < 37{,}000 \text{ kg} \\ t &= 2.6M^{0.167} \quad \text{for } M > 37{,}000 \text{ kg} \end{aligned} \tag{5-2}$$

where

$t$ is fireball duration (sec)
$M$ is mass released (kg)

The fireball is assumed to be constant over this duration, although in reality some growth and decay may occur. Films taken of actual BLEVEs indicate that a fireball rises. This is not addressed in most models; in other words, the fireball is assumed to be a sphere resting on the ground.

The surface emissive power of a fireball is usually assumed to be in the range 150–300 kW/m$^2$. Values for LPG of 270 kW/m$^2$ for releases below 125 tons and 200 kW/m$^2$ for larger releases have been used in the UK. Experimental measurements of average surface emissive power of butane fireballs have been reported as 300–350 kW/m$^2$ (Spouge, 1999). Uncertainty in determining the surface emissive power accounts for much of the inaccuracy in fireball modeling.

In all cases, the fraction of heat energy radiated, $F_r$, lies between 0.2 and 0.4. For more detailed modeling, a correlation giving $F_r$ based on initial vapor pressure of the fluid at ambient temperature (Roberts, 1982) can be used:

$$F_r = 0.27P^{0.32} \tag{5-3}$$

where

$F_r$ is fraction of heat energy radiated

$P$ is vapor pressure of fluid (MPa abs)

Roberts makes clear that $P$ is intended to be the vapor pressure when the failure occurs. In a BLEVE, this might be the relief valve setting $P_0$, whereas in a fireball resulting from an impact failure, it will be the vapor pressure at ambient temperature, as is used in FLARE (described in Appendix C). For a fireball following a release of gas (as opposed to liquefied gas), $P$ should be the storage pressure.

Since the fireball surface is similar to that of jet fires, a simple approach for estimating the distances to various thermal radiation levels would be to determine them as functions of the fireball radius, refer to Table 5-4. Note that the difference between exposures to fireballs and jet flames is that the short term nature of the exposure is such that a "dosage" approach is more appropriate than the steady state flux approach used for jet flames.

## 5.8. Liquid or Pool Fires

Pool fires begin with the release of a flammable material from process equipment or storage. If the material is liquid, stored at a temperature below its normal boiling point, the liquid will collect in a pool. The geometry of the pool will be dictated by the surroundings. If the liquid is stored under pressure above its normal boiling point, then a fraction of the liquid will flash into vapor, with the unflashed liquid remaining to form a pool in the vicinity of the release.

To determine the impact of a pool fire on adjacent equipment, a series of calculations are required as shown in Figure 5-6.

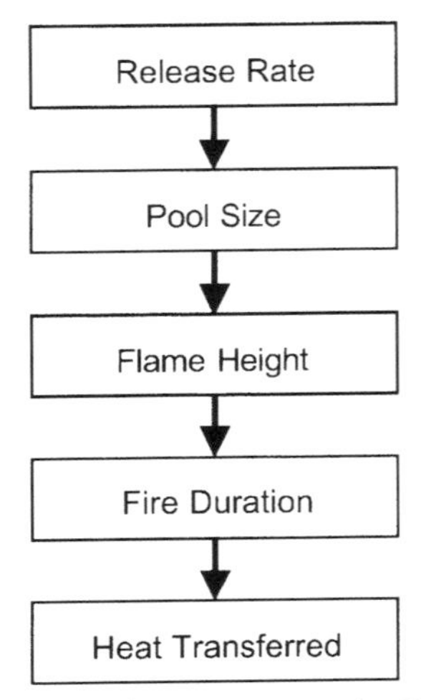


**Figure 5.6.** Evaluation Process for Pool Fire

### 5.8.1. *Release Rate*

Release rate is required to determine the size of the pool and fire duration. The release rate can be calculated from other CCPS books such as *Guidelines for Consequence Analysis of Chemical Releases* (CCPS, 1999), or can be defined as small, medium, or large as shown in Table 5-1.

### 5.8.2. *Pool Size*

Once sustained combustion is achieved, liquid fires quickly reach steady-state burning with a near constant mass-burning rate. As such, the heat release rate for the fire becomes a function of the liquid surface area exposed to air.

A liquid fuel spill may either be confined or unconfined. A confined spill is limited by physical boundaries (e.g., a diked area) and results in a pool of liquid with a depth that is greater than would be obtained if the fuel spilled unconfined. An unconfined spill will tend to have thin fuel depths (typically less than 5 mm), which will result in slower burning rates of the fuel.

The spill area, $A_s$, for a confined pool fire is defined by the physical boundaries. For an unconfined spill of greater than 95 l that has reached a static, maximum size, the following rules can be used to estimate the area:

$$A_s = 360 \cdot V \tag{5-4}$$

where

$V$ is the volume of fuel spilled ($m^3$)

$A_s$ is the area of liquid spill ($m^2$).

The above equation corresponds to a spill depth of 2.8 mm.

The larger the area of the pool, the larger the fire, however the duration of the fire will be significantly decreased. The damage resulting from a fire is a function of time and distance. Thus, for any given spill volume, greater damage may result from a smaller but deeper pool fire because it will burn longer.

For an unconfined spill where fuel continues to flow after ignition, the pool fire will eventually reach a steady-state size. The steady-state pool diameter $D$ will be determined based on a balance between the volumetric flow rate of fuel and the volumetric burning rate of fuel. This may be expressed as:

$$D_{ss} = \left( \frac{4\dot{V}_{Leak} \cdot \rho}{\pi \cdot \dot{m}''} \right)^{1/2} \tag{5-5}$$

where

$D_{ss}$ is steady-state diameter of the fire (m)
$\dot{V}_{Leak}$ is volumetric flow rate ($m^3/s$)
$\dot{m}''$ is mass burning rate per unit area ($kg/m^2$ s)
$\rho$ is density of fuel ($kg/m^3$)

Equation (5-5) assumes that the burning rate is constant. More detailed pool burning geometry models are available (Spouge, 1999). Circular pools are normally assumed; where dikes lead to rectangular shapes, an equivalent diameter is used in the calculation.

### 5.8.3. Flame Height

When determining flame height, an assumption that the flame is a solid gray emitter with a well-defined cylindrical shape is made for ease of calculation. The cylinder may be straight or tilted as a result of wind. For pool fires, the flame height above the fire source can be determined by (Heskestad, 1981; 1983):

$$H = 0.23\dot{Q}^{2/5} - 1.02D \tag{5-6}$$

where

$H$ is flame height (m)
$\dot{Q}$ is heat release rate (kW)
$D$ is effective fire diameter (m) [see Equation (5-17)]

The effect of wind and tilt on flame geometry has been addressed in the *SPFE Handbook* (Beyler, 2002). The angle of tilt can be determined from the following equation and used to calculate the vertical and horizontal components of the flame length (see Figure 5-7):

$$\cos\theta = 1 \qquad \text{for } u^* \le 1$$

$$\cos\theta = 1/\sqrt{u^*} \qquad \text{for } u^* > 1$$

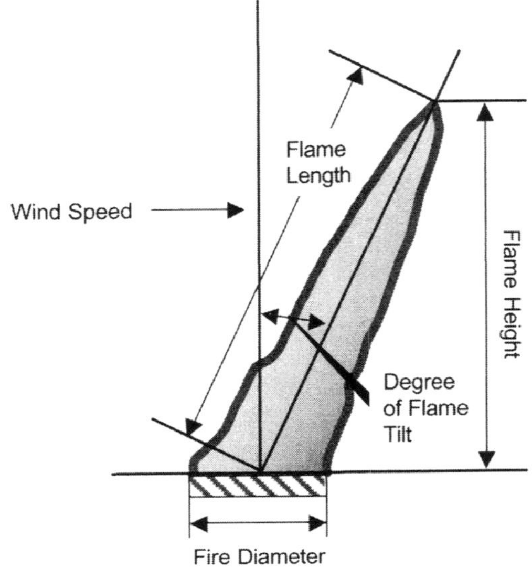


**Figure 5-7.** Flame Inclinations due to Wind (SFPE, 2002)

where $u^*$ is the nondimensional wind velocity given by

$$u^* = \frac{u_w}{(g\dot{m}''D / \rho_v)^{1/3}} \tag{5-7}$$

where

$D$ is effective fire diameter (m)
$u_w$ is the wind speed measured at a height of 1.6 m (m/s)
$g$ is the acceleration of gravity
$\dot{m}''$ is the mass burning rate (kg/m$^2$s)
$\rho_v$ is vapor density at the boiling point of the liquid (kg/m$^3$)

Additional discussion on the effect of wind and tilt of flame can be found in the *SPFE Handbook* (Beyler, 2002).

### *5.8.4. Duration of Burning Pools*

Once a mass-burning rate of the fire is established, the duration of the fire (burn time) can be calculated as:

$$t_b = \frac{m_f}{\dot{m}''A} = \frac{V\rho}{\dot{m}''A} \tag{5-8}$$

where

$t_b$ is burn time (s)
$m_f$ is mass of fuel spilled or contained in the pool (kg)

$V$ is volume of fuel ($m^3$)
$\dot{m}''$ is mass burning rate per unit area ($kg/m^2$ s)
$A$ is fire area ($m^2$)
$\rho$ is density ($kg/m^3$)

Storage tanks can be treated as a confined pool fire. For confined pools that have a significant level of material, Table 5-2 shows the burning rate in inches per hour for a variety of materials. When first ignited, the fire spreads rapidly across the full extent of the hydrocarbon pool and proceeds to consume the liquid at a characteristic burning rate (Spouge, 1999).

Burning rates for hydrocarbon fuels on land are given in Table 5-3 (Mudan & Croce, 1988). More extensive data were developed by Shell Research (Spouge, 1999).

**Table 5-2**
*Burning Rates of Ordinary Petroleum Fuel Liquids*

| | Burning Rate | |
|---|---|---|
| **Liquid Fuel Type** | **Depth Burned Per Hour (in)** | **Depth Burned Per Hour (cm)** |
| Motor gasoline | 6–12 | 15–30 |
| Aviation gasoline | 10–12 | 25–30 |
| Kerosene (No. 1, FO) | 5–8 | 13–20 |
| Diesel oil | 6–9 | 15–23 |
| Fuel oil (Heavy, No. 5) | 5–7 | 13–18 |

**Table 5-3**
*Hydrocarbon Burning Rates on Land*

| **Material** | **Mass Burning Rate ($kg/m^2s$)** | **Burning Velocity (mm/s)** |
|---|---|---|
| Gasoline | 0.05 | 0.07 |
| Kerosene | 0.06 | 0.07 |
| Hexane | 0.08 | 0.11 |
| Butane | 0.08 | 0.13 |
| LNG | 0.09 | 0.20 |
| LPG | 0.11 | 0.21 |

## 5.8.5. Heat Transfer

### 5.8.5.1. Flame Temperature

The buoyant gas flow above the fire, including any flames, is typically referred to as the fire plume (see Figure 5-8). Flame temperatures typically range from 900°C to 1200°C, and will vary with the type of fuel, ambient conditions, and oxygen availability. Temperature variations result from the amount of soot particles within the flame (which absorb energy and allow for convective or radiative heat transfer) (Drysdale, 1998). In general, the sootier the flame, the cooler its temperature.

## 5.8.6. Convective Heat Transfer above the Plume

In well-developed fires, the convective heat fraction is typically measured at more than about 65% of the total heat release rate (Heskestad, 2002). This heat is carried away by the plume above the flames. Prediction of plume velocity and temperatures above the flames serve as the basis for convective heat transfer calculations where overhead equipment exists. Widely used fire plume theory assumes a point source origin, and uniformity throughout the plume relative to air density, air entrainment, velocity profile, and buoyancy.

The temperature above the flaming portion of the fire plume can be estimated using the following:

$$T_{CL} = T_{\infty} + 19.7Q^{2/3}(z - z_0)^{-5/3} \tag{5-9}$$

where

$T_{CL}$ is center line gas temperature (K) above the flaming region at a height, z

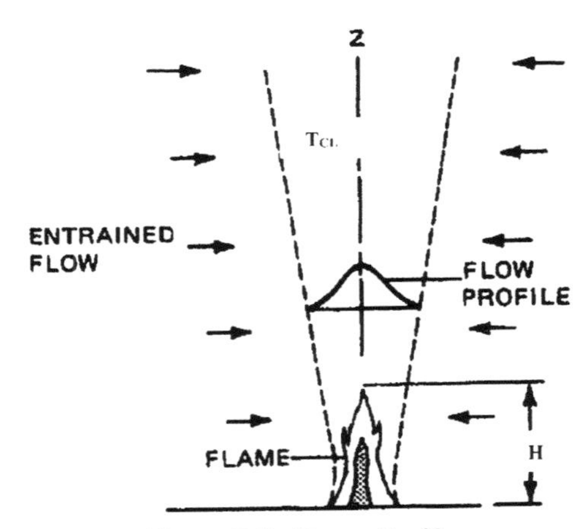


**Figure 5-8.** Plume Profile

$T_\infty$ is ambient temperature (K)
$Q$ is heat release rate (kW)
$z$ is height above the top of the combustible (m) at which $T_{CL}$ is measured
$z_0$ is height of virtual origin above or below the top of the combustible (m)

The virtual origin is calculated as:

$$z_0 = -1.02 + 0.083\left(\frac{\dot{Q}^{2/5}}{D}\right) \tag{5-10}$$

where

$z_0$ is virtual location of the point source origin
$\dot{Q}$ is heat release rate (kW)
$D$ is diameter of fire (m)

### 5.8.6.1. *Flame Radiation to External Targets*

Several methods have been described for prediction of radiation from pool fires (SFPE, 1999). The primary methods are based on correlations developed from experimental data.

Shokri and Beyler correlated experimental data of flame radiation to external targets in terms of an average effective emissive power of the flame (Shokri and Beyler, 1989). The flame is assumed to be a cylindrical, black body radiator with an average emissive power, diameter ($D$), and height ($H_f$), see Figure 5-9.

The radiant heat flux to a target ($q''$) relative to a source outside the flame is expressed as:

$$q'' = EF_{12} \tag{5-11}$$

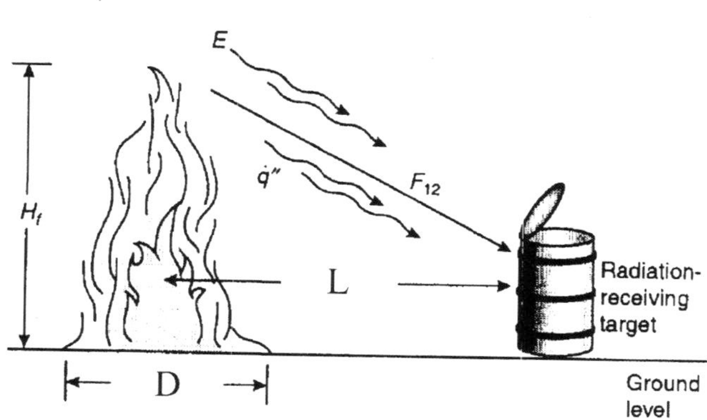


**Figure 5-9.** Shokri and Beyler Representation

where

$q''$ is radiant heat flux ($kW/m^2$)

$E$ is emissive power ($kW/m^2$)

$F_{12}$ is radiation view (configuration) factor between the target and the flame ($0 < F_{12} < 1$)

$E$ may be expressed in terms of the diameter ($D$) of the fire:

$$E = 58(10^{-0.00823D}) \quad (kW/m^2, m) \tag{5-12}$$

This emissive power is assumed to be over the whole flame surface area, and is significantly less than the emissive powers that can be calculated from point source measurements. Increasing the pool diameter reduces the emissive power due to the increasing black smoke outside the flame and the resulting obscuration effect on the luminous flame.

The calculation of the view factor ($F_{12}$) from basic principles would involve lengthy algebraic equations. Figures 5-10 through 5-14 (SFPE, 2002) provide precalculated view factors based on several configuration and aspect ratios of the fire and target location. The view factor ($F_{12}$) can be determined using the flame dimensions, distance to the target ($L$), and height of the target.

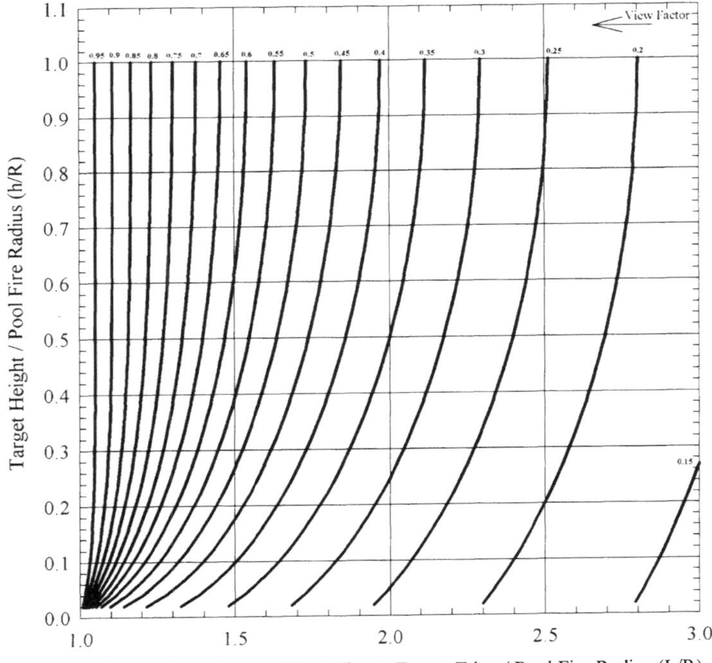


**Figure 5-10.** Maximum Configuration Factor for a Flame Height to Pool Fire Radius Ratio $H_f/R_p=2$

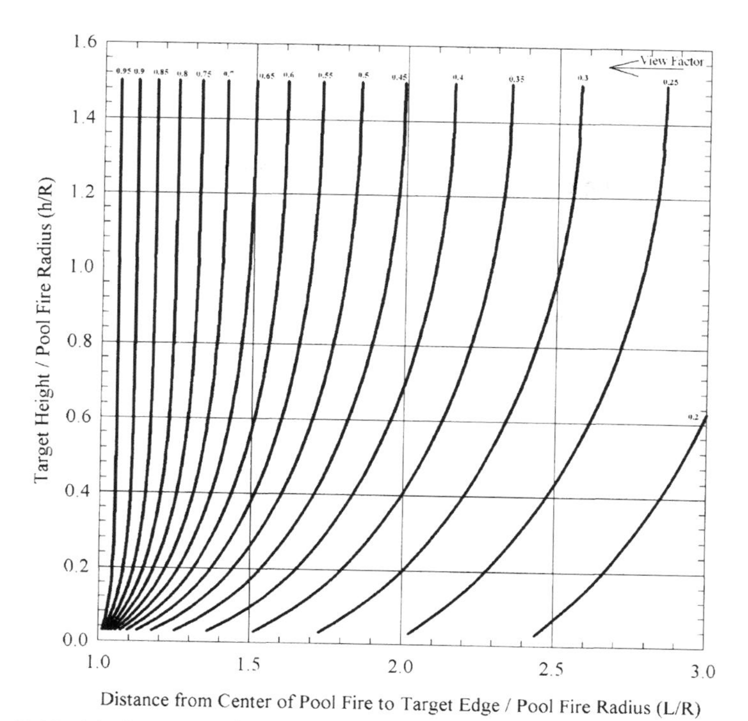


**Figure 5-11.** Maximum Configuration Factor for a Flame Height to Pool Fire Radius Ratio $H_f/R_p=3$

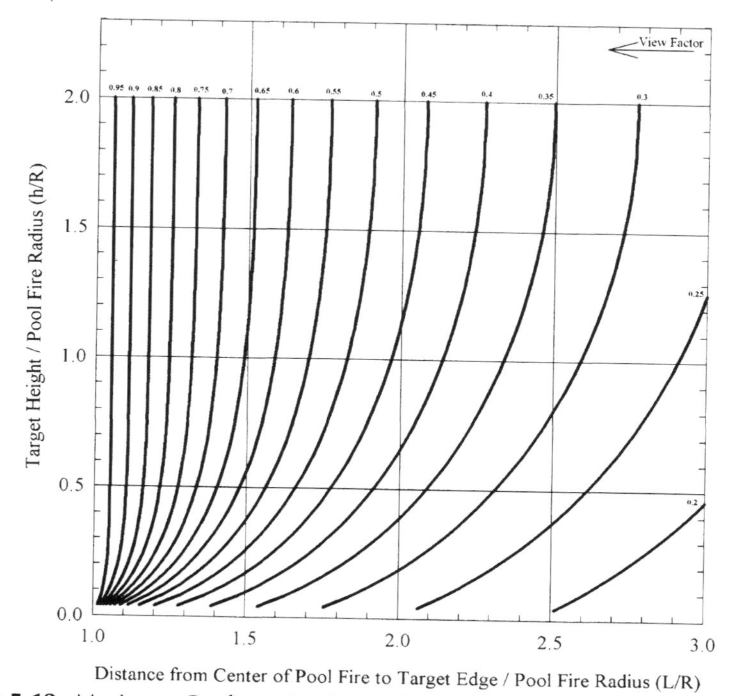


**Figure 5-12.** Maximum Configuration Factor for a Flame Height to Pool Fire Radius Ratio $H_f/R_p=4$

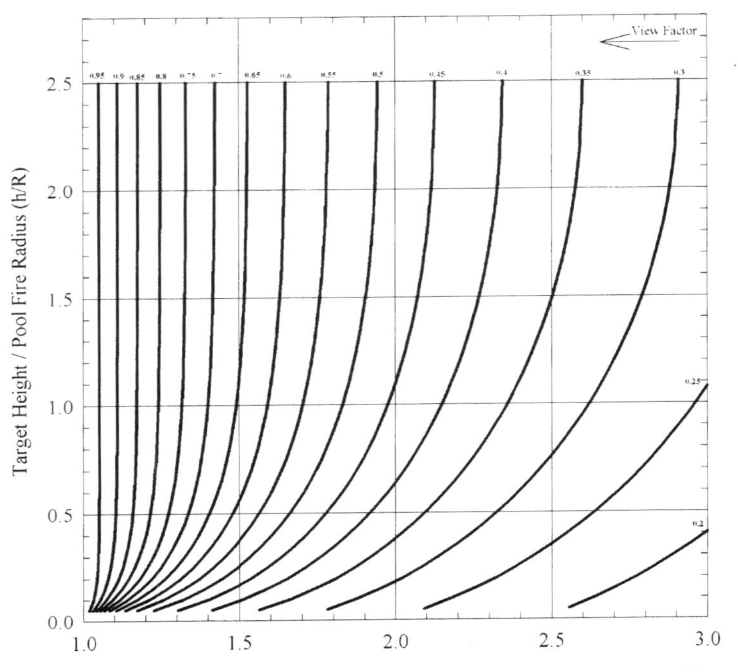


**Figure 5-13.** Maximum Configuration Factor for a Flame Height to Pool Fire Radius Ratio $H_f/R_p=5$

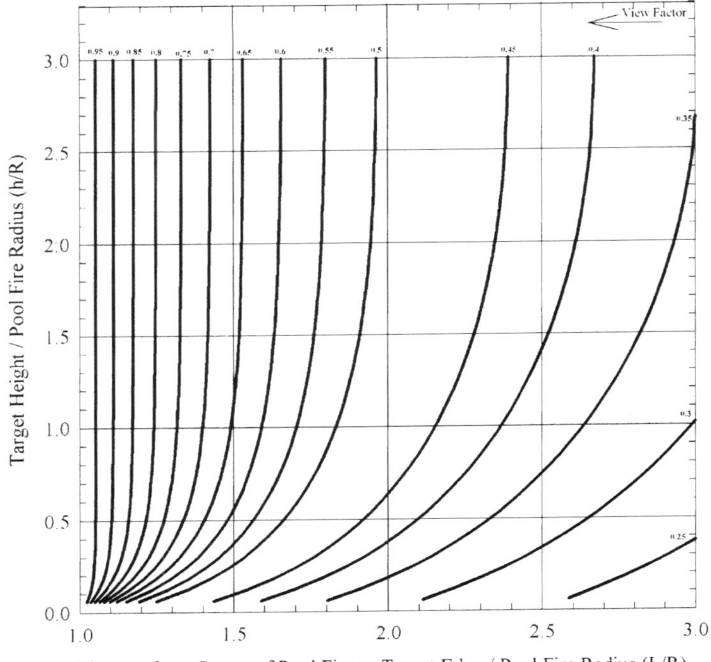


**Figure 5-14.** Maximum Configuration Factor for a Flame Height to Pool Fire Radius Ratio $H_f/R_p=6$

Utilizing the Shokri and Beyler method to estimate the incident flux on a target involves the following steps:

1. Calculate the heat release rate of the fire, $\dot{Q}$.
2. Determine the diameter and flame height of the fire (if the fire is noncircular, use the effective diameter). The flame height is calculated using Equation (5-6).
3. Determine the view (configuration) factor (use Figures 5-10–5-14).
4. Calculate the emissive power of the flame using Equation (5-12).
5. Calculate the incident heat flux to the target using Equation (5-11).

## 5.9. Gas and Jet Fires

Jet flames occur in the processing industry as accidental gas releases or intentional release and burning of gas at flares. Jet fires are typically characterized as high pressure releases of gas from limited size openings. High pressure and small openings (e.g., due to a small leak in a vessel or broken drain valve) can lead to very high gas velocities, approaching sonic.

The fire hazards associated with jet fires have been developed in greater detail in other references (Beyler, 2002; SINTEF, 1997). This section provides an overview of assessing a jet fire hazard based on the most straightforward approach. The primary focus is to first quantify the physical size of the resulting fire, its heat release rate, and then to determine the heat transfer from the jet to adjacent targets. The heat release rate from a jet fire is governed by the mass flow rate of the fuel being released. The discharge of fuel is dependent on the nature of the release (e.g., the pressure, size, shape of the discharge/nozzle opening and properties of the fuel). Consequently, the specific scenario will determine the flow rate of fuel.

A series of calculations are required to determine the impact from a jet fire. These steps are shown in Figure 5-15.

### *5.9.1. Estimating Discharge Rates*

The initial release rate of hydrocarbon gas through a hole to the atmosphere depends on the pressure inside the equipment, the hole shape/size, and the molecular weight of the gas.

For a small hole in the containment, there are two possible release conditions:

- Adiabatic if the pressure drop across the orifice is large.
- Isothermal if the pressure drop is small.

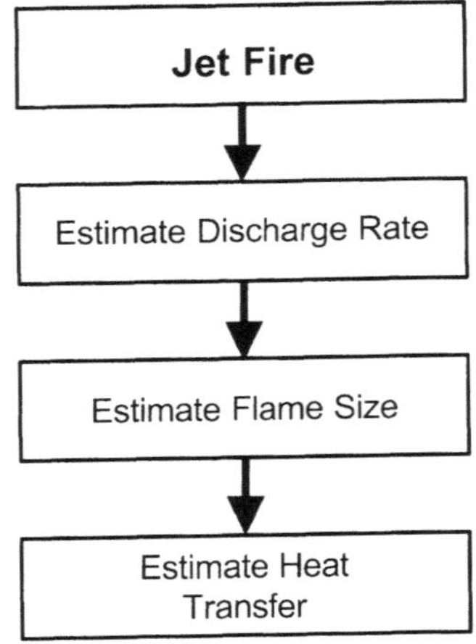


**Figure 5-15.** Steps in Jet Fire Calculations

The adiabatic case is the most common for accident conditions. The process is treated as an isentropic free expansion of an ideal gas using the equation of state:

$$Pv^k = \text{constant} \tag{5-13}$$

where

- $P$ is pressure
- $v$ is the specific volume of the gas
- $k$ is the isentropic expansion factor which is equal to $\gamma$, the ratio of specific heats for pure isentropy; but in practice pure isentropy is not achieved, hence $k$ is less than $\gamma$

Equation (5-14) is combined with Bernoulli's equation. Assuming flow on a horizontal axis and using a coefficient of discharge to account for friction at the orifice, the mass flow rate of an ideal gas through a thin hole in the containment wall is:

$$M = C_d \times \rho_{\text{ambient}} \times A_h \sqrt{\frac{2 \times P_{\text{process}}}{\rho_{\text{process}}} \times \frac{k}{k-1} \times \left[1 - \left(\frac{P_{\text{ambient}}}{P_{\text{process}}}\right)^{(k-1)/k}\right]} \tag{5-14}$$

where

- $M$ is mass flow rate (kg/s)
- $P$ is pressure (Pa)
- $C_d$ is coefficient of discharge , typically 0.85 for gas releases
- $A_h$ is area of hole ($m^2$)
- $\rho$ is density of the gas ($kg/m^3$)

If the pressure ratio is above a critical value given below, the exiting mass flow is limited to a critical maximum value. This is sonic or choked flow:

$$\left(\frac{P_{\text{ambient}}}{P_{\text{process}}}\right)_{\text{critical}} = \left(\frac{2}{k+1}\right)^{k/(k-1)} \tag{5-15}$$

and

$$M_{\text{max}} = C_{\text{d}} \times A_{\text{h}} \times \sqrt{P_{\text{process}} \times \rho_{\text{process}} \times k \times \left(\frac{2}{k+1}\right)^{(k+1)/(k-1)}} \tag{5-16}$$

### 5.9.2. *Jet Flame Size*

Once the fuel flow rate is determined, the heat release rate is calculated as:

$$\dot{Q} = \dot{m} \cdot \Delta h_{\text{c}} = \dot{m}'' A \Delta h_{\text{c}} \tag{5-17}$$

where

$\dot{Q}$ is heat release rate (kW)
$\Delta h_c$ is heat of combustion (kJ/kg)
$\dot{m}$ is mass flow rate (kg/s)
$\dot{m}''$ is the mass burning rate (kg/m²s)
$A$ is area (m²)

In relatively still air, the flame length, $L$ (m), of most jet flames can be estimated as:

$$L = 0.2 \cdot \dot{Q}^{2/5} \tag{5-18}$$

The base of the jet flame is usually not attached to the release point, due to the high velocity and richness of the fuel near the heat source. This lift-off distance has been measured on flares to be 20% of jet length. This effect is important in reducing the predicted radiation level on the leak source, which might otherwise cause a small leak to escalate to a full-bore failure.

A jet flame in the presence of a crosswind has been addressed (Brzustowski et al., 1975a; Gollahalli et al., 1975; Kalghatki, 1983). Generally, the flame can be considered as the frustum of a cone, with increasing flame diameters further from the nozzle. Calculation methods and the effects of wind on the jet flame diameter and lift-off distance are presented in the literature (Beyler, 2002; SINTEF, 1997).

When a jet fire impinges on an object, its shape may be very distorted compared to the free-field shapes modeled. If the jet fire impinges perpendicularly on a flat object such as a fire-wall or deck, it will produce a thin circular flame over the object's surface.

Large jet fires inside congested units may impinge on many items of process equipment simultaneously, producing a highly distorted flame shape. A modified fireball model may then give a better representation than a free-field jet fire model.

For a jet fire close to obstacles but not impinging on them, the flame shape will be similar to free-field predictions, but the radiation field will be distorted. A simple judgmental representation of the effects of obstacles on the radiation zones is described below:

- *Solid obstacles* such as walls on enclosed spaces are considered impervious to thermal radiation, and are assumed not to affect the radiation zones outside them.
- *Partial obstacles* such as process equipment or decks on open-sided modules are approximated by reducing the radiation zones as follows:

| | | |
|---|---|---|
| 37.5 kW/m$^2$ | becomes | 12.5 kW/m$^2$ |
| 12.5 kW/m$^2$ | becomes | 5 kW/m$^2$ |
| 5 kW/m$^2$ | becomes | 2 kW/m$^2$ |

These reductions are only applied more than 10 m from the edge of the obstructed zone.

When a jet fire has decayed to a pressure of 10 psig the fire is assumed to have effectively ceased. This pressure is close to the transition pressure from sonic to subsonic flow. When a jet fire event has decayed to this level, its magnitude and exposure potential are considered to have reached a threshold level below which no significant further damage can occur (i.e., no escalation potential) and active fire fighting measures can effectively bring the fire under control.

### 5.9.3. Heat Transfer

The point source model assumes that the fire can be represented as a point that is radiating to a target at a distance, $R$, from the point. The model is most appropriate for calculating incident heat fluxes to targets where fluxes are in the range from 0 to 5 kW/m$^2$ (SFPE, 1999). The point source model has been shown to be accurate for calculating the incident heat flux from a jet flame to a target outside the flame (Beyler, 2002). The literature contains more refined line or cylinder models (Beyler, 2002; SINTEF, 1997).

To calculate the incident heat flux per unit surface area on a target, $q''$, at a distance $R$ from the point source, the following equations apply:

$$q'' = \frac{Q_r \cos\theta}{4\pi R^2} \quad (5\text{-}19)$$

where

$$Q_r = x_r Q \quad (5\text{-}20)$$

where

$Q$ is heat release rate (kW)
$Q_r$ is the heat radiated from the fire (kW)
$R$ is distance from point source to target (m)
$x_r$ is the radiative fraction
$q''$ is the incident heat flux per unit surface area of a target (kW/m$^2$)

The radiative fraction is the fraction of heat that is radiated from the fire plume (SFPE, 1999); it can be estimated using:

$$x_r = 0.21 - 0.0034D \tag{5-21}$$

where

$x_r$ is the radiative fraction
$D$ is the diameter of the fire (m)

For pool fires greater than 50 m, the radiative fraction for 50 m should be used. As the fire diameter increases, the radiative fraction decreases.

The radiative fraction, $x_r$, generally ranges from 0.2 to 0.4. This range reflects differences in fuel properties, with values of 0.2 for hydrocarbon fuels with one carbon atom (e.g., $C_1$ for methane) to values of 0.4 for hydrocarbons with five or more carbon atoms (Brzustowski, 1971). Fuels such as propane and butane ($C_3$) have reported radiative fractions of 0.3.

Data by McCaffrey (1995) show that the radiative fraction decreases for jet flames. Based on laboratory-scale tests, as the jet velocity increases the radiative fraction decreases. Although the aerodynamics of the flow have a significant effect on the radiation from the flame, the experimental data show that a radiative fraction of 0.4 represents a conservative maximum value for most fuels and conditions. If a more detailed or less conservative analysis is desired, Beyler (2002) presents experimental data for radiative fractions for the conditions discussed above.

Utilizing the point source model to estimate the incident flux on a target involves the following steps:

1. Calculate the heat release rate of the fire, $Q$
2. Determine the diameter of the fire (if the fire is noncircular, use the effective diameter).
3. Determine the location of the point source, centered over the fire at a height equal to one half of the flame height [Equation (5-6)]. For jet fires, use Equation (5-18).
4. Calculate the distance between the point source and the target, $R$.
5. Determine the radiative output from the fire, Equations (5-20) and (5-21).
6. Calculate the incident heat flux from Equation (5-19).

The impingement of a flame jet on a target provides high heat fluxes, more than that which would occur from a pool fire. The high heat fluxes are a result of the combination of large convective and radiative components. Maximum heat fluxes of 250 to 300 kW/m$^2$ have been measured for large-scale jet flames (Cowley and Prichard, 1990) with fluxes averaged over the impingement area of 200 kW/m$^2$ for sonic natural gas jets and 150 kW/m$^2$ for two-phase LPG jets.

To calculate the radiant heat contours for gas or liquid jet fire, approximated multiplying factors have been developed as illustrated in Table 5-4 (Spouge, 1999). The factors do not account for the affects of objects on radiant heat. The multiplying factor for each heat contour is applied to the flame length.

### 5.9.4. Radiative Exposure

A simple, conservative technique for equating the imposed heat flux that will cause a failure is to calculate the heat transfer. This approach assumes no thermal gradients within the target material (i.e., the target is at a uniform temperature—it is the same throughout the material but can change with time). The incident heat flux from a fire onto the surface of a target will increase the surface temperature of the target until steady-state conditions are achieved. Steady-state conditions will occur when the temperature of the target reaches a level at which the incident heat flux equals the heat losses from the target.

The energy balance for the target is approximated as:

$$q'' = q''_{rerad} + q''_{conv} = \sigma(T_s^4 - T_\infty^4) + h(T_s - T_a) \quad (5\text{-}22)$$

where

$q''$ is incident heat flux to the target (kW/m$^2$)
$q''_{rerad}$ is heat flux reradiated from the target to the surroundings (kW/m$^2$)
$q''_{conv}$ is heat flux convected from the target to the surroundings (kW/m$^2$)
$\sigma$ is Stefan-Boltzmann constant ($5.67 \times 10^{-11}$ kW/m$^2$K$^4$)
$T_s$ is surface temperature of the target (K)
$T_\infty$ is temperature of the surroundings (K)
$T_a$ is ambient temperature of the air around the target (K)
$h$ is heat transfer coefficient (0.015 kW/m$^2$K can be used as a first estimate)

**Table 5-4**
*Estimated Distance to Radiation Level*

| Radiation Level (kW/m$^2$) | Multiplying Factor |
|---|---|
| 250 | 1.0 (objects impinged by flame) |
| 37.5 | 1.2 |
| 12.5 | 1.45 |
| 4.7 | 1.75 |

This energy balance also assumes that the absorbtivity and emissivity of the target surface are nominally equal and for simplicity are considered as one. The view factor associated with the re-radiation from the target to the surroundings has been assumed to be one for simplicity. This assumption will lead to higher estimated temperatures, particularly as the fire encompasses more of the field of view of the target.

The heat transfer coefficient for convective cooling is dependent on the configuration of the target and the orientation and velocity of the air around the target. Details on calculating heat transfer coefficients can be found in most heat transfer textbooks (Incropera and DeWitt, 1995). A value of 0.015 $kW/m^2K$ can be used as a reasonable estimate.

Solving Equation (5-22) for the surface temperature, $T_s$, is most easily done with mathematical software, which can iteratively solve the nonlinear equation. Alternately, the temperature can be iteratively solved via the following method:

1. Estimate a value for $T_s$.
2. Solve the right side of Equation (5-22) (call this value *RX*).
3. Compare *RX* to the incident heat flux, $q''$.
   - If *RX* is within an acceptable variance of $q''$ (e.g., 5%), then $T_s$ is solved.
   - If $RX > q''$ then guess a new $T_s$ (lower than the previous value) and repeat Step 2.
   - If $RX < q''$ then guess a new $T_s$ (greater than the previous value) and repeat Step 2.

In some cases, convective cooling of the target can be neglected. For example, at high surface temperatures, radiative cooling of the target will dominate, and the convective heat loss will be minimized. Assuming that the convective heat loss can be neglected, Equation (5-22) can be solved explicitly for $T_s$:

$$T_s = \left[\frac{q''}{\sigma} + T_\infty^4\right]^{1/4} \qquad (5\text{-}23)$$

Eliminating the convective heat transfer term will lead to higher estimates of the surface temperature. Given the various assumptions used in the simplified equations above [Equations (5-22) and (5-23)], Equation (5-23) will tend to provide the more conservative estimate of the surface temperature.

More sophisticated, computerized heat transfer calculation techniques are available where specific scenarios require detailed analysis (Idling et al., 1977; Paulsson, 1983). These techniques involve the use of multidimensional finite element programs.

## 5.10. Solid Fires

While solid fires generally do not have the same impact as flammable material fires in process units, the hazards from solids are important in several respects. Class A materials may be the source of ignition for hazards having a greater combustible loading or posing a greater threat in terms of impact and Class A or D solids may pose a threat due to inherent reactivity or use in a process. For more information on solid fires, refer to *SFPE Handbook* (Beyer, 2002). Radiant heat from solid fires can be calculated similarly to that of pool fires.

## 5.11. Fire Impact to Personnel, Structures, and Equipment

Once the fire characteristics are calculated, an assessment of the impact of the fire needs to be completed. Fires produce four major outputs: gases, flame, heat, and smoke. The materials involved in the fire will determine the combination of these four outputs. For example, crude oil will produce a very dark thick smoke cloud, and ethylene does not produce much smoke, but does have a very large flame.

The outputs can create consequences to personnel, structures, and equipment. There are different approaches for assessing consequences. Criteria can be developed that will allow the analyst to compare the results of the calculations to predetermined criteria. The results either meet or exceed the criteria. These criteria can be established based on a conservative approach of assuming a steady-state condition. Another approach is the use of more sophisticated heat transfer techniques, where a time-dependent onset of critical criteria is modeled.

Data and procedures presented in this section can be used in either approach. Time-independent approximations of failure criteria are presented to provide first-order estimates of fire consequences. Time-dependent criteria are also presented where specific scenarios warrant more detailed analysis. Most of the thermal criteria is presented in terms of heat flux, although some temperature criteria are also presented. A conservative methodology is presented to translate heat flux from a fire to surface temperature on a material target.

### *5.11.1. Impact to Personnel*

#### *5.11.1.1. Thermal Radiation*

When there is a line-of-sight between a person and the flame, the main impact is thermal radiation. The primary potential effects of thermal radiation are:

- Burns to exposed skin.
- Ignition or melting of clothing.

Burns are classified in increasing degrees of severity:

- *First degree*—superficial burns giving a red, dry skin (similar to mild sunburn).
- *Second degree*—burns more than 0.1 mm deep, affecting the epidermis and forming blisters.
- *Third degree*—burns more than 2 mm deep, affecting the dermis and nerve endings, resulting in a dry skin that has no feeling (major blistering).

The time to damage human skin increases logarithmically with the increase in skin temperature. Skin damage begins at about 45°C (113°F) and becomes virtually instantaneous at 72°C (162°F). Complex methods involving the use of thermal skin property are detailed in the literature (SFPE, 2000).

Figure 5-16 (SFPE, 2002) shows a method for predicting first and second degree burns based on heat flux and time. By knowing the exposure flux, one can predict the time to injury. No safety factor is included in this calculation; the estimate is for bare skin, unprotected by clothing.

Other personnel impact criteria are reported in the literature. Table 5-5 reports on data from several sources.

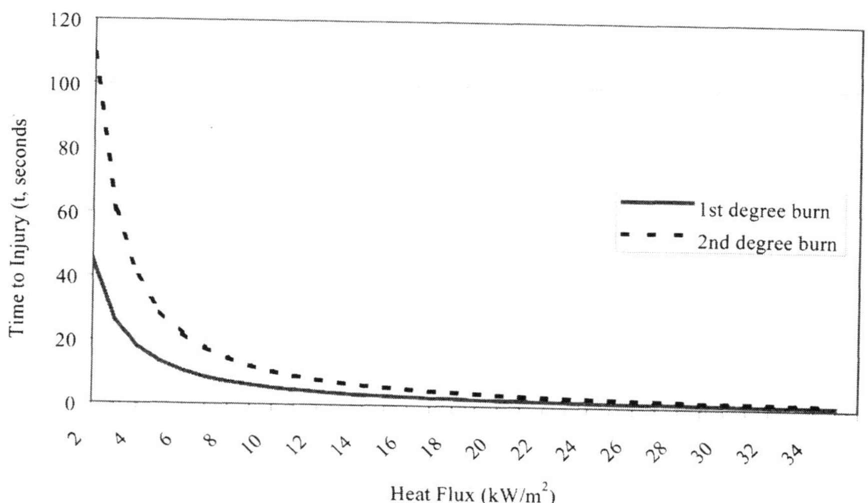


**Figure 5-16.** Prediction of First- and Second-Degree Burns

**Table 5-5**

*Estimated Effects of Heat on Personnel*

| Incident Flux ($kW/m^2$) | Impact |
|---|---|
| 37.5 | 100% lethality in 1 minute (Barry, 2002) |
| 25 | 1% lethality in 10 seconds (Barry, 2002) |
| 15.8 | 100% lethality in 1 minute, significant injury in 10 seconds (Barry, 2002) |
| 12.5 | 1% lethality in 1 minute; first degree burns in 10 seconds (Barry, 2002) |
| 6.3 | Emergency actions lasting a minute can be performed by personnel without shielding, but with appropriate clothing (API RP 521) |
| 4.7 | Emergency actions lasting several minutes can be performed by personnel without shielding, but with appropriate clothing (API RP 521) |

#### *5.11.1.2. Smoke and Gases*

Smoke is composed of combustion gases, soot (solid carbon particles), and unburnt fuel. For outdoor fires, the impact of smoke is usually a secondary consideration after the heat transfer. In many circumstances, the immediate thermal threat from the fire plume (jet, pool, or flash fire) overwhelms the smoke threat, particularly for personnel in close proximity to the event. There may be circumstances where personnel are in a downwind smoke plume where there is no immediate thermal threat. As a rule-of-thumb, all people within a smoke plume may be immediately or nearly immediately affected and at risk from a life safety standpoint (be it from lack of visibility or by toxic products).

Estimates of visibility in smoke plumes have been made and are on the order of less than 1 meter near the source (i.e., high smoke concentration). Generic smoke dilution factors for large plumes have also been estimated and are presented in the CMPT Handbook (Spouge, 1999). Outdoor smoke plume models can be used to estimate the specific areas of smoke involvement.

Where personnel take refuge in a building from an outdoor fire, smoke infiltration into a building may be a concern. A methodology is available to estimate this impact based on the concentration of the material present outside of the building, the building ventilation rate, and the time of exposure (SINTEF, 1997; Lees, 1996).

The component of the smoke that has the quickest impact to people is carbon monoxide (CO). Table 5-6 illustrates the impact of carbon monoxide concentrations on personnel (Spouge, 1999).

Smoke infiltration may result from natural leakage openings (e.g., around ventilation ducting/grills), open doors, or breached fire barriers. The integrity of

**Table 5-6**
*Effects of Carbon Monoxide*

| Carbon Monoxide Concentration (ppm) | Effects |
|---|---|
| 1500 | Headache in 15 minutes, collapse in 30 minutes. Death in 1 hour. |
| 2000 | Headache in 10 min, collapse in 20 minutes. Death in 45 minutes. |
| 3000 | Maximum "safe" exposure limit for 5 min. Danger of collapse in 10 minutes. |
| 6000 | Headache, dizziness in 1–2 minutes. Danger of death in 10–15 minutes. |
| 12,800 | Immediate effect, unconsciousness in 2–3 breaths. Danger of death in 1–3 minutes. |

closures will be a factor; gaps in dampers and stuffing materials around pipes/ducts are sources of natural leakage. Natural air leakage on the order of 0.01 to up to 1.0 air changes per hour have been estimated, based on the integrity of enclosure construction. Wind may also be a factor. Based on estimated or empirical air leakage rates, the rate of change of internal smoke conditions in an enclosure due to smoke infiltration may be estimated as suggested in the CMPT Handbook (Spouge, 1999):

$$\frac{dc_i}{d_t} = \frac{(c_0 - c_i)R}{3600} \qquad (5\text{-}24)$$

where

$c_i$ is internal concentration (ppm)
$c_0$ is outside concentration (ppm)
$R$ is air change rate (per hour)
$dc_i/d_t$ is rate of change of internal concentration (ppm per second)

### 5.11.2. Impact to Structures

Steel, aluminum, concrete, and other materials that form part of a process or building frame are subject to structural failure when exposed to fire. Bare metal elements are particularly susceptible to damage. A structural member undergoes any combination of three basic types of stress: compression, tension, and shear. The time to failure of the structural member will depend on the amount and type of heat flux (i.e., radiation, convection, or conduction), and the nature of the exposure (one-sided flame impingement, flame immersion, etc.). Cooling effects from suppression systems and effects of passive fire protection will reduce the impact.

When evaluating the impact to structures, standard fire test exposures can be utilized to determine the onset of critical temperatures or the impact mitigation strategies. Hydrocarbon fire time–temperature exposures have been developed to simulate the rapid temperature rise to approximately 2000°F experienced with liquid hydrocarbon fuel fires.

The new CCPS *Guidelines for Investigating Chemical Process Incidents* provides the following guidance: "Three recommended references on metallurgy aspects of investigations are: *Assessment of Fire and Explosion Damage to Chemical Plant Equipment/Analyzing Explosions and Pressure Vessel Ruptures*, by D. McIntyre, *Defects and Failures in Pressure Vessels and Piping,* by H. Thielsch, and *Understanding How Components Fail*, by D. Wulpi."

The following section from the book may also be of interest:

> Professional arson investigators have developed highly effective methods of deducing facts from a systematic study of burn, char, and melt patterns. Typical examples include:
>
> - Most woods will burn at a steady rate of 1.5 inches/hr (3.6 cm/hr).
> - Hydraulic fluids usually exhibit a consistent response of smoke color, flame color, autoignition temperature, and a whitish residue.
>
> Not all evidence is simple to diagnose. Steel weakens at approximately 1,100°F (575°C). Steel exposed to 1,500°F (816°C) for a short period can begin to fail and show the same degree of damage as steel exposed to a lower temperature for a longer period of time.
>
> Thus, a sag pattern can be a relatively reliable indicator that the steel was exposed to a temperature of at least 1,100°F (575°C), but, the maximum temperature above 1,100°F (575°C) cannot be accurately determined without additional evidence.

Also note the following temperatures of interest to process safety incident investigation teams (Perry and Green, 1997; NFPA 422M; NFPA 1997; and Avallone and Baumeister, 1996).

- Paint begins to soften 204°C (400°F)
- Zinc primer paint discolors to tan 232°C (450°F)
- Zinc primer discolors to brown 260°C (500°F)
- Normal paints discolor 310°C (600°F)
- Zinc primer paint scorches to black 371°C (700°F)
- Lube oil autoignites 421°C (790°F)
- Stainless steel begins to discolor 427-482°C (800-900°F)
- Plywood autoignites 482°C (900°F)
- Vinyl coating on wire autoignites 482°C (900°F)
- Rubber hoses autoignite 510°C (950°F)
- Aluminum alloys melt 610–660°C (1,125–1,215°F)

- Glass melts 750–850°C (1,400–1,600°F)
- Brass melts (instrument gauges) 900–1025°C (1,650–1,880°F)
- Copper melts 1,083°C (1,980°F)
- Cast iron melts 1,150–1,250°C (2,100–2,200°F)
- Carbon steel melts 1,520°C (2,760°F)
- Stainless steel melts 1,400-1,532°C (2,550-2,790°F)

ASTM E 1529 *Standard Test Methods for Determining Effects of Large Hydrocarbon Pool Fires on Structural Members and Assemblies* and Underwriters Laboratories Inc. 1709 *Standard for Rapid Rise Fire Tests of Protection Materials for Structural Steel* are two tests which are used to evaluate the performance of structures, equipment, and protective materials to hydrocarbon fires (see Figure 5-17).

Simple, conservative approaches are available to determine the possibility of structural failure. Several rules-of-thumb have been published and are summarized in Table 5-7 (Spouge, 1999).

Knowing the heat flux from a fire and temperatures, the time to structural failure can be estimated. A somewhat more detailed approach is to evaluate the heat transfer to the structural element and compare the resulting temperature to critical failure temperatures. Failure of a structural metal element occurs

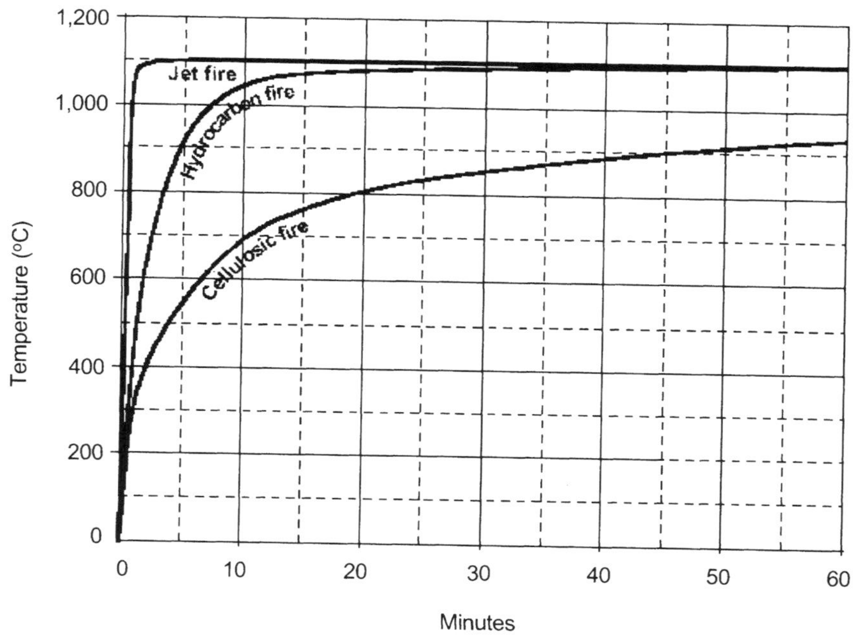


**Figure 5-17.** Time–Temperature Curve for Fire Tests

**Table 5-7**
*Estimated Failure Times of Steel Elements*

| | Time to Failure (minutes) | | |
|---|---|---|---|
| **Component** | **Impinging Jet Fire** | **Impinging Pool Fire** | **Nonimpinging Fire (37.5 kW/m² exposure)** |
| 25 mm steel pipe | 5 | 10 | No failure |
| 7 mm steel plate (flame impingement on one side only) | 2 | 4 | 13 |
| 305 mm web × 127 mm flange steel beam | 3–4 | 4 | 13 |

when the element is unable to support the imposed load. Elevated temperatures negatively affect the yield strength of metals. This varies with the type of metal and alloy. The yield strength for steel at 600°C (1,112°F) is reduced by about 50%; for aluminum, the 50% reduction of yield strength occurs at about 150°C (302°F). The failure is also a function of the designed factor of safety. The American Institute of Steel Construction (AISC, 1978) limits the maximum permissible design stress to 60% of the yield strength. At 538°C (1,000°F), the yield strength is about 60%. As such, typical failure criteria used for structural steel calculations is 500–550°C (932–1,022°F). A failure temperature of 150°C (302°F) for aluminum is a conservative criterion.

The time it takes for the steel member to reach this temperature depends on:

- *The degree of heating*—whether impingement from a jet flame or a cooler diffusive flame, or radiant or convective heating from outside the flame.
- *The nature of the exposure*—whether the member is engulfed or heated from one side only, and whether heated over its full length or over a small section.
- *The geometry of steel*—the mass of steel and the exposed surface area.
- *The mitigation (if any)*—from cooling water or of passive fire protection on steel.

The time to failure for a closed-in empty pipe at ambient pressure can be expressed (Spouge, 1999) as:

$$t_f = \frac{T_f - T_a}{qK - q_0} C_p \frac{M}{S} \tag{5-25}$$

where

$t_f$ is time to failure (sec)
$T_f$ is failure temperature (K)
$T_a$ is ambient temperature (typically 283 K)
$q$ is heat flux from flame ($kW/m^2$)
$q_0$ is output heat flux (typically 15 $kW/m^2$)
$K$ is heat exposure factor (range of 0.2–1)
$C_p$ is specific heat capacity of steel (typically 0.46 kJ/kgK)
$M/S$ is mass per unit surface area of member ($kg/m^2$)

Some conclusions can be drawn directly from this equation:

- Larger members (such as large pipes and thick plates) have longer failure times, since they have greater mass per unit surface area, and hence take longer to heat up.
- Failure may not occur at all if the flame heat flux is below about 45 $kW/m^2$. Thus, failure is only likely within or very close to the flame.

If piping or vessels containing fluid under pressure are heated in a fire, the temperature of the pipe and fluid will rise. The impacts are:

- Reduction in strength of the pipe or vessel, which may ultimately fail under its internal pressure; failure of a pipe is defined here as resulting in a loss of containment (i.e., a leak).
- Introduction of thermal stresses due to variations in the heating.
- Thermal expansion of a section of pipe restrained between two brackets, which may lead to buckling.
- Reduction in strength of pipe supports, which may allow the pipe to buckle under its own weight
- Possible failure of flange seals.

The presence of fluid inside the pipe may have a large influence on the time to failure because of internal convective cooling. There are two scenarios:

1. *If the pipe is isolated by valves,* then the heating of the pipe will heat the fluid inside and set up natural convection currents. The heat transfer coefficient is then low, and depends on the fluid temperature, which increases as the closed-in volume is heated
2. *If the pipe fluid continues to flow,* then forced convection with the moving fluid providing very effective cooling to the pipe. The heat transfer coefficient is then high, and depends on the flow velocity.

The heat transfer coefficients in either situation can be determined from rather complicated empirical formulas, but in general, two conclusions emerge:

1. *If a gas pipe is closed-in*, internal convection provides only a modest cooling effect, with little effect on the rate of temperature rise. However, for liquids, local boiling may occur that causes the internal pressure to increase and may rupture the pipe if not relieved.
2. *If the pipe fluid continues to flow*, internal convection is often sufficient to prevent failure, with the temperature reaching equilibrium below the failure temperature. However, this is not invariably the case.

Water from water spray or deluge sprinklers is highly effective when applied on steel decks for pool fires, and can be expected to prevent failure. In jet fires, it has relatively little effect on the failure time.

Passive protection can be used to increase the time to structural failure. For example, intumescent mastic coatings of less than 1 inch thickness have been shown to provide up to 4 hours of fire resistance when applied to steel columns. Cementitious materials have been shown to provide 1–4 hours fire resistance for thicknesses of 2.5–6.3 cm (1–2.5 in). For additional information on passive fire protection, see Chapter 7.

Experiments on gas jet fires impinging on steel tubular members by Shell/British Gas (*Offshore Research Focus*, 1980) evaluated two types of passive fire protection:

1. *A cementitious coating, 34 mm thick*—this kept the temperature to 100°C (212°F) for approximately 45 minutes, corresponding to evaporation of water from the coating. Subsequently the temperature rose to approximately 200°C (392°F) after 1 hour.
2. *An intumescent epoxy coating, 14 mm thick*—this kept the temperature rise to approximately 6°C (43°F) per minute, reaching 370°C (698°F) after 1 hour.

Simple calculations of concrete structural fire resistance are not readily available. Above 300°C (572°F), the strength of concrete is reduced by 25% and should be discounted structurally (Spouge, 1999). Experiments on concrete slabs with hydrocarbon fires indicate that after 2 hours the outside 130 mm exceeds 300°C (572°F). Normally this concrete will spall away.

If the concrete member is designed to have 4-hour fire resistance in a cellulosic fire test, as recommended in design guidance by the Institute of Structural Engineers and the Concrete Society, then 3-hour fire resistance would be expected in hydrocarbon pool fires, provided that spalled material remained in place (Spouge, 1999).

Fiberglass reinforced plastic (FRP) is used in composite systems. It is particularly important in the process industry because of its corrosion resistance and light weight. The epoxy resin in the FRP matrix begins to melt at approximately 150°C (302°F). This may be used as a first-order failure criterion. Failure of empty 5-mm thick FRP pipes in 2–6 minutes has been reported (SINTEF,

1997). Liquid-filled pipes and tanks may not fail as readily because of the heat-sink effect from the liquid.

No general rules-of-thumb are available for plastic tanks/containers; failure criteria should be based on available test data (SINTEF, 1997). FRP is used offshore for fire water lines in order to reduce weight. Wet pipe FRP systems may be effective in a fire scenario because water is flowing, allowing the heat to be carried away. If the pipe is dry, failure may rapidly occur.

### 5.11.3. Thermal and Nonthermal Impact on Electrical and Electronic Equipment

A heat flux of 25 kW/m$^2$ has been published as a general rule-of-thumb for damage to process equipment (Barry, 2002). Clearly, this excludes electrical and electronic equipment, which may fail to operate at much lower heat fluxes and resulting temperatures. For example, data on the thermal impact of fire on electrical and electronic equipment have been summarized for U.S. Navy applications (Scheffey et al., 1990). The following limits were derived from a literature evaluation:

- 50°C (122°F) for faults in operating electronic equipment.
- 150°C (302°F) for permanent damage to nonoperating equipment.
- 250°C (482°F) for failure of standard Polyvinyl Chloride (PVC) cable.

Additional data on damageability is available in the literature (Bryan, 1986).

Combustion products can affect sensitive electronic equipment. For example, hydrogen chloride (HCl) is formed by the combustion of PVC cables. Corrosion due to combusted PVC cable can be a substantial problem. This may result in increased contact resistance of electronic components. Condensed acids may result in the formation of electrolytic cells on surfaces. Certain wire and cable insulation, particularly silicone rubber, can be degraded on exposure to HCl. A methodology for classifying contamination levels and ease of restoration is presented in the *SFPE Handbook*
(SPFE, 2002). There is evidence that relatively prompt (48 hours) salvage procedures can greatly mitigate the effects of HCl exposure to electronic equipment.

### 5.11.4. Impact on the Environment

Impact on the environment may result from both unwanted fires, improper control of fire effluent or improper use of suppression system agents. Environmental considerations impact decisions on whether to provide protection for a hazard, and whether this protection should be provided automatically or manually. Scenarios to be considered include uncontrolled fires, potential hazardous situations, firefighting training, and fixed or mobile vehicle suppression system discharge testing.

One option for reducing the effects on the environment is the use of remote impounding of the water runoff and any hazardous material. Remote impounding accomplishes two objectives: it removes hazardous material that could burn or, if burning, allows it to burn at a safe location. Secondly, water and other spilled material are contained so that the environment is not endangered.

Consider a large warehouse containing flammable liquids as an example of the trade-offs to consider for a hydrocarbon fuel hazard (Scheffey, 2002). The building may be unprotected or protected using water-only sprinklers designed for less hazardous conditions. If a fire spreads uncontrolled in this situation, large amounts of smoke will be discharged to the air, and liquid/chemical contents may be discharged to the ground, natural bodies of water, or down sanitary sewers to a wastewater treatment plant. The fire will likely be fought manually.

A rough estimate of firefighting water that may be used is 15 to 50 times the minimum anticipated agent required for suppression (Rasbash, 1986; Scheffey and Williams, 1991). The effluent resulting from such an incident can easily be several orders of magnitude greater than the scenario where a properly designed suppression system is installed (e.g., a fixed automatic foam system). The impact of effluent from the manual firefighting attack of an uncontrolled fire can be dramatic (Donkelaar, 1995). Containment is an obvious mitigation strategy along with a fire suppression system designed to control a fire to manageable proportions.

Environmental considerations also impact on the selection of firefighting agents. Total flooding gaseous agents used in enclosures should be environmentally acceptable. Users of foam agents should be aware of potential use and discharge limitations. For example, foam from system acceptance or maintenance tests may ultimately discharge to wastewater treatment plants or sensitive natural bodies of water. This may result in an unacceptable environmental impact. Companies need to consider environmental regulations and impact of discharge of fire protection systems and runoff.

## 5.12. Examples

### *5.12.1. Example—Warehouse Pool Fire (Indoor)*

**Problem**

A 55-gal (208-l) drum of diesel fuel is suddenly ruptured during a warehouse accident. The fuel is released quickly across the concrete slab and is ignited when it comes in contact with a piece of equipment. Physical barriers limit the fuel spill to an area of 40 m$^2$. Determine the size of the resulting fire and the potential for damage to adjacent steel structure and personnel. The closest equipment to the edge of the pool is 5 m away at ground level. Assume a steel failure at 500°C (932°F).

**Solution**

The heat release rate for the fire is calculated using Equation (5-17) as:

$$\dot{Q} = \dot{m}'' A \Delta h_c$$

The heat of combustion of diesel (assume JP-5) is found to be 43.0 MJ/kg, the density is 810 kg/m$^3$ and the maximum mass burning rate per unit area for a pool, $\dot{m}''_{max}$, is 0.054 kg/m$^2$s.

The resulting heat release rate is:

$$\dot{Q} = (0.054 \text{ kg/m}^2\text{s})(40 \text{ m}^2)(43 \text{ MJ/kg}) = 92.88 \text{ KW}$$

Based on the given area of the pool, the effective diameter of the pool, $D$, is calculated as:

$$D = \left(\frac{4A}{\pi}\right)^{1/2}$$

where $A$ is the area of the pool. The effective diameter, $D = 7.1$ m.

The flame height is calculated per Equation (5-6) as:

$$H_f = 0.23\dot{Q}^{2/5} - 1.02D = 0.23(92{,}880 \text{ KW})^{2/5} - 1.02(7.1 \text{ m}) = 15.1 \text{ m}$$

If it is assumed that the fuel spill burns at the maximum rate for the duration of the fire, the burn time, $t_b$ [Equation (5-8)] for the fuel spill fire will be:

$$t_b = \frac{m_f}{\dot{m}''A} = \frac{V \cdot \rho}{\dot{m}''A} = \frac{0.208 \text{ m}^3 \cdot 810 \text{ kg/m}^3}{0.054 \text{ kg/m}^2\text{s} \cdot 40 \text{ m}^2} = 78 \text{ s} = 1.3 \text{ min}$$

This predicted burn time of 1.3 minutes is most likely shorter than will actually occur. In reality, there will be additional time associated with the growth period of the fire and the pool fire may take minutes before reaching a steady-state burning rate. This time also does not account for secondary materials igniting and burning.

The impact of the fire on equipment 5 m away is determined by first calculating the incident heat flux to the equipment. Using the method of Shokri and Beyler, the incident flux [Equation (5-11)] is calculated as:

$$q'' = E \cdot F_{12}$$

where $E$ is the emissive power of the fire and is calculated based on the pool diameter ($D = 7.1$ m) per the following correlation [Equation (5-12)]:

$$E = 58(10^{-0.00823D}) = 50.7 \text{ KW/m}^2$$

The view factor, $F_{12}$, is determined using the following ratios:

- The ratio of the flame height to pool fire radius, $H/R$, equals 4.2 (15.1/3.56).

- The ratio of the target height to pool fire radius, $H_{target}/R$, equals 0.56 (2/3.57).
- The ratio of the distance between the center of the pool fire and the target to the pool fire radius, $L/R$, equals 2.4 (8.57/3.57).

From Figure 5-12, the view factor is approximately 0.3. The incident heat flux to the target equipment is thus:

$$q'' = E \cdot F_{12} = (50.7 \text{ kW/m}^2)(0.3) = 15.2 \text{ kW/m}^2$$

A conservative estimate of the resulting surface temperature is obtained by assuming no convective heat losses from the target structures. Equation 5-23 is used to calculate the surface temperature of the equipment due to an incident radiative heat flux from the fire and accounting for only radiation losses from the target.

$$T_s = \left[\frac{q''}{\sigma} + T_\infty^4\right] = \left[\frac{15.2 \text{ kW/m}^2}{5.67 \times 10^{-11} \text{ kW/m}^2 \text{ K}^4} + 298^4\right]^{1/4} = 724 \text{ K} = 452^\circ\text{C}$$

The resulting surface temperature of 452°C is below the failure criteria temperature for steel of 500 to 550°C. Therefore, it can be concluded that all steel structures 5 m or more from the fire will not be damaged. The relatively short duration of the pool fire will also help limit the extent of damage. If equipment is constructed of aluminum, the target temperature of 452°C would indicate that the equipment would be damaged, since the failure criteria for aluminum is about 150°C. Electrical and electronic equipment would also be in jeopardy of damage.

Assuming that people are impeded by obstructions and are only able to move to a maximum distance of 5 m away from the fire, what is the potential for injury? Using Figure 5-16, a person within 20 m of the fire would experience first- and second-degree burns within 3 and 6 seconds, respectively.

The times to first and second degree burns are quite fast, indicating that people would need to escape to much further distances to be safe.

### *5.12.2. Example—Process Jet Fire*

**Problem**

A 10-cm diameter natural gas pipe is accidentally ruptured by a forklift. The accident leads to the ignition of the gas and a vertical jet fire. Determine the impact of the jet fire on structural steel components that are 10 m above the fire and on 2-m-high process equipment that is 5 m away at ground level. The pipe rupture occurs at ground level and the gas release rate has been determined to be 0.8 kg/s [though application-specific, Sprouge (1999) provides guidance for calculating potential release rates for various scenarios].

**Solution**

The heat release rate, $Q$, of the resulting fire is calculated [Equation (5-17)] as:

$$Q = \dot{m}\Delta h_c = (0.8 \text{ kg/s})(50 \text{ MJ/kg}) = 40 \text{ MW}$$

Assuming relatively still air, the flame height, $L$ (m), is first calculated [Equation (5-18)] as:

$$H = 0.2Q^{2/3} = 0.2(40{,}000 \text{ kW})^{0.4} = 13.9 \text{ m}$$

The ratio of $L/D$ (flame length to the gas exit diameter) is calculated and found to be 139, which is less than 200. Therefore, the jet flame is buoyancy-controlled and the calculated flame length of 13.9 m is appropriate.

With a flame length of 13.9 m, the jet flame will impinge on the steel structure overhead. Consequently, the steel will see high convective and radiative heat fluxes on the order of 200 kW/m$^2$. Since the structure will be exposed to direct flame impingement, the expected failure time would be 3–4 minutes (Table 5-7) or less due to the high heat flux from the jet fire, depending on the type of steel structure and design factor of safety.

To estimate the impact of the jet fire on process equipment located 20 m from the source, the point source model can be used to determine the incident heat flux from the jet flame to the equipment. The incident heat flux per unit surface area of target, $q''$ is calculated as follows.

The heat radiated from the fire, $Q_r$, is calculated from Equation (5-20) using a radiative fraction, $x_r$, of 0.2 for methane (Brzustowski, 1971):

$$Q_r = x_r Q = 0.2 \cdot 40{,}000 \text{ KW} = 8{,}000 \text{ KW}$$

The distance from the center of the flame to the top of the process equipment, $R$, is calculated assuming the arrangement shown in Figure 5-18. Selecting the top of the process equipment as the target will lead to higher incident fluxes, as this location is closest to the point source representation of the fire (i.e., mid-height of the flame).

$$R^2 = X^2 + (0.5L + H_{pipe} - H_{target})^2 = 5^2 + (0.5 \cdot 13.9 + 1 - 2)^2 = 60.4 \text{ m}^2$$

Based on the geometry,

$$\cos\theta = \frac{X}{0.5L + H_{pipe} - H_{target}} = \frac{5}{0.5 \cdot 13.9 + 1 - 2} = 0.84$$

$$q'' = \frac{Q_r \cos\theta}{4\pi R} = \frac{8{,}000 \text{ kW} \cdot 0.84}{4\pi 60.4 \text{ m}^2} = 8.9 \text{ kW/m}^2$$

The consequence of the 8.9 kW/m$^2$ incident heat flux on nearby equipment is determined by estimating the surface temperature of the equipment.

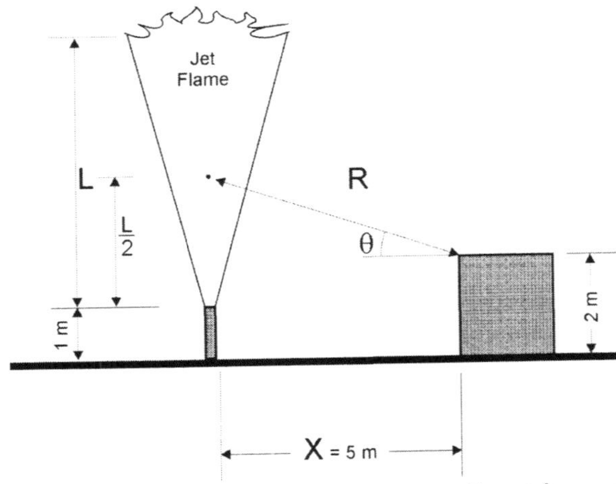


**Figure 5-18.** Schematic of Radiant Exposure to a Target from a Jet Fire

As a conservatively high estimate of temperature, Equation (5-23) can be used (this equation neglects convective cooling of the equipment):

$$T_s = \left[\frac{q''}{\sigma} + T_\infty^4\right] = \left[\frac{8.9 \text{ kW/m}^2}{5.67 \times 10^{-11} \text{ kW/m}^2 \text{ K}^4} + 298^4\right]^{1/4} = 637 \text{ K} = 364°\text{C}$$

A surface temperature of 364°C (687°F) is not expected to cause structural damage to steel equipment, based on a failure criteria of 500 to 550°C (932–1022°F). However, one can assume that electronic equipment and seals that are housed or in contact with a steel enclosure may see similarly high temperatures. Since some electronic equipment may fail at temperatures of 50 to 250°C (122–482°F) it is reasonable to assume that such equipment would be at least compromised if not failing when exposed to this radiation jet fire.

### *5.12.3. Example—Storage Tank Fire*

**Problem**

A full 24.4-m (80-ft) diameter floating roof gasoline storage tank suffers the double accident of the sinking of the roof and the ignition of the gasoline. Determine the radiant exposure to personnel on top of an adjacent tank located 15.2 m (50 ft) away and the time available for evacuation. Also determine the effect of the fire on communications equipment located 76.2 m (250 ft) away at a height of 4.6 m (15 ft) above the lip of the tank (see Figure 5-19).

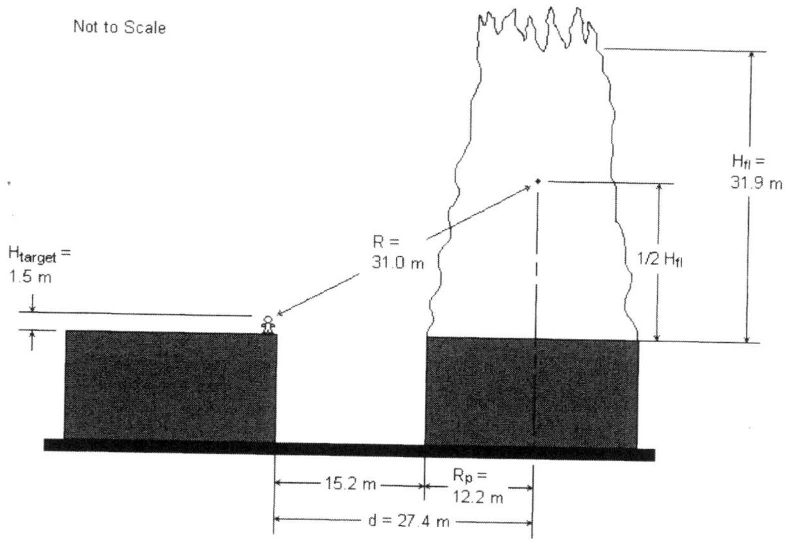


**Figure 5-19.** Schematic of Tank Fire Exposure to Personnel

**Solution**

The mass-burning rate per unit area may be estimated for a gasoline pool fire from the burning rate listed in Table 5-2, assuming motor gasoline, 23 cm/hr (6.3e-05 m/s), and the density of gasoline, 740 kg/m$^3$, as:

$$\dot{m}'' = (6.3\text{e-}05 \text{ m/s})(740 \text{ kg/m}^3) = 0.047 \text{ kg/m}^2\text{s}$$

The heat release rate, $\dot{Q}$, of the resulting fire is calculated [Equation (5-17)] as:

$$Q = \dot{m}A_s\Delta h_c = (0.047 \text{ kg/m}^2\text{s})(467 \text{ m}^2)(43.7 \text{ MJ/kg}) = 956.5 \text{ MW}$$

Assuming relatively still air, the flame height, $H_f$, is calculated [Equation (5-6)] as:

$$H_f = 0.23\dot{Q}^{2/5} - 1.02D = 0.23(956{,}500 \text{ kW})^{2/5} - 1.02(24.4 \text{ m}) = 31.9 \text{ m}$$

For large gasoline pool fires, the radiant fraction of heat released can be calculated [Equation (5-21)] as:

$$\chi_r = 0.21 - 0.0034D = 0.21 - 0.0034(24.4 \text{ m}) = 0.13$$

To estimate the impact of the pool fire on the nearby personnel, start with the point source model. Assuming a head height target, use a target point at a height of 5 ft (1.5m). The distance from the point source representation of the flame to the target point is calculated as:

$$R^2 = d^2 + (\tfrac{1}{2}H_{fl} - H_{target})^2 = (27.4\text{ m})^2 + [\tfrac{1}{2}(31.9\text{ m}) - (1.5\text{ m})]^2 = 960\text{ m}^2$$

and

$$R = 31.0\text{ m}$$

Based on geometry, the cosine of the angle between the point source representation of the flame, and the target point is calculated as:

$$\cos\theta = \frac{d}{R} = \frac{27.4\text{ m}}{31.0\text{ m}} = 0.89$$

The incident heat flux per unit area to the personnel can now be calculated [Equations (5-19) and (5-20)] as:

$$q'' = \frac{\chi_r \dot{Q}\cos\theta}{4\pi R^2} \frac{(0.13)(956{,}500\text{ kW})(0.89)}{4\pi(960\text{ m}^2)} = 96\text{ kW/m}^2$$

Since the point source estimated incident heat flux is much greater than 5 kW/m$^2$, the limit for applicability, it is clear that the people are too close to the fire for the point source to be a valid technique. Therefore, an estimation using the method of Shokri and Beyler is warranted. The emissive power of the fire is calculated [Equation (5-12)] as:

$$E = 58(10^{-0.00823D}) = 58(10^{-0.00823(24.4\text{ m})}) = 37\text{ kW/m}^2$$

The view factor between the flame and target is estimated using the geometry and one of the Figures 5-10 through 5-14. The flame height to pool fire radius is:

$$H_f/R = (31.9\text{m})/(12.2\text{m}) = 2.6.$$

which is approximately 3. Therefore, use Figure 5-11, for $H_f/R_p = 3$.

The ratios of target height to pool fire radius, $H_{target}/R_p$, and ratio of distance from center of pool fire to target edge to pool fire radius, $d/R_p$, are likewise estimated as 0.13 and 2.25, respectively.

From Figure 5-11, the view factor, $F_{12}$, associated with the above values is approximately 0.23.

The incident heat flux to the target is calculated [Equation (5-11)] as:

$$q'' = EF_{12} = (37\text{ kW/m}^2)(0.23) = 8.6\text{ kW/m}^2$$

From Figure 5-16, an incident heat flux of 8.6 kW/m$^2$ will produce first-degree burns on unprotected personnel within 10 seconds and second-degree burns within 17 seconds.

### 5.12.4. Example—Flowing Pool Fire

**Problem**

A 5,000-gallon gasoline tanker truck rolls over and catches on fire. The tank is penetrated and releases fuel at a rate of 0.032 $m^3/s$ (500 gallons per minute). Assuming the ground is level and the resulting spill forms a growing circular pool, determine the impact the fire will have on a fiberglass process line that runs parallel to the road at a distance of 30.5 m (100 feet) away and 0.9 m (3 feet) off the ground (see Figure 5-20).

**Solution**

Before the effects of the fire on the fiberglass process line can be determined, the size of the fire needs to be calculated. The first step is to determine the steady-state size of the spill recognizing that as fuel is spilling onto the ground, the fire is consuming it. From Table 5-3, the mass-burning rate of gasoline is 0.05 $kg/m^2s$. The density of gasoline is 74 $kg/m^3$. The steady-state pool diameter is calculated [Equation (5-5)] as:

$$D_{ss} = \left(\frac{4V_{\text{leak}}\rho}{\pi \dot{m}''}\right)^{1/2} = \left(\frac{4(0.032\ \text{m}^3/\text{s})(740\ \text{kg}/\text{m}^3)}{\pi(0.05\ \text{kg}/\text{m}^2\text{s})}\right)^{1/2} = 24.4\ \text{m}$$

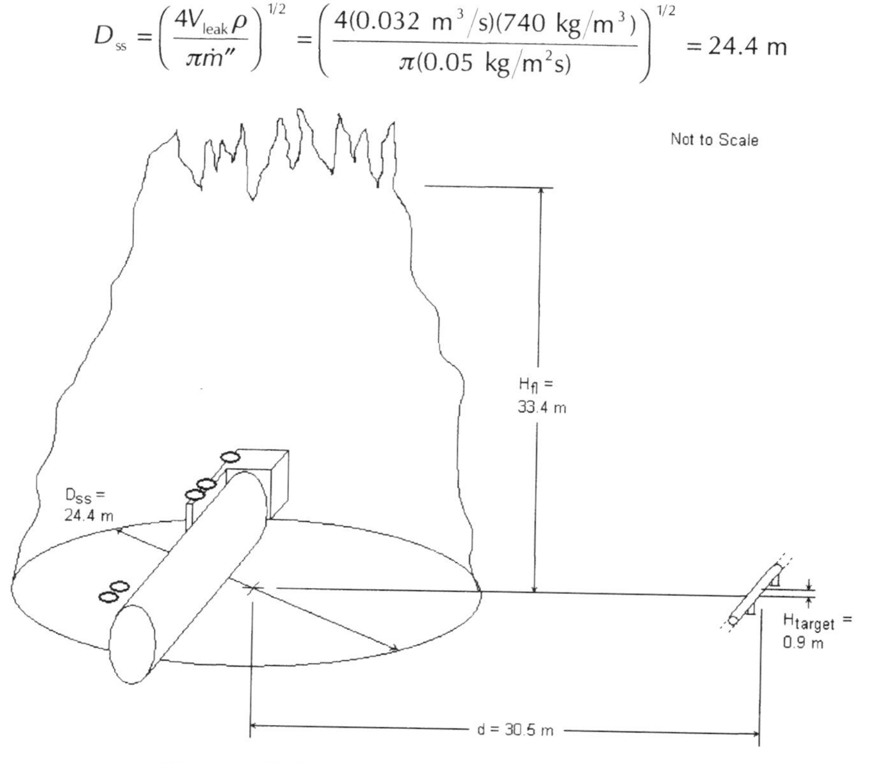


**Figure 5-20.** Schematic of Tanker Roll-Over Incident

For the assumed circular shape of the spill, the fire occupies an area of 467 $m^2$ centered on the leak.

The heat release rate, $\dot{Q}$, of the fire is calculated [Equation (5-17)] as:

$$Q = \dot{m}''A_s\Delta h_c = (0.05\ \text{kg/m}^2\text{s})(467\ \text{m}^2)(43.7\ \text{MJ/kg}) = 1{,}020\ \text{MW}$$

The flame height can be calculated [Equation (5-6)] as:

$$H_f = 0.23\dot{Q}^{2/5} - 1.02D = 0.23(1{,}020{,}000\ \text{kW})^{2/5} - 1.02(24.4\ \text{m}) = 33.4\ \text{m}$$

An estimation of incident heat flux can be calculated using the method of Shokri and Beyler. The emissive power of the fire is calculated [Equation (5-12)] as:

$$E = 58(10^{-0.00823D}) = 58(10^{-0.00823(24.4\text{m})}) = 37\ \text{kW/m}^2$$

The view factor between the flame and target is estimated using the geometry and one of the Figures 5-10 through 5-14. The flame height to pool fire radius is:

$$H_f/R_p = (33.4\text{m})/(12.2\text{m}) = 2.7$$

which is approximately 3. Therefore, use Figure 5-11, for $H_f/R_p = 3$.

The ratios of target height to pool fire radius, $H_{target}/R_p$, and ratio of distance from center of pool fire to target edge (30.5 m) to pool fire radius, $d/R_p$, are likewise estimated as 0.08 and 2.5, respectively.

From Figure 5-11, the view factor, $F_{12}$, associated with the above values is approximately 0.2.

The incident heat flux to the target is calculated [Equation (5-11)] as:

$$q'' = EF_{12} = (37\ \text{kW/m}^2)(0.2) = 7.2\ \text{kW/m}^2$$

The steady-state temperature of the fiberglass process line can be estimated with a lumped mass approach by iteratively solving Equation (5-22) for $T_s$. Assume the ambient and initial target temperature to be 20°C (293 K) and a nominal heat transfer coefficient of 0.015 kW/m$^2$K.

$$q'' = \sigma(T_s^4 - T_\infty^4) + h(T_s - T_\infty)$$

$$(7.2\ \text{kW/m}^2) = \sigma(T_s^4 - 293\ \text{K}^4) + (0.015\ \text{kW/m}^2\text{K})(T_s - 293\ \text{K})$$

Solving for $T_s$ yields an estimated steady-state temperature of 249°C (480°F). Because this temperature exceeds the first-order failure criterion for fiberglass of 150°C and the fire burns for 10 minutes, which is longer than the reported 2–6 minutes failure time for empty fiberglass pipes, the process line will be compromised.

# 6

# FIRE RISK ASSESSMENT

A fire risk assessment is a tool for making decisions on fire protection issues. Typically, inherently safer design (CCPS, 1996) is applied prior to the fire risk assessment, however the results of the fire risk assessment may indicate that further review of the design should be conducted.

The objective of this chapter is to provide an understanding of how and why a fire risk assessment is performed. Completing a fire risk assessment provides management the information needed to make risk-based decisions.

## 6.1. Fire Risk Assessment Overview

A fire risk assessment is an integral part of a company's overall risk management system (RMS) and should be integrated with other risk assessments. Performing a fire risk assessment is not required by regulations, but is used by some companies. This is especially so with increased acceptance of a performance-based approach to fire protection. The authorities having jurisdiction (AHJ) are progressively recognizing a fire risk assessment as an equivalent to codes and standards for determining the need for and adequacy of fire protection. There are certain key principles that apply to fire risk assessments, including:

- A fire risk assessment should begin early in the design process.
- A fire risk assessment includes conducting a thorough hazard identification.
- Fire risk assessments should be revised as new information becomes available and the design evolves.
- Fire risk assessments should be used in the identification of prevention, control, and mitigation measures.

Fire risk assessments are also useful tools for reviewing existing facilities, particularly:

- When changes are contemplated within or adjacent to an existing facility
- As a retrospective on past changes, especially if multiple changes have occurred over the years
- With changes in materials or processes

A fire risk assessment should be documented to provide a clear overall picture of the possible fire hazards and the role safety systems play in hazard control and mitigation. Also, a fire risk assessment should be maintained evergreen during the lifecycle of the facility to ensure ongoing management of fire hazards.

## 6.2. Fire Risk Assessment Methodology

The methodology for conducting a fire risk assessment is illustrated in Figure 6-1. Each step in the methodology is described further in this section.

The benefit of applying a fire risk assessment as a decision tool will vary between companies and projects. In many cases, a formalized fire risk assessment is not necessary because the solution is relatively obvious based on design standards and appropriate fire protection is either applied or not applied. Fire risk assessments may be necessary:

- For very large projects
- When the fire hazards are not well understood
- When the fire impact is so large that decisions on what is appropriate protection is difficult
- When the cost of fire protection is significant relative to the cost of the fire risk assessment

The methodology outlined in this chapter follows that in *Guidelines for Chemical Process Quantitative Risk Analysis* (CCPS, 2000). NFPA 550 *Guide to the Fire Safety Concept Tree* provides another example of fire risk assessment. There are three keys to a successful fire risk assessment:

- It is important that experienced and knowledgeable personnel perform the assessment. A fire risk assessment should be performed by a fire protection engineer who can apply common sense, understand if the results are realistic, and transfer the results into a performance-based approach to fire protection.
- In any risk assessment, the fire protection engineer will need to make certain assumptions. The assumptions can be in the development of a

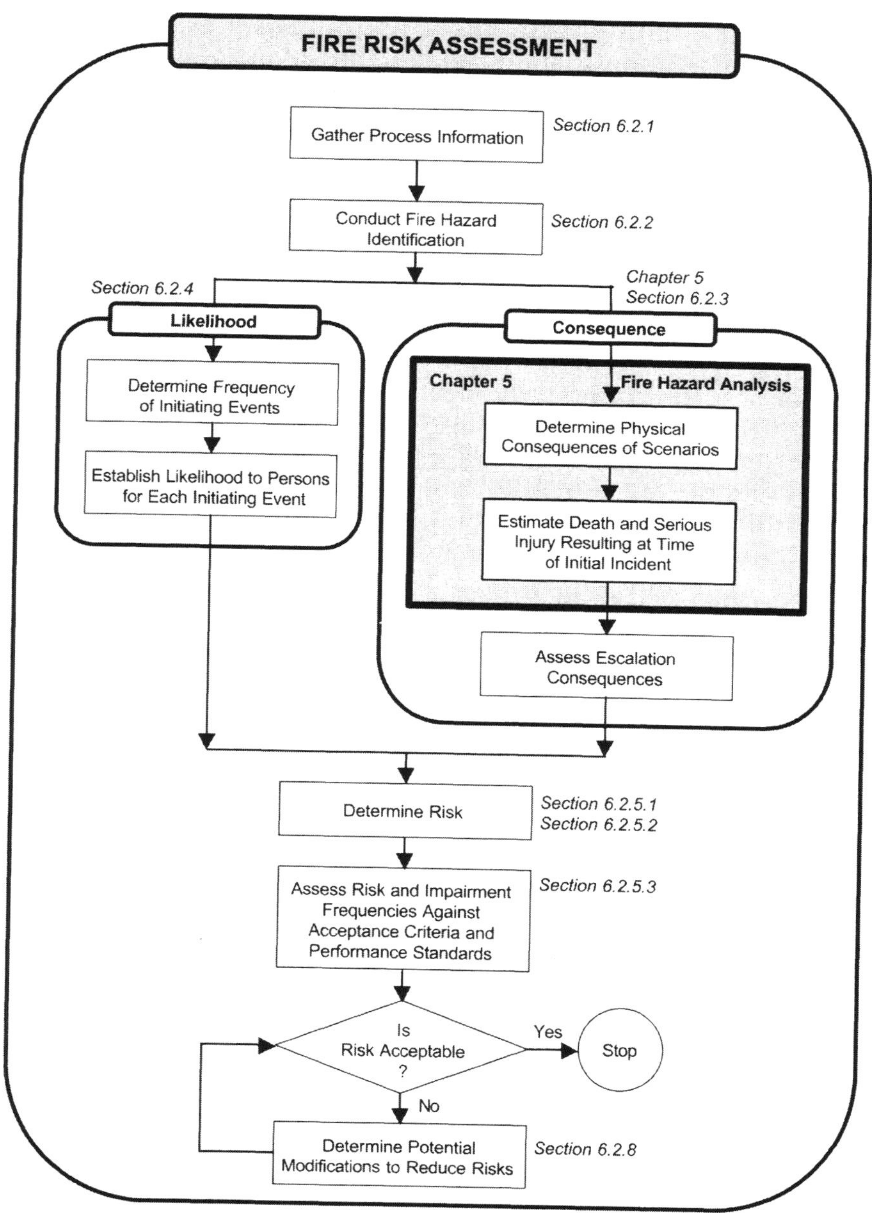


**Figure 6-1.** Fire Risk Assessment Methodology

rule set, and use of judgment data. All assumptions in a risk assessment should be documented and the justification for data selection provided. Performing a sensitivity analysis on the assumptions may be warranted.

- Documentation is critical to understanding the results and report long after the fire risk assessment is completed. The fire protection engineer should ensure that the fire risk assessment is thoroughly and completely documented.

### 6.2.1. *Process Information*

The fire risk assessment process begins by collecting basic information about the design. For new projects there may be limited information available; however a list of hazards can be developed, the potential fire impact can be assessed, and preliminary risk can be calculated. The level of detail of the fire risk assessment is only as good as the level of detail of the design information. The information needed to perform a fire risk assessment could include:

- Operating and maintenance philosophies
- Plot plans and layout drawings
- Piping and instrumentation diagrams (P&IDs)
- Process flow diagrams (PFDs)
- Hazardous materials data
- Equipment lists
- Process data sheets

### 6.2.2. *Fire Hazard Identification*

Identifying and analyzing fire hazards and scenarios is the next step in a fire risk assessment. The hazard identification should be structured, systematic, auditable, and address all fire hazards, including nonprocess fires. The result of the hazard identification is a list of potential fire hazards that may occur at the facility, for example, jet, pool, flash, BLEVE, electrical, or Class A fires. This list should also include the location where each fire could occur. Hazard identification techniques used to identify potential hazards are shown in Table 6-1.

Each identified hazard will have a range of possible scenarios; it may not be reasonable to evaluate every scenario. Therefore, representative fire scenarios should be chosen to cover a range of foreseeable scenarios. The scenarios to evaluate are those where the initial release and ignition characteristics are likely to cause the most extensive damage, loss of production, and the greatest risk to personnel. The fire scenarios selected should have a sufficient inventory that will burn long enough to cause failure of equipment and/or the structure.

**Table 6-1**
*Hazard Identification Methods*

| Hazard Identification Methods |
|---|
| ▪ Checklist |
| ▪ Hazard Identification (HAZID) |
| ▪ What-If? |
| ▪ Hazard and Operability Studies (HAZOP) |
| ▪ Other methods |

Smaller fire scenarios that can escalate and less severe but very likely scenarios should also be considered.

### 6.2.3. *Fire Hazard Analysis*

Fire hazard analysis (FHA) is the process to determine the size, severity, and duration of a scenario and its impact on personnel, equipment, operations, and the environment. Chapter 5 provided details of performing an FHA. The following paragraphs provide an overview of the FHA process. For example, one scenario could be a seal failure where the material being released is ignited and a fire results. In assessing consequences, several questions must be considered:

- *What is the severity of the initiating event?*
- *Can the event escalate?*
- *What is the impact of the event on personnel, property, the supply chain, customers, and the public?*

In most cases, each scenario will have a variety of conditions that need to be evaluated. These include the size of the release, orientation of release, temperature and pressure of operation, and weather conditions.

#### 6.2.3.1. *Escalation*

It is important to look beyond the initiating event to determine the potential for fire to spread to adjacent areas. If the fire is not detected early and quickly controlled, then the fire can escalate and involve other equipment and units. For escalation to occur, the fire must impact adjacent equipment by either radiant heat or flame impingement. In most risk assessments, escalation is taken into account by establishing a rule set. If any of the conditions within the rule set are exceeded, then escalation is assumed to occur. Typically, rule-sets may include:

- Radiant heat exceeding 37.5 kW/m$^2$ for 10 minutes
- Jet fire flame impingement on vessels or structural steel that are not protected for 15 minutes

- Pool fire that travels into an area not previously on fire

In assessing the impact of radiant heat on equipment, evacuation routes, and support structures, the potential for escalation must be examined.

#### 6.2.3.2. *Estimating Impact*

Fire hazard analysis (FHA) is a method for estimating fire impact to personnel, structures, and equipment and is discussed in Chapter 5, Section 5.13. Computer models used in fire consequence analysis range from spreadsheets that use simple equations to computational fluid dynamics (CFD) modeling that can take a day to evaluate one scenario. In the conceptual or design-stage, simple models can be used. As the design details increase, the complexity of the consequence analysis also needs to increase. Appendix C describes computer programs that can be used to model fires in industrial facilities.

#### 6.2.3.3. *Consequence-Only Decision Making*

In some cases, after completing the consequence portion of the analysis, the impact of the consequences is deemed so severe that the company may decide to provide fire protection that will provide mitigation without completing the likelihood analysis. **It is important to take the time to analyze the consequences (*conduct an FHA*) and determine if reasonable mitigation measures can be applied before continuing with the fire risk assessment.** Credit for additional mitigation measures can be taken in the fire risk assessment.

### 6.2.4. *Likelihood*

A consequence-only basis for installing fire protection equipment may not be cost effective. In reality, the likelihood of the fire scenario should be taken into consideration. There are several issues that need to be considered:

- The frequency of the initiating event (loss of containment)
- Probability of ignition
- Postrelease events (escalation), their different consequences, and their related frequencies that can occur after the flammable material is released

#### 6.2.4.1. *Initiating Event Frequency Analysis*

The initiating frequency estimate is derived from the cause of the fire scenario. This may initially be obtained from historical data (or more specific data if available) and modified where necessary to take account of any site-specific considerations that may affect the frequency. See Section 6.2.4.6 for additional discussion on failure rate data. Any assumptions made during this process should be documented with the basis for data selection.

The flammable release frequencies may be estimated by counting all relevant system components whose failure could result in a flammable release and multiplying by failure rate data appropriate to the type, standard or design, use, and operating conditions.

Under certain circumstances, it may be appropriate to examine the sequence of events that may lead to the initiating event. Techniques such as fault tree analysis or event trees may be used to estimate the frequency of these events.

Fault tree analysis is based on a graphical, logical description of the failure mechanisms of a system. Before construction of a fault tree can begin, a specific definition of the top event is required: for example the release of propylene from a refrigeration system. A detailed understanding of the operation of the system, its component parts, and the role of operators and possible human errors is required. Refer to *Guidelines for Hazard Evaluation* (CCPS, 1992) and *Guidelines for Chemical Process Quantitative Risk Assessment* (CCPS, 2000).

### *6.2.4.2. Postrelease Frequency Analysis*

Event trees are used to perform postrelease frequency analysis. Event trees are pictorial representations of logic models or truth tables. Their foundation is based on logic theory. The frequency of *n* outcomes is defined as the product of the initiating event frequency and all succeeding conditional event probabilities leading to that outcome. The process is similar to fault tree analysis, but in reverse.

A complete understanding of the system under consideration and of the mechanisms that lead to all the hazardous outcomes is required. This may be in the form of a time sequence of instructions, control actions, or in the sequence of physical events that lead to hazardous consequences.

The starting point in event tree analysis is the initiating event. The quantitative evaluation of the event tree requires condition probabilities. These may be based on reliability data, historical records, experience, or from fault trees.

The outputs of an event tree from a post-release frequency analysis are a number of outcomes ranging from more to less hazardous. An event tree highlights failure routes for which no protective system can intervene and where additional protective systems/mitigative action may be contemplated. The quantitative output is the frequency of each event outcome. These outcomes (which might specify BLEVE, flash fire, pool fire, jet fire) are used to determine individual and societal risk.

The event tree technique is a relatively simple approach and can be used in various levels of detail. The level to use for a particular task can be selected based on the importance of the event or the amount of information available (CCPS, 2001).

### 6.2.4.3. *Ignition Probability*

Major flammable releases may be ignited at varying distances from the release source. In a fire risk assessment, it is necessary to identify ignition sources that may be reached by any cloud of flammable concentration. Ignition can occur immediately (due to the energy of the failure event, immediate contact with a hot surface or a release above a material's autoignition temperature) or can be delayed until the cloud, or pool encounters an ignition source. These sources may include open flames, hot surfaces, sparks or mechanical friction. Ignition may also occur due to human activities. Typical sources of ignition that may be found in process areas include:

- Flares
- Boilers
- Fired heaters
- Static electricity
- Vehicle traffic
- Electrical motors
- Hot work—welding and cutting
- Hot surfaces
- Lighting
- Overhead high voltage lines
- Mechanical—sparks, friction, impact, vibration, etc.
- Chemical reactions

Given the presence of a flammable mixture, the *probability of ignition* is typically modeled as a function of two components. The first is the probability the ignition source will be present. The second is the probability that, given the ignition source is present, it actually ignites the material. The second factor is much harder to estimate as it is a function of minimum energy required to ignite the flammable material and the ignition energy of the source. Table 6-2 provides generic ignition probabilities based on size of release (Cox et al., 1990).

### 6.2.4.4. *Weather Conditions*

Weather conditions have a major effect on the mechanisms and extent a release spreads. Meteorological data are readily available from the National Oceanic and Atmosphere Administration (NOAA) and from airports in the vicinity of the facility. These data include wind direction, several wind speeds, and several atmospheric stability categories. A typical 16-point wind rose is

**Table 6-2**
*Generic Ignition Probability*

| Release Rate Category | Release Rate lbs/s (kg/s) | Gas Leak (probability) | Oil Leak (probability) |
|---|---|---|---|
| Tiny | <1.1 (<0.5) | 0.005 | 0.03 |
| Small | 1.1–11 (0.5–5) | 0.04 | 0.04 |
| Medium | 11–55 (5–25) | 0.10 | 0.06 |
| Large | 55–441 (25–200) | 0.30 | 0.08 |
| Massive | >441 (>200) | 0.50 | 0.10 |

shown in Figure 6-2, which shows the percentage frequency of the wind blowing in each direction, wind speeds in each direction, and the frequency of the calms. Calms may be treated quantitatively as wind speed of 0.5 m/s. Atmospheric stability is important because it affects the degree of mixing with air, and thus dispersion results (CCPS, 1987).

The degree of meteorological data required for performing the analysis depends on the accuracy of the results desired. A single weather condition (combination of atmospheric stability and wind speed) can be used, however, it is usually impossible to isolate a single average condition that adequately represents all weather conditions. Many risk analyses use at least two weather conditions: one stable (e.g., 2 m/s, stability F) and the other characteristic of average conditions (e.g., 5 m/s, stability D).

Additional information on atmospheric conditions and stability factors can be found in *Guidelines for Consequence Analysis* (CCPS, 2000).

### 6.2.4.5. *Population Density*

Knowledge of the distribution and density of people is necessary to assess the impact of radiant heat and smoke from fires. This allows an estimate to be made of the risk to which the population in and around the facility may be exposed. Extensive population data is necessary where an estimate of societal risk is required. Where only an estimate of individual risk is desired, extensive population data may not be required. However, it is still necessary to determine the location of the people whose individual risk is being estimated.

The population distribution is often defined as population density. Sources of population data for an area are census reports, detailed maps, aerial photographs, and site inspections. Special attention must be made to various

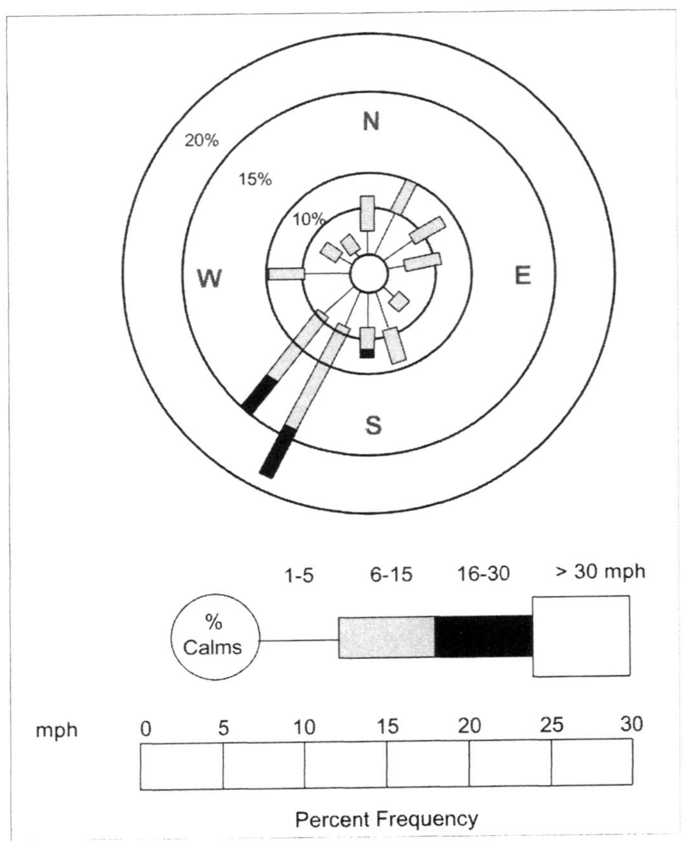


**Figure 6-2.** Example wind rose

types of population and day/night variations (i.e., residential or industrial), and concentrations of people such as hospitals, churches or schools. If day/night population variations are significant, then separate sets of calculations using meteorological data developed for daytime and nighttime conditions should be carried out.

### 6.2.4.6. *Failure Rate Data*

It is important to distinguish the types of time-related and demand-related equipment failure rate data that can be found and used in a fire risk assessment. Basically, four types of data and corresponding data sources provide both time-related and demand-related failure data as shown in Table 6-3 and discussed in the following sections.

**Table 6-3**
*Failure Rate Data*

| Data Type | Data Source |
| --- | --- |
| Facility-specific data | Internal sources |
| Generic data | External data resources |
| Predicted data (generic quality) | Estimation techniques |
| Expert opinions (crude estimates) | Internal expert opinion |

#### 6.1.4.6.1. FACILITY-SPECIFIC DATA

Failure rate data generated from collecting information on equipment failure experience at a facility are referred to as facility-specific or field failure rate data. Facility-specific data contain failure rates specific to equipment (e.g., a certain valve or pump in use at a facility by manufacturer, make, model, and serial number) and are cataloged accordingly. The collection of facility-specific data from internal operations for use in a risk analysis is desirable because such data reflect the practices, environmental factors, and other reliability influences specific to the equipment under study. The ideal situation is to have valid historical data from identical equipment, in the identical application, functioning under the identical operating and maintenance conditions. Where these are not available, but data on similar equipment are, then they may be used with appropriate judgment.

In order for facility-specific data to be usable, they must be maintained for a sufficient period of time to accurately portray component failure tendencies, and must be stored in a format which is readily accessible and retrievable (CCPS, 2000).

#### 6.1.4.6.2. GENERIC DATA

Although facility-specific failure rate data on the specific equipment under study are preferred, often the only way to assemble sufficient data to satisfy study needs is to construct a "generic" data set. This data set is built using inputs from all of the facilities within a company, various facilities within the industry, literature sources, or commercial databases.

Generic data are much less specific and detailed than the data derived from facility-specific failure rate data. At best, generic data can claim to approximate facility-specific failure rate data. Generic failure rate data are not limited to a specific manufacturer, make, model, and serial number. For example, a generic database cannot be used to differentiate the future performance of two new pumps, each designed for the same service but provided by different manufacturers. Generic failure rates would most likely be the same for both pumps. While losing specificity, a generic database can draw on a much larger pool of facilities for input and can overcome the lack of operating exposure and failure mode bias encountered in working with facility-specific data.

A substantial amount of generic failure rate data applicable to process equipment have been identified by CCPS (CCPS, 2000).

#### 6.1.4.6.3. PREDICTED DATA

Increasing attention is being given to developing methods to predict failure rate data for process equipment and systems. Such methods are beginning to appear in published literature. These methods include correlations, factored estimation procedures, and analogies to predict equipment failure rates. They are desirable because they offer efficient means of providing equipment failure rate data for risk assessments, and they can be conveniently incorporated into computer software.

These methods include estimating failure rate data using models or correlations developed from an engineering or scientific analysis of the influences on the reliability of particular types, classes, or groups of equipment. For example, Thomas provides a factor-based technique for estimating the probability of catastrophic leakage from a pipe or pressure vessel. Factors include size and shape influences, weld zones, facility age, and other quality factors (CCPS, 2000).

#### 6.1.4.6.4. EXPERT OPINION

Where the expert opinion equipment failure rate data is derived using a group of individuals with experience with similar equipment, operating under similar conditions, it may be more accurate than generic data.

The Delphi-type approach is one method that can be used to improve the accuracy of expert opinion estimates generated by a group of experts. The data contained in the Institute of Electrical and Electronic Engineer's (IEEE) Std. 500, *Guide to the Collection and Presentation of Electrical Electronic, Sensing Component and Mechanical Equipment, Reliability for Nuclear-Power Generating Station* was compiled using this approach (IEEE, 1985). If facility-specific data are unavailable, this reference is a good source of failure rate data for electrical, electronic, and mechanical components.

#### 6.1.4.6.5. ASSUMPTIONS

In any risk assessment, the analysts will need to make some assumptions. These are important as the validity of the risk analysis is dependent on the validity of the assumptions. The assumptions can be in the development of a rule set, use of judgment data or other information not readily available. All assumptions in a risk assessment should be documented and the justification for selecting those data provided. In some cases, performing a sensitivity analysis on each of the assumptions may be warranted.

### 6.2.5. *Risk*

Risk is defined as a measure of human injury, economic loss, or environmental damage in terms of both the likelihood and severity of the consequences. It is important to recognize that risk is an estimate. It can not be exactly measured or, if calculated, is not an accurate number. There will be a level of uncertainty with any risk estimate and management should understand that risk estimates have some uncertainty.

There are three commonly used ways of calculating risk estimates: risk indices, individual risk, and societal risk.

#### 6.2.5.1. *Risk Indices*

*Risk indices* are single numbers or a tabulation of numbers that are correlated to the magnitude of the risk to people. Some risk indices are relative values with no specific units. The limitations on the use of indices are that they may not be an absolute criteria for accepting or rejecting the risk. Risk indices also do not communicate the same information as individual or societal risk measures. An example of risk indices is a risk ranking matrix. Table 6-4 (modified from CCPS, 1992) shows how severity and likelihood are combined to obtain risk indices. An example risk matrix is shown in Figure 6-3 (RRS, 2002).

**Table 6-4**
*Risk Indices*

| Risk Indices | Severity–Likelihood |
|---|---|
| 1 | 1–1, 1–2, 2–1 |
| 2 | 1–3, 2–2, 3–1 |
| 3 | 2–3, 3–2 |
| 4 | 1–4, 3–3, 4–1 |
| 5 | 2–4,3–4, 4–2, 4–3, 4–4 |

| LIKELIHOOD OF OCCURRENCE *WITH SAFEGUARDS* / SEVERITY OF CONSEQUENCES *WITHOUT SAFEGUARDS* | FREQUENT 1 This incident has occurred at this facility and/or is reasonably likely to occur at any time. | OCCASIONAL 2 This incident is likely to occur at this facility within the next 15 years. | SELDOM 3 This incident has occurred at a similar facility and may reasonably occur at this facility within the next 30 years. | UNLIKELY 4 Given current practices and procedures, this incident is not likely to occur at this facility. |
|---|---|---|---|---|
| **MAJOR** 1<br>**SAFETY** - Fatality or permanently disabling injury.<br>**OPERABILITY** - Major or total destruction to process areas above $10MM; plant downtime in excess of 30 days.<br>**ENVIRONMENTAL & COMMUNITY IMPACT** - One or more severe injuries; significant release with serious long-term off-site impact. | 1 | 1 | 2 | 4 |
| **SERIOUS** 2<br>**SAFETY** - Severe injury.<br>**OPERABILITY** - Major damage ($1MM to $10MM) to process areas with up to 30 days plant downtime.<br>**ENVIRONMENTAL & COMMUNITY IMPACT** - One or more injuries or possible evacuation; significant release with serious environmental impact. | 1 | 2 | 3 | 5 |
| **MINOR** 3<br>**SAFETY** - Single injury, not severe, possible lost time.<br>**OPERABILITY** - Some equipment damage ($100k to $1MM) with possible downtime.<br>**ENVIRONMENTAL & COMMUNITY IMPACT** - Odor or noise complaint from the public. Release that results in some Agency notification or violation. | 2 | 3 | 4 | 5 |
| **INCIDENTAL** 4<br>**SAFETY** - Minor injury or no injury.<br>**OPERABILITY** - Minimal equipment damage (less than $100k) with negligible plant downtime.<br>**ENVIRONMENT & COMMUNITY IMPACT** - No impact off site. Environmental recordable event with no agency notification. | 4 | 5 | 5 | 5 |

***LEGEND:***
1 = Very High Risk; Additional Consideration Required
2 = High Risk; Additional Consideration Required
3 = Moderate Risk; Additional Consideration Recommended
4 = Possible Risk; Additional Consideration at Discretion of Team
5 = Negligible Risk; Additional Consideration Not Required

**Figure 6-3.** Risk Ranking Matrix (RRS, 2002)

### 6.2.5.2. *Individual Risk*

*Individual risk* can be a single number or a set of risk estimates for various individuals or geographic locations. In general, individual risk considers the risk to an individual who may be in the effected zone of an incident. The risk includes the nature of the injury or damage, likelihood of the injury or damage occurring, and the time period over which the injury or damage might occur. For

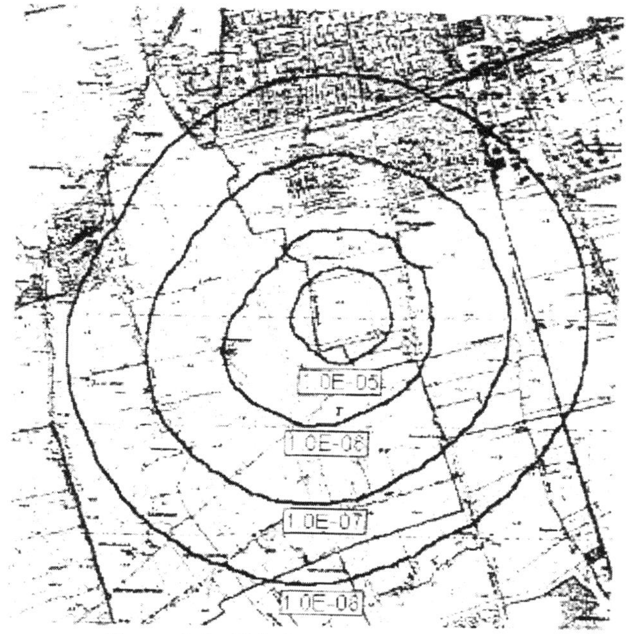


**Figure 6-4.** Risk Contours (CCPS, 2000)

example, a fatality from a fire can be expressed as $1.0 \times 10^{-4}$ fatalities per year from a fire in the facility.

There are different ways to calculate individual risk (CCPS, 2000). The most common are given below:

- *Individual risk contours* illustrate the geographical distribution of individual risk as shown in Figure 6-4. The risk contours show the expected frequency of an event capable of causing the specified level of harm at a specified location, regardless of whether or not anyone is present at that location to suffer that harm. Thus, individual risk contour maps are generated by calculating individual risk at every geographic location, assuming that somebody will be present and subject to the risk 100% of the time.
- *Maximum individual risk* is the individual risk to the person(s) exposed to the highest risk in an exposed population. This is often the operator working at the unit being analyzed, but might also be the person in the general population of highest risk. Maximum individual risk can be determined from risk contours by locating the person most at risk and determining what the individual risk is at that point. The most meaningful means of estimating the individual risk associated with the most exposed person is to track that person's movement throughout his/her

working day and sum the value on a time-exposed weighted basis. This is also a means of setting a basis for restricting a person's exposure in recognized high risk areas. Alternatively, it can be determined by calculating individual risk at every geographical location where people are present and searching the results for the maximum value.

- *Average individual risk* (exposed population) is the individual risk averaged over the population that is exposed to risk from the facility (e.g., all of the operators in a building, or those people within the largest incident effect zone). This risk measure is only useful if the risk is relatively uniformly distributed over the population, and can be extremely misleading if risk is not evenly distributed.

The calculation of individual risk at a geographical location near a facility assumes that the contributions of all incident outcome cases are additive (IChemE, 1985). Thus, the total individual risk at each point is equal to the sum of the individual risks, at a specific point, for all fire scenarios that can impact that point.

$$IR_{x,y} = \sum_{i=1}^{n} IR_{x,y,i} \tag{6-1}$$

where

$IR_{x,y}$ is the total individual risk of fatality at geographical location $x,y$ (chances of fatality per year, or $yr^{-1}$)

$IR_{x,y,i}$ is the individual risk of fatality at geographical location $x,y$ from incident outcome case $i$ (chances of fatality per year, or $yr^{-1}$)

$n$ is the total number of incident outcome cases considered in the analysis

$$IR_{x,y,i} = f_i P_{f,i} \tag{6-2}$$

where

$f_i$ is frequency of incident outcome case $i$, from frequency analysis ($yr^{-1}$)

$P_{f,i}$ is probability that incident outcome case $i$ will result in a fatality at location $x,y$ from the consequence and effort models

### 6.2.5.3. *Societal Risk*

*Societal risks* are single number measures, tabular sets of numbers, or graphical summaries that estimate risk to a group of people located in the effected zone of an incident. Since major incidents have the potential to affect many people, societal risk is a measure of risk to a group of people. It is most often expressed in terms of the frequency distribution of multiple casualty events, such as the F-N curve shown in Figure 6-5. The calculation of societal risk requires the same frequency and consequence information as individual risk. Additionally,

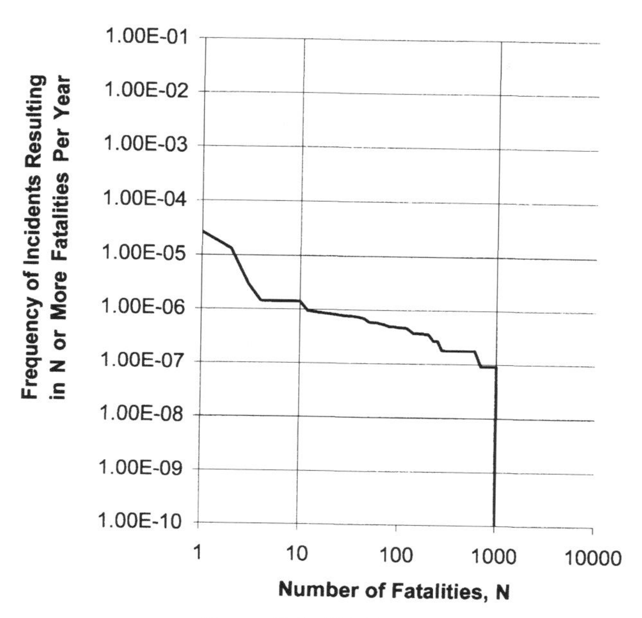


**Figure 6-5.** FN Curve

societal risk estimation requires a definition of the population at risk around the facility. This definition can include the population type (e.g., residential, industrial, school), the likelihood of people being present, or mitigation factors.

Individual and societal risks are different presentations of the same underlying combinations of incident frequency and consequences. Both of these measures may be of importance in assessing the benefits of risk reduction measures or in judging the acceptability of a facility in absolute terms. In general, it is impossible to derive one from the other. The underlying frequency and consequence information are the same, but individual and societal risk estimates can only be calculated directly from that basic data. A high societal risk does not necessarily mean any one person has a high individual risk and vise-versa. These are two very different criteria and should be treated separately.

### 6.2.6. *Other Risks*

#### *6.2.6.1. Financial Impact*

There are different costs associated with a fire at the facility. There is direct cost for damage and loss of equipment, product or materials, business interruption, and increased cost of insurance because the facility now has a history of fire, etc. There is also the cost of possible loss of market share that is not covered by insurance. These costs have been identified and discussed in Chapter 3.

In determining the risk from a financial perspective, a company should add the financial factors they determine important and arrive at a total loss the company is willing to accept. This becomes the criteria used in the risk assessment for financial impact.

#### 6.2.6.2. *Public Disruption/Corporate Image*

When a fire occurs at a facility, it is generally noticed by the public very quickly. Local news organizations are quick to arrive and communicate to the public via television and radio. Damage to reputation can impact public opinion and directly impact the value of a company (the price of stock).

#### 6.2.6.3. *Environmental Impact*

Fires result in impacts on the environment: air pollution, firefighting water runoff, and hazardous and nonhazardous waste.

- *Air pollution*—generally, local environmental agencies are quick to arrive at industrial fires and start monitoring to determine if hazardous materials are in the air that could cause harm to the public or the environment. Depending on the materials on fire, there is a potential for hazardous materials to travel offsite. The Emergency Response Plan must identify potential pollutants and appropriate responses for the public.
- *Water runoff*—the Emergency Response Plan must address where water runoff is to be directed and the potential materials that the runoff can carry. This information will be requested by local environmental agencies.
- *Hazardous and Nonhazardous Waste*—after a fire is extinguished, the clean-up begins. There may be both hazardous and nonhazardous wastes that need disposal. Asbestos wastes may be an issue in older facilities.

Environmental impact is part of a facility's risk tolerance criteria. First, there is the direct cost of clean-up, monitoring costs, and environmental costs. There could be potential fines from local, state, or federal authorities. Finally, there may be potential civil actions that could result in millions of dollars of damage.

### 6.2.7. Risk Tolerance

After the risk is calculated, the results may be compared to either governmental or company criteria to determine if the risk is at an accepted level. If it is, then additional fire protection is not required and the level of fire protection used in the risk calculation is adequate.

If the level of risk does not meet the risk criteria, then additional protection is required. The options for reducing the risk are identified and selected and the analysis is recalculated to determine the impact on the risk. Some options provide significant risk reduction; others have very little impact on the risk.

#### 6.2.7.1. Cost Benefit Analysis

One useful tool of risk assessment is to compare the risk before and after prevention or mitigation to determine the difference in risk. A cost benefit analysis can be completed that determines the cost of the mitigation versus the amount of risk reduction. All costs need to be calculated to determine a cost per year. These costs would include fire damage, injury or fatality, insurance cost increases, loss of profits, etc. The cost of the mitigation, including capital and maintenance costs, needs to be determined.

Cost of Mitigation per Year / Costs of Loss per Year = Ratio

If the ratio is less than 1, then the cost exceeds the benefit. If the ratio is greater than 1, then the cost to provide the mitigation is justified based on reduced risk.

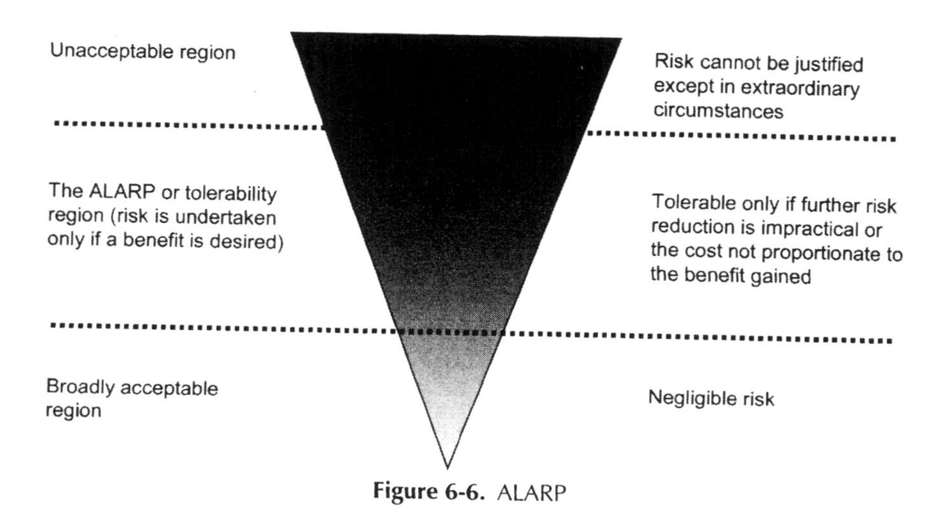


**Figure 6-6.** ALARP

### 6.2.7.2. ALARP

ALARP (as low as reasonably practicable) is a concept that was developed in the United Kingdom, by the Health and Safety Executive (HSE, 2001). It is accepted by regulatory bodies elsewhere and has gained a wide acceptance within the petroleum and hydrocarbon processing industry. This concept is illustrated in Figure 6-6 (modified from HSE, 2001). The figure shows three regions. The top area is where the risk is clearly not acceptable and action must be taken to reduce the risk. The bottom region is where the risk is clearly acceptable and no further action is required. The middle region is the ALARP area where a company must demonstrate that all practical mitigation has been applied and additional mitigation is not reasonable.

There have been attempts by companies to assign quantitative risk numbers to these regions. It is important to note that different values should be considered for workers vs. for the public near the facility. Generally, the risk to the public is one order of magnitude lower than for workers. The workers accept the risk, whereas the public wants a lower risk. **Each company should establish their own risk tolerance values.** This may be used to assess financial and business risk.

**Table 6-5**
*Prioritizing Risk Reduction in a Fire Protection Strategy*

| Method | Definition |
|---|---|
| *Elimination* | The removal or minimization of a hazard through inherently safer design. |
| *Prevention* | The minimization of the likelihood of an event through the installation, integrity, and application of prevention systems to address each hazard through design practices. |
| *Detection and Control* | The limitation of the severity of incidents by early detection of the incident and the operation of safety systems, including fire protection systems through fire and gas detection and process isolation. |
| *Mitigation* | The protection of people and physical assets from the effects of any remaining credible hazards through fire protection systems. |
| *Emergency Response* | The protection of people from the effects of an escalating or catastrophic incident. |

### 6.2.8. Risk Reduction Measures

#### 6.2.8.1. Hierarchy of Risk Reduction

A common aspect of any fire protection strategy is defining how hazards are managed and describing the order of priority for managing those fire hazards. Table 6-5 identifies the priority of how hazards should be managed.

A systematic process should be used to eliminate or minimize hazards:

- For every identified hazard or hazardous activity, determine if it can be designed out of the process
- For every cause, examine ways to make failure inherently less likely to occur through the inherent strength, reliability, longevity, and simplicity of the design
- Examine the severity of hazards for opportunities to minimize and limit damage potential
- Study the immediate impact of the event on the facility and people to see if changes to the layout or the way people operate it can reduce their exposure

#### 6.2.8.2. Risk Reduction Strategies

Measures to reduce the impact of fire include active and passive systems. Active systems include automatic sprinkler, water deluge, water mist, gaseous agent, dry chemical, foam, and standpipe handle systems. Passive protection is provided by fire resistive construction, including spray-applied or cementitious fireproofing of steel, concrete/masonry construction, and water-filled steel columns. Chapter 7 provides details on the design of fire protection systems.

The *CMPT Handbook* notes that fully spray-protected steel plates and beams can be expected to survive pool fires, but rapid failure will still occur for jet fires (Spouge, 1999). The *SINTEF Handbook* notes that there are no commonly accepted formulations or correlations quantitatively describing the interaction between water spray and fires (SINTEF, 1997).

Fire resistant coatings or structural materials can increase the time to onset of structural failure. Much of the data available relates to standardized testing, although there are now numerous analytical techniques to assess performance (Milke, 2002). The important consideration when using either test or analytical methods to evaluate passive fire protection in the process industry is to use a hydrocarbon pool fire exposure. Standard time-temperature fire exposures are used for evaluation of general building construction. Materials for use in the process industry should be evaluated using the higher incident flux "hydrocarbon fire" time-temperature exposure (ASTM E1529, 2000; UL 1709, 1991).

Care must be taken in attempting to correlate the results between standard time–temperature data and corresponding protection with a hydrocarbon exposure. The hydrocarbon exposure flux is on the order of 158–200 kW/m$^2$ ASTM is 158, MC is 200, achieved in the first five minutes of the test. The standard time-temperature curve achieves a flux of about 100 kW/m$^2$ at 60 minutes into the test. Some materials that withstand the standard time–temperature exposure fail rapidly when exposed to the hydrocarbon threat (Scheffey et al., 1993).

### 6.2.9. Reassessment of Risk

In the fire risk assessment, it is important to reevaluate the risk once options for mitigation are determined. The amount of risk reduction should be calculated for each option or combination of options. Often, the results indicate that some options do not provide much, if any, risk reduction. The use of cost benefit analysis can help management in deciding which option to select. Facilities that depend exclusively on the local fire department for fire protection should complete a fire hazard analysis to determine the appropriate fire protection.

# 7

# FIRE PROTECTION FUNDAMENTALS

While Chapters 7 and 8 could be written as a single chapter, its length would make it very difficult for practical use. Chapter 7 is intended to describe a fire protection system in general terms, while Chapter 8 concentrates on where fire protection systems may be applied. Some overlap is provided where necessary for clarity. Chapter 7 provides:

- General information on design of fire protection systems that applies to all systems.
- A basic understanding of fire protection systems (both passive and active).
- General design information on fire protection commonly used in industry.
- References for more detailed information.
- Advantages and disadvantages of the different systems.

Chapter 8 provides guidance on where fire protection may be necessary and suggested design criteria. For example, in process structures, Chapter 8 suggests either fireproofing or water spray to provide structural steel protection against fire with specific guidance on duration of fireproofing or density of water spray. Chapter 8 does not suggest types of materials, how it should be installed or designed, or installation requirements. This information is contained in Chapter 7.

This chapter provides the fundamentals of design for passive and active fire protection systems. A passive system should be used wherever possible, as this is an inherently safer approach than an active system.

The recommended design criteria presented in this chapter are based on generally accepted codes and standards and individual company standards.

**Where values are given for application densities, etc., these should be regarded as typical rather than definitive. The optimum amount of fire protection is affected by a number of factors and greater or lesser values may be acceptable or warranted.**

## 7.1. General Design Criteria

This section addresses general criteria that apply to most fire protection system designs and provides information for use when considering all aspects of fire protection design.

### *7.1.1. Automatic versus Manual Activation*

Active fire protection systems can be installed to provide the desired mitigation by either *automatic* or *manual* activation.

- *Automatic activation* is a system where fire protection devices are integrated with a detection system designed to automatically activate upon sensing fire. The advantage of automatic activation is minimizing the response time delay and, thus, significantly reducing the chance of a larger fire developing. One disadvantage is that when faults occur in the detection or logic associated with the system, there is the potential for unwanted activations which can result in damage to the protected equipment, shutdown of production, and require the fire protection system to be recharged. A second disadvantage is that if there is a failure of the detection system or the activation mechanism, the system does not operate. This is one reason why automatic systems also have a manual activation option.
- *Manual activation* requires a person to activate the system by pushing a button or opening a valve in response to either an observation of a fire or a signal from a detection system. The advantages of manual systems are lower cost, both installation and maintenance, and less complexity. The disadvantage is that manual systems are liable to result in significant delays in activation due to the reliance upon human action. Such delays inevitably result in growth of the fire and associated damage. Delays associated with manual systems may be exacerbated by the growing tendency for having fewer operators in the field.

Consider these points when deciding to use automatic versus manual activation:

- *Automatic Activation Systems*
    - Increased cost of initial installation due to inclusion of the activation system

- Additional maintenance is required
- Damage that could occur due to automatic activation

- *Manual Activation Systems*
  - Additional emergency response capability may be required because the fire may be larger and escalation is more likely
  - Surveillance capability
  - Availability of personnel
  - If a delay in activation occurs, the fire may be larger and longer, resulting in more property damage (cost)

### 7.1.2. Isolation

Process fires will continue and may escalate until the flow of fuel is stopped, the fuel is fully consumed, or the fire is extinguished. Isolation valves are used to reduce or isolate inventories of flammable gases or liquids.

Isolation valves may be located near the property line, the edge of a process unit, or the liquid outlet of a vessel. Valves should be installed on all hazardous materials lines entering or leaving the facility to ensure the facility can be isolated in the event of a spill or fire. Similarly, valves should be located at or near the battery-limit of each unit or outside dike walls for the same reasons and for safety and ease of access.

These battery-limit, unit isolation valves should be clearly identified and installed in easily accessible locations that are a safe distance from potential fire sources. Generally, 25–50 ft (8–15 m) provides an acceptable separation distance. These isolation valves can also be used for turnaround purposes and should be located at or near ground level.

Equipment such as pumps, compressors, tanks, and vessels associated with large inventories of flammable gas or liquid (>5000 gallons) should be provided with equipment emergency isolation valves to stop the flow of material if a leak occurs. For example, the decision to add emergency isolation valves to the inlet and outlet of a compressor is dependent on the flammability of the gas, its pressure, and the quantity of gas in the associated piping and vessels.

Isolation valves can be operated in the following four modes:

- *Remote automatic*—based on a sensing element, logic solver, and output signal to close the valve. In most cases, these automatic valves are designed to fail safe. Automatic isolation can be associated with the automatic activation of a shutdown system for equipment or a process unit.

- *Local automatic*—some isolation valves may be automatic, but locally actuated. In this case, the actuation device may be a fusible-link or fusible-tubing arrangement.
- *Remote manual*—unit or equipment isolation valves are sometimes in elevated locations. These valves are activated manually from a remote location.
- *Local manual*—manual unit and equipment isolation valves may be operated by a handle or by an actuator. Typically, these single action actuators are powered by air and are of a spring-to-close design.

All isolation valves should be clearly identified by sign or color. For additional information, see Section 8.1.10.

### *7.1.3. Depressurization*

When the shell of a vessel is exposed to extreme heat on the outside and the inside is in contact with vapor, metal temperatures may reach levels where tensile strength will be reduced such that rupture may occur even though the pressure does not exceed the set point of the pressure relief valve [for ordinary carbon steel, this temperature is ~900°F (482°C)]. Vapor depressuring provides fire protection for process units by reducing the contained pressure at such a rate that:

- There is a significant reduction in the chance of rupture of any steel pressure vessel exposed to fire.
- The driving force behind pressure jet flames is rapidly decreased.
- The leakage rate of liquid spills decreases, allowing containment of pool fires.

Depressurization may also be used as a protection during startup, shutdown, and short-term upsets by reducing the pressure to prevent a pressure relief valve discharge. However, vapor depressuring should not be used as a permanent operating control in lieu of correcting the causes of a process upset. There are several advantages of vapor depressuring, including the following:

- An auto-refrigeration effect may be produced in a pressure vessel, which provides cooling of liquid contained in the vessel.
- The need for a liquid blowdown system is eliminated. By retaining liquid in the vessel as a heat sink, any increase in temperature of the wetted shell is minimized.

It should be noted that vapor depressuring may not be practical when the vessel design pressure is less than 100 psig (690 kPa) because valves and piping can become unreasonably large and costly or when the vapor depressuring load governs the size of pressure relief and flare headers. Refer to API RP 521,

*Guide for Pressure-Relieving and Depressuring Systems* (API, 1997) and *Guideline for Protection of Pressurized Systems Exposed to Fire* (Scanpower, 2002).

#### 7.1.3.1. Design

Good engineering judgment is required when investigating the application of vapor depressuring. Vapor depressuring design considerations include the following:

- Vessels should be depressured to at least 50% of the design pressure within approximately 15 minutes for vessels exposed to pool fires with wall thicknesses of 1 in (2.5 cm) or greater. Jet flame impingement requires more rapid rates.
- Vessels with thinner walls usually require a greater depressurization rate.
- The depressurization rate percentage is based on the wall thickness, initial metal temperature, and the rate of heat input from the fire (API RP 521).
- Bare pressure vessels in a process unit protected by vapor depressuring usually do not require protection by fixed water spray systems.

Some companies limit the application to facilities operating over 250 psig (1724 kPa), while others depressure all light hydrocarbon processes.

#### 7.1.3.2. Depressuring Flow Rate

To calculate the vapor flow rate needed to accomplish depressuring, the maximum expected operating pressure of the vessels under consideration should be used as the initial pressure, and 50% of the design pressure or 100 psig, whichever is lower, after a maximum of 15 minutes as the final pressure (API RP 521). When estimating flows for sizing depressuring valves and piping, it is important to consider the:

- Effect of the initial depressuring rate on the closed pressure relief system and flare
- Average depressuring rate used to determine depressuring time
- Variations in temperature, pressure, vapor composition, and possible liquid entrainment into the relieving system during the depressuring period

#### 7.1.3.3. Depressuring Valves

Vapor depressuring valves may be electric motor-operated gate valves, pneumatically operated control valves, or manually operated valves. Pneumatically operated valves should open on air failure, with provision to maintain pressure on the diaphragm in the event of instrument air failure. A back-up air supply should be provided; for example, an air tank provides a simple solution. When pneumatically operated control valves are used, they should be of the tight-shutoff type. Locked-open block valves should be considered to facilitate the testing and maintenance of the depressuring valve. Features of a typical vapor depressuring valve include:

- A vapor depressuring valve is usually installed in a bypass line around the vessel pressure relief valve that handles the largest relief load in the process train.
- A vapor depressuring valve should discharge into the relief valve header, not to the atmosphere.
- Normally locked-open block valves should be considered upstream and downstream of each depressuring valve to permit testing and maintenance.
- The actuator and power supply for any depressuring valve with a double action or an energize to open actuator that may be exposed to fire should be fireproofed for a minimum of 15 minutes, per the UL 1709 high rise fire test. Fireproofing of the actuator and power supply is not required with the fail-open design.

In sizing depressuring valves, it should be assumed that heater burners are shut-off, reboilers are shutdown, and normal flow in the vessel has ceased. Vapor depressuring valves should be designed such that the initial, instantaneous depressuring flow rates do not exceed the capacity of the closed pressure relief system and the flare.

### 7.1.4. Approved/Listed Equipment

In the context of fire protection system design, approved or listed equipment for fire service is generally understood to mean acceptable to the authority having jurisdiction (AHJ). The AHJ includes organizations or individuals that are responsible for accepting equipment, materials, an installation, or a procedure. In other words, the AHJ could be a fire marshal, insurance company, owner, or some other organization.

Listed equipment is typically defined as materials that have been tested to a standard by an organization that is acceptable to the AHJ. The listing organization, such as Underwriters Laboratories (UL) or Factory Mutual (FM), maintains a periodic audit of listed equipment or materials to ensure that the equipment or material meets appropriate designated standards or has been tested and found suitable for a specified purpose. The testing organization then publishes a list of equipment that has passed the evaluation process. Hence, the fire protection community uses the term "listed" to define equipment that has been specifically evaluated for use in fire protection applications.

The use of listed equipment is important for demonstrating that fire protection system components can function in accordance with the requirements of a particular design standard. In addition, listed equipment has a level of reliability that can be traced back to a standardized test protocol. This is not to suggest that listed equipment is the best equipment for every application. The higher capacity and higher pressure rated equipment (pumps, valves, etc). necessary for optimum protection in processing facilities is not always available as listed. Listing is an expensive process for manufacturers and the market for this larger equipment is relatively small and not economically justifiable. Such equipment is available in the required capacity/pressure rating that meets the same specification as the smaller listed equipment and considered to be equivalent. Every effort should be made to purchase such equivalent equipment.

### *7.1.5. Qualification and Competence of Personnel*

The competence of personnel who select, design, install, test, and maintain fire protection systems is ultimately the factor that controls the effectiveness and reliability of fire protection systems. Many of the personnel described in the following sections are subject to jurisdictional requirements for licensing.

#### *7.1.5.1. Fire Protection Engineer*

The company or facility should make use of the services of an engineer knowledgeable and trained in fire protection. Ideally, a registered fire protection engineer should be available to review fire protection designs. Fire safety, loss prevention, or process safety engineers should assist in the analysis of hazards, selection of protection system specifications, approval of the system, and acceptance testing.

#### *7.1.5.2. Facility Personnel Who Perform Fire Protection Inspections*

Facility personnel who perform inspections on fixed fire protection systems must be knowledgeable of the systems and have received training on the inspection protocols. Facility personnel may be operations personnel, maintenance personnel, security personnel, fire protection personnel, or others assigned by management.

#### *7.1.5.3. Facility Personnel Who Perform Testing and Maintenance*

Facility personnel who perform testing and maintenance on fire protection equipment and systems must be trained, experienced, and knowledgeable in the systems and the protocols for testing and maintenance. Knowledge can include work history, educational experience, craft certification, manufacturer certification, field verification, and job assessment and testing. Facility personnel may be pump mechanics, pipe fitters, instrument technicians, electrical technicians, millwrights, fire protection personnel, or other qualified personnel assigned by management. The facility should have National Institute for Certification in Engineering Technologies (NICET) certified personnel on staff or contract with a company providing that level of competency. For additional information on certification, refer to NICET's website at http://www.nicet.org .

#### *7.1.5.4. Fire Protection Service Companies*

Many facilities outsource all or part of the design, installation, and testing and maintenance functions to engineering and fire protection service companies. These companies perform a variety of services including design of fixed systems, flow test of water systems, operational test of equipment, maintenance and repair systems, and design and installation of new systems.

It is important that the outside service company meet the facility pre-qualification criteria. As part of the selection process, it must be determined if the service company is knowledgeable and experienced in the types of fire protection systems they will be designing, installing, or maintaining.

The service company should be able to provide evidence that personnel performing design, installation, and maintenance are qualified to perform the work.

Fire protection service companies should have a NICET certified person on staff. A registered fire protection engineer should be available to the service company, especially if design is being performed.

### *7.1.6. Life Safety*

Life safety addresses construction, protection and occupancy features in buildings necessary to minimize danger to life from fire, including smoke, fumes, or panic. NFPA 101, The Life Safety Code, has been adopted by many jurisdictions. The requirements of NFPA 101 are not intended for property protection, but are designed for life safety. There may be some life safety benefits that are achieved from the property protection features provided, or vice versa.

#### *7.1.6.1. Buildings*

The term *building* can include office, laboratory, control building, warehouse, maintenance, and open/closed process structures.

The life safety features that apply begin with the classification of the building or structure under evaluation. Classification of a building is a combination of hazard of contents and occupancy.

Hazard of contents classifications are low, ordinary, or high depending on the amount and type of fuel available to burn.

Occupancy classifications fall into the following categories:

- Assembly
- Educational
- Health care
- Detention and correctional
- Residential
- Mercantile
- Business
- Industrial
- Storage
- Mixed occupancies

In many cases, buildings are classified as mixed occupancy, which means that within a single building there may be several occupancy types, depending on the different activities carried on in it.

Most of the buildings at chemical, petrochemical, and hydrocarbon processing facilities will fall into the *Industrial* classification. Industrial buildings and structures are classified in NFPA 101 as follows:

- *General Industrial Occupancy*—ordinary and low hazards in buildings of conventional design suitable for various types of processes.
- *Special Purpose Industrial Occupancy*—ordinary and low hazards in buildings designed for, and suitable only for, particular types of operations involving a low density of employee population with much of the area occupied by vessels, equipment, piping or machinery.
- *High Hazard Industrial Occupancy*—high hazard materials or processes, or high hazard contents. High hazard occupancy is one with significant fire or explosion hazards involving flammable liquids or gases, liquefied flammable gases, heated combustible liquids, or combustible dusts. Incidental high hazard operations within low or ordinary occupancies should be the basis for the overall occupancy classification.
- *Open Structures*—structures supporting equipment or operations not enclosed within building walls, such as those found in refineries, chemical processing, or power plants. Roofs or canopies without enclosing walls should not be considered an enclosure.

For the processing industries, most process structures fit into the *special purpose industrial, high hazard industrial,* or *open structures* classification. See NFPA 101 for more information.

Once the occupancy and hazard of contents classifications have been defined, then appropriate life safety design features can be identified. These design features include exit access, travel distance, number of exits, exit width, emergency lighting, etc.

#### 7.1.6.2. Exits and Arrangements

For open ground level processing areas, the provision of emergency evacuation routes is relatively simple. There are no specific codes and standards. However, by common practice of equipment and vessel location and piping arrangement, reasonably straight pathways should be provided through the processing area to its perimeter. These pathways should allow egress in at least two different directions and be free of trip hazards, have adequate illumination, and avoid proximity to potential hazards, such as spill accumulation areas.

Elevated process structures require particular attention to egress to permit personnel to escape safely and quickly in an emergency. Some important definitions are:

- *Common Path of Travel*—that portion of the route to an exit access that must be traversed before two separate and distinct paths of travel to two separate exits are available.
- *Dead-End*—a potential corridor, area, or path of travel that has no opportunity for exit at its end, forcing a person to retrace the path traveled to again reach a choice of egress routes to an exit.

Common paths of travel and dead-ends in general industrial and special purpose industrial occupancies cannot exceed 50 ft (15 m) and are prohibited in high hazard industrial occupancies. An exception allows dead-end corridors in general industrial and special purpose industrial occupancies fully protected by automatic fixed fire protection to be up to 100 ft (30 m).

In general, at least two means of egress are required from every level, story, or section of a process structure. There are limited exceptions allowing a single means of egress for certain structures, defined as low hazard and for small areas, as follows:

- Structures that do not handle flammable or combustible materials (considered low or ordinary hazard industrial special purpose occupancies).
- If the travel distance to the exit or stair does not exceed:
  - 50 ft (15 m)
  - 100 ft (30 m) (with automatic fixed fire protection)
- Levels, floors, or platforms with an area less than 200 ft$^2$ (18 m$^2$) if the travel distance to the exit or stair does not exceed 25 ft (8 m).
- If occupied by not more than three persons.

The required two means of egress from each level should be separated from each other as much as possible. Based on the maximum overall diagonal dimension of the area or level, the stairway entrances should be separated by the following minimum distances:

- *With* automatic fixed fire protection, 1/2× diagonal dimension.
- *Without* automatic fixed fire protection, 1/3× diagonal dimension.

Properly designed ladders are recognized and allowed as the primary and, possibly only, means of access to and egress from certain locations. In common practice, these are locations accessed only by people capable of using a ladder, not frequently attended, and not occupied by more than three people. These may include:

- Towers and columns
- Tops of horizontal vessels
- Tops of smaller vertical cylindrical tanks
- Platforms mounted on the sides of vessels or equipment
- Platforms around machinery

Ladders are permitted as a second means of egress from areas not considered a high hazard, such as from boiler rooms or similar utility or equipment spaces that are normally not occupied by more than three people who are all capable of using a ladder.

### *7.1.6.3. Enclosed Process Structures*

In special purpose industrial and high hazard industrial occupancies where unprotected vertical openings exist, (graded flooring is an example of unprotected vertical openings), every floor level must have direct access to enclosed stairs or other exits protected against obstruction by any fire or smoke in the open areas connected by the vertical openings.

For enclosed process structures, any inside stairs that serve as an exit or exit component require enclosure to provide personnel protection from smoke and fire during emergency egress. The stair enclosures serving as an exit way must have the following fire-resistive construction:

- *3 stories or less*—at least 1-hour fire resistance rating
- *4 stories or more*—at least a 2-hour fire resistance rating

For enclosed process structures, open outside stairs may be used for an exit, but require protective separation from the interior of the building by walls with the same fire resistance rating as required for enclosed stairs. This construction should extend vertically from the ground to a point 10 ft (3 m) above the topmost landing of the stairs or to the roofline, whichever is lower, and to a point not less than 10 ft (3 m) horizontally.

## 7.2. Fire Control

There are different types of fire and many different firefighting agents for combating them. An understanding of how these different types of firefighting agents are used in fire protection is important because their effectiveness can vary widely when applied to different types of fires.

### *7.2.1. Type of Fires*

There are innumerable situations where gases, liquids, and hazardous chemicals are produced, stored, or used in a process that, if released, could potentially result in a fire. It is important to analyze all materials and processes associated with a particular process including production, manufacturing, storage, or treatment facilities. Each process requires analysis of the potential for fire.

Hydrocarbon fires are a principal concern in many processing facilities. There are many different types of hydrocarbon fires. The mode of burning depends on characteristics of the material released, temperature and pressure of the released material, ambient conditions, and time to ignition. Types of hydrocarbon fires include:

- Jet fires
- Unconfined vapor cloud fires or flash fires
- Pool fires (two-dimensional fires)
- Running liquid fires (three-dimensional fires)
- Boiling liquid expanding vapor explosions (BLEVE) or fire balls

Figure 7-1 illustrates two- and three-dimensional fires. Other fires that can occur in specific areas within a processing facility include:

- Solid material fires, e.g., fires involving wood, paper, dust, plastic, etc.
- Warehouse fires
- Electrical equipment fires, e.g., transformer fires

**Figure 7-1.** Two- and three-dimensional fires

- Fires involving oxygen, e.g., systems for oxygen addition to a Fluid Catalytic Cracking (FCC) unit
- Fires involving combustible metals, e.g., sodium
- Fires involving pyrophoric materials, e.g., aluminum alkyls

### 7.2.2. General Control Methods

Fire protection systems should be designed to accomplish a combination of the following objectives:

- Extinguishment of fire
- Control of burning
- Exposure protection
- Prevention of fire

#### 7.2.2.1. Extinguishment of Fire

Fire protection systems achieve extinguishment of fire by a number of methods, principally:

- Reducing the heat release rate of a fire and preventing flashback by cooling—this reduction of the heat release rate and cooling usually occurs by direct and sufficient application of cooling medium through or into the fire plume and onto the burning fuel surface.
- Separating fuel vapors from oxygen (smothering).
- Inhibiting chemical chain reaction.

For example, extinguishment of fire by water is accomplished by any or a combination of cooling, smothering from produced steam, emulsification of some liquids, and dilution.

#### 7.2.2.2. Control of Burning

Fire protection systems achieve control of burning by limiting the size of a fire by:

- Distribution of extinguishing agent to absorb heat released of decrease the heat release rate
- Providing exposure protection to adjacent combustibles
- Containment

Control of burning systems operates until one of the following occurs:

- Agent supply is exhausted
- Burning fuel is consumed
- Flow of fuel is stopped
- Leaking fuel is extinguished

The control of burning principle may be applied where combustible materials are not susceptible to complete extinguishment or where complete extinguishment is not considered desirable.

For example, control of fires is accomplished by application of water to a burning material to control fire spread. Likewise, it is not advisable to completely extinguish a fire involving flowing gaseous fuel until the source of the fuel is stopped, otherwise the release of unburned gas will continue and may be reignited by another source of ignition, resulting in a larger fire or perhaps an explosion.

#### 7.2.2.3. *Exposure Protection*

Fire protection systems achieve exposure protection by absorption of heat through application of extinguishing agents to structures or equipment exposed to a fire. The application of some extinguishing agents removes or reduces the heat transferred to the structures or equipment from the exposing fire, as well as limits the surface temperature of exposed structures and equipment to a level that will minimize damage and prevent failure.

Exposure protection systems provide protection by the application of water to structures and equipment for the anticipated duration of the exposure fire. Water spray curtains are generally less effective than direct application due to unfavorable conditions such as wind, thermal updrafts, and inadequate drainage. Extinguishing agents such as $CO_2$ or dry chemical agents are not able to provide this type of cooling.

#### 7.2.2.4. *Prevention of Fire*

Fire protection systems achieve prevention of fire by operating until flammable vapor, gases, or hazardous materials dissolve, dilute, disperse, or cool.

### 7.2.3. *Firefighting Agents*

Before fire protection systems can be adequately discussed, it is important to understand the agents available for fire protection and their particular application. These include:

- Water
- Foam
- Carbon Dioxide
- Dry Chemical
- Clean Agents

**WARNING : Some firefighting agents are not compatible with certain chemicals. For example, multipurpose, dry chemical is not compatible with some oxidizers. See the MSDS for the involved materials being protected.**

Table 7-1 highlights the advantages and limitations of the various extinguishing agents.

*7.2.3.1. Water*

Water is not suitable for electrical (Class C) fires, but is effective on all combustible (Class A) and many flammable liquid (Class B) fires. Water can be applied by hose streams, monitors, sprinklers, water spray systems, or as water curtains for such purposes as:

- Fire extinguishment
- Fire control

**Table 7-1**
*Advantages and Limitations of Various Extinguishing Agents*

| Agent | Type Extinguishment | Advantages | Limitation |
|---|---|---|---|
| Water | Cooling<br>Smothering<br>Dilution<br>Exposure | Available<br>Very low cost | Not for Class C electrical fires<br>Freezes at 32°F (0°C)<br>Reactive with some material, e.g., sodium, magnesium<br>Cannot extinguish low flash point materials |
| Foam | Smothering | Best for Class B Pool Fires (Two-dimensional fires) | Not for electrical fires<br>Foam blanket may break-up<br>Not applicable for LPG |
| $CO_2$ | Smothering<br>Reduction<br>Some cooling | Nonreactive<br>No residue<br>Class C | Reduces $O_2$ level<br>Toxic to people (asphyxiant)<br>Not applicable for oxidizers |
| Dry Chemical | Chain breaking | Class B & C | Fire reflashes if not completely extinguished or hot surfaces are present (especially flammable/combustible liquids) |
| Clean Agent | Chain breaking<br>Inerting | Good for Class A, B, C | Not for outdoors<br>May produce toxic gases |

- Cooling of equipment, tanks, or structures
- Dispersion of flammable or toxic vapors

Water extinguishes fire primarily by cooling. This cooling is most effective when the water is applied as fog or spray. One gal (3.8 l) of water absorbs about 6,000 BTU (1512 k/cal) when vaporized to steam. Even when complete extinguishment is not achieved (such as during fires involving low flash point liquids), much of the heat is absorbed, and damage is vastly reduced. **In the process industry, the prime purpose of applying water streams in a fire situation is to provide cooling and containment, not extinguishment.** Water can also be used as a mist, which has different extinguishing characteristics than water streams.

A secondary means of extinguishment results when steam, created by evaporation of water to about 1,700 times the original volume, limits combustion by displacing oxygen in enclosed fire areas. Water can also exclude air when applied over heavier than water liquids.

#### *7.2.3.2. Foam*

Foams for fire protection are an aggregate of air-filled bubbles that float on the surface of a flammable liquid. These foams are made from an aerated water solution and a small percentage of foam liquid concentrate. They are used principally to form a cohesive floating blanket on the liquid surface that extinguishes the fire by excluding air, thus smothering and cooling the fuel. Foam application also helps prevent reignition by averting formation of combustible mixtures of vapor and air. Low expansion foams are suited particularly for extinguishing two-dimensional (pool) flammable liquid fires that involve spills, or storage tanks where the foam forms a vapor-sealing blanket that secures the surface after extinguishment. High expansion foams are good for contained or indoor three-dimensional fires. High expansion foams are also used for control of LNG spill fires by absorbing heat energy. When a fire involves polar or water-miscible liquids, only foams compatible with such liquids should be used.

##### 7.2.3.2.1. FLUOROPROTEIN FOAMS

Fluoroprotein foam is available as concentrates for proportioning with water to a concentration of 3% or 6%. The manufacturer should be consulted for the correct concentrate to be used in a particular system. Proportioners, the devices that meter the concentration, must be designed and set for the percent of foam liquid concentrate to be used.

Fluoroprotein foam is produced through the turbulent mixing of atmospheric air into the foam solution (water and foam liquid concentrate). This turbulence is typically produced by air introduced into the solution by venturi action (foam maker).

#### 7.2.3.2.2. AQUEOUS FILM FORMING FOAM (AFFF)

The air foams generated by AFFF solutions possess a low viscosity and have fast spreading, leveling and self-sealing properties. Like other foams, AFFF acts as a surface barrier to limit fuel vaporization and exclude air. To ensure extinguishment, the foam blanket should cover the entire surface of the spill or fire.

The foam produced with Aqueous Film Forming Foam (AFFF) concentrate is dry-chemical compatible. Protein and fluoroprotein foam concentrates and AFFF concentrates are incompatible and should not be mixed; although foams separately generated with these concentrates are compatible and can be applied to a fire in sequence or simultaneously. AFFF is available in various liquid concentrate percentages.

AFFF concentrates should not be used to combat fires involving water-miscible and polar liquids unless they are the "alcohol-resistant" type. A high concentration and or application rate may be required for extinguishing these liquids. The manufacturer should be consulted regarding the suitability of a foam concentrate for a particular liquid. These concentrates have components that chemically react with polar solvents to form a plastic-like film. AFFF concentrates can be used in conventional foam-making devices. Greater concentrations and application rates are required when combating water-miscible and polar liquids than are required when combating hydrocarbon fires. The foam manufacturer should be consulted for instructions.

AFFF is suitable for subsurface injection into tanks containing ordinary hydrocarbons; it is not suitable for subsurface injection into storage tanks containing water-miscible liquids or polar solvents, or for oil with a viscosity of 2000 or more Saybolt Standard Units (SSU).

#### 7.2.3.2.3. HIGH-EXPANSION FOAM

High-expansion foam is made from a synthetic foam concentrate diluted with water as recommended by the manufacturer. The foam equipment operates by passage of air or other gas through a screen or net that is wetted by the foam solution. The foam may also be made in an air-aspirating device. Foam expansion ratios of 100:1 to 1,000:1 can be obtained. This foam is particularly well suited for confined Class A or Class B fires in buildings, warehouse pits, etc., displacing the vapor, heat and smoke. Because of the low pressure developed by the generating equipment and the low weight of the foam, high expansion foam may not be effective on outdoor fires due to wind, thermal currents, etc. Manual firefighting can then become impossible due to lack of visibility.

### *7.2.3.3. Halon*

Halogenated extinguishing agents are hydrocarbons where one or more hydrogen atom is replaced by fluorine, chlorine, bromine, or iodine atoms. The substituted atom is not only rendered nonflammable, but it acts as a very efficient chain-breaking agent for many of the reactions that occur during combustion. These extinguishing agents are simply known as *halon*.

The Montreal Protocol of September 1987 restricts the production of certain halon extinguishing agents, most notably Halon 1301. Existing halon fire protection systems may remain in service. However, the dwindling supply of halon continues to drive up the cost of recharging existing systems.

#### *7.2.3.4. Carbon Dioxide*

Carbon dioxide ($CO_2$) is a noncorrosive, electrically nonconductive, residue-free gas that will not freeze or deteriorate with age. For fire extinguishment or inerting purposes, $CO_2$ is stored in liquid form for low pressure systems and gaseous form for high pressure systems. In both designs, the motive force for discharge is provided by the inherent pressure of the $CO_2$. Its pressure will decrease with temperature, however, this is not significant until the atmospheric temperature is below 0°F (–18°C). $CO_2$ may be applied for fire extinguishment through three different media: hand hoses with $CO_2$ supplied cylinders, total flooding fixed systems, and local application fixed systems. $CO_2$ fixed fire extinguishing systems are used extensively for protecting highly valuable equipment, occupancies, or enclosures susceptible to fire, smoke or water damage. They are also used where an electrically nonconductive, nonresidue forming extinguishing agent is needed.

Carbon dioxide gas is an asphyxiant, a potent respiratory stimulant, and both a stimulant and depressant of the central nervous system. Fatalities have occurred after people have entered enclosures where air has been largely displaced by $CO_2$. Therefore, fixed, automatic $CO_2$ systems require a time delay pre-evacuation alarm period (often 30 seconds), warning signs, and an alarm signal incorporated into the system design to allow sufficient time for personnel evacuation prior to $CO_2$ release. Verification of the oxygen level must be made prior to reentry.

#### *7.2.3.5. Dry Chemicals*

There are several types of dry chemical extinguishing agents. The three most popular are sodium bicarbonate, potassium bicarbonate (Purple K), and monoammonium phosphate.

##### 7.2.3.5.1. SODIUM BICARBONATE

Sodium bicarbonate- and borax bicarbonate-based dry chemicals were among the first dry agents used in portable fire extinguishers. About 1960, the sodium bicarbonate dry chemical was modified to make it compatible with protein-based, low-expansion foam to create a dual agent extinguishing system. Shortly thereafter, the more effective Purple K based extinguishers replaced sodium bicarbonate extinguishers.

#### 7.2.3.5.2. POTASSIUM BICARBONATE (PURPLE K)

Potassium bicarbonate has greater extinguishing capability on Class B fires than sodium bicarbonate and is usually foam compatible. This characteristic permits emergency response personnel to fight fire more effectively.

#### 7.2.3.5.3. MONOAMMONIUM PHOSPHATE

Monoammonium phosphate, or a mixture of monoammonium and diammonium phosphate, a multipurpose dry chemical, is the only dry chemical that is effective on Class A combustible, non–deep-seated fires as well as Class B flammable liquid and gas fires. It is more effective on Class B fires than sodium bicarbonate, but less effective than potassium bicarbonate. It may be compatible with AFFF foams, depending on the manufacturer's process and its listing. This agent has disadvantages when used on a Class C fire. Its securing action, due to agent decomposition, results in the formation of a sticky, tar-like residue, which solidifies on cooling. This not only leads to major clean-up problems, but of greater importance, could damage delicate instruments, equipment, and electrical wiring. Furthermore, monoammonium phosphate is somewhat corrosive to metals.

### 7.2.3.6. *Clean Agents*

Clean agents are nonconductive, volatile, or gaseous fire extinguishants that do not leave a residue upon evaporation. Clean agents fall within two categories: halocarbons and inert gases. Typical halocarbons include hydrofluorocarbons (HFCs), hydrochlorofluorocarbons (HCFCs), perfluorocarbons (PFCs or FCs), and fluoroiodocarbons (FICs). Typical inert gases include argon, nitrogen, carbon dioxide, or combinations of these agents.

Clean agent systems can also be used for explosion prevention and suppression where flammable materials are confined. Clean agent fire extinguishing systems are used primarily to protect hazards that are in enclosures or equipment enclosures. Some typical hazards that could be protected by clean agents are:

- Electrical and electronic hazards
- Subfloors and other concealed spaces
- Flammable and combustible liquids
- Telecommunication equipment hazards

Clean agents *should not* be used on fires involving the following:

- Materials capable of rapid oxidation, e.g., cellulose nitrate and gunpowder
- Reactive metals, e.g., lithium, sodium, potassium, magnesium, titanium, zirconium, uranium, and plutonium

- Metal hydrides
- Chemicals capable of decomposition, e.g., organic peroxides and hydrazine

Some clean agents may not be appropriate for all applications as they can produce materials toxic to people, especially in inerting concentrations.

## 7.3. Passive Protection Systems

Passive protection systems discussed in this section are:

- Spacing and layout
- Fireproofing
- Containment and drainage
- Diking
- Fire walls
- Electrical area classification
- Static electricity, lightning, and stray current protection

### *7.3.1. Spacing and Layout*

This section discusses general layout and spacing guidelines as they apply to fire protection issues. For complete discussion of facility siting and layout, refer to *Guidelines for Facility Siting and Layout* (CCPS, 2003b).

The purpose of layout and spacing is to design a workplace that will minimize personnel injuries, overall property damage, and related business interruption resulting from potential toxic releases, fires, and explosions. Areas to address during layout and spacing include both those that will minimize the incident size and those that will minimize the incident impact. The magnitude of a potential incident may be reduced by:

- Minimizing or eliminating piping runs through unrelated units by locating vulnerable, interconnecting piping and equipment adjacent to one another.
- Keeping fuel separate from potential sources of ignition.
- Separating high consequence areas from high risk operations.
- Removing pumps from underneath fin fans.

Separation distances are typically determined through one of two methods: utilizing spacing tables or calculating potential fire distances. Company spacing tables are generally developed based on past practice, learnings from incidents, regulatory data, and engineering experience. *Guidelines for Facility Siting and Layout* (CCPS, 2003b) contains spacing tables that have been developed based on a review of various major refining and petrochemical company spacing tables, insurance guidelines, historical spacing guidance, regulations, and engineering experience. Table 7-2 (CCPS, 2003b) provides an overview of the spacing tables. Although spacing tables may not provide an exact, analytical answer, they are a means to quickly determine layout and spacing requirements, while taking advantage of the significant experience contained in the tables. When spacing tables are used, care should be exercised to assure that the spacing table is applicable for the process being used and the hazard of concern.

It is important to note that the distances in the spacing tables are based on potential fire consequences. Explosion and toxic concerns may require greater spacing. The separation distances cited are suggested minimum distances. Increased spacing may be warranted based on site-specific hazards and concerns. Applicable codes, standards, and local regulations should be researched.

If the spacing table is not applicable to the process, then an alternative method should be used. One alternative to the spacing tables is to develop spacing distances for the site's specific layout and process parameters through fire and explosion consequence modeling. Given the large number of equipment pieces involved in a site layout, this can be a time-consuming endeavor. Basic steps when taking this approach are to:

- Identify the hazards inherent in the process unit.
- Identify the release of flammable materials that could result from accidents involving the hazards.
- Calculate the fire and explosion impacts on exposed process equipment, populations, facilities, and adjacent areas.

**Table 7-2**
*Spacing Table Overview*

| Tables Contained in *Guidelines for Facility Siting and Layout* | Subject |
|---|---|
| Separation Distances for Equipment | Primarily spacing between pieces of equipment in the same unit and spacing between that equipment and the edge of the unit |
| Tank Spacing to Other Areas and Equipment | Primarily separation distances between storage tanks and site/unit boundaries |
| Tank to Tank Spacing | Separation distances between various types of storage tanks |
| Minimum Spacing Requirements for Onsite Buildings | Primarily distances between buildings and major site features such as process and property boundaries, tanks storage, loading racks, and utilities |

- Identify the opportunities to mitigate the consequences of incidents in terms of magnitude and impact.

A hazard analysis may be performed to review reductions in the suggested separation distances. Increased risk can be mitigated by providing additional safeguards, such as fireproofing, automatic water-spray systems, emergency shutdown systems, or additional firefighting equipment.

Certain siting and layout guidelines apply to the entire site, as described below:

- Firefighting access should be provided from at least two directions and should not require crossing an adjacent unit. Access ways should be provided at least every 200 ft (61 m) and be at least 20 ft (6 m) wide and not pass under pipe ways. These access ways serve as firebreaks and permit firefighting from two directions with 100 ft (30 m) lengths of hose.
- The classification of electrical equipment should be considered relative to the surrounding equipment and the area electrical classification plan.

In addition to fire protection considerations, access for maintenance activities must be provided above, below, and between equipment. Locate equipment subject to frequent maintenance and cleaning close to unit boundaries for ease of access. Lifting arrangements for pumps, heavy valves, and other equipment should be considered in spacing and layout of the unit.

#### *7.3.1.1. Fire Barriers*

Fire barriers should be considered when the spacing recommended can not be met and hazards are not easily mitigated with active fire protection systems. Barriers, such as walls, partitions, and floors, provide physical separation of spaces and materials. The effectiveness of a fire barrier is dependant on its fire resistance, materials of construction, and the number of penetrations. Inattention to the integrity of penetrations is one of the primary reasons fire barriers fail to provide proper protection. Factors to consider in the design and placement of fire barriers include:

- Types, quantity, density, and location of combustible materials
- Location and configuration of plant equipment
- Consequence of fire exposure on adjacent plant equipment
- Location of fire detection and suppression systems

Fire barriers have been used to separate:

- Adjacent turbine generators beneath the underside of the operating floor
- Battery rooms from adjacent areas

- Cable spreading room(s) and cable tunnel(s) from adjacent areas
- Control room, computer room, or combined control/computer room from adjacent areas
- Emergency diesel generators from each other and from adjacent areas
- Fire water pumps from adjacent areas
- Fossil fuel-fired auxiliary boiler(s) from adjacent areas
- Fuel oil pumping, fuel oil heating facilities, or both, used for continuous firing of the boiler from adjacent areas
- Large turbines from compressors
- Main fire water pump(s) from reserve fire water pump(s) when these pumps provide the only source of fire protection water
- Maintenance shop(s) from adjacent areas
- Office buildings from adjacent areas
- Process areas with potential for runaway reactions
- Rooms with major concentrations of electrical equipment, such as switchgear room and relay room, from adjacent areas
- Storage areas for flammable and combustible liquid tanks and containers from adjacent areas
- Storage of reactive materials, such as peroxide catalysts
- Telecommunication rooms from adjacent areas
- Warehouses from adjacent areas

Any opening in a fire barrier is a potential weak point. It is advisable to minimize the number of openings. All openings in fire barriers should be provided with fire door assemblies, fire dampers, penetration seals (fire stops), or other approved means having a fire protection rating consistent with the designated fire resistance rating of the barrier.

### *7.3.2. Fireproofing*

This section discusses basic design guidelines for fireproofing or passive fire protection in areas where flammable liquids and gases are processed, handled, and stored. API 2218, *Fireproofing Practices in Petroleum and Petrochemical Processing Plants* (API, 1999) can be referenced for additional information.

Fireproofing is a fire resistant material or system that is applied to a surface to delay heat transfer to that surface. Fireproofing, a form of passive fire protection, protects against intense and prolonged heat exposure that can cause the weakening of steel and eventual collapse of unprotected equipment, vessels, and supports and lead to the spread of burning liquids and substantial loss of property. The primary purpose is to improve the capability of equipment/structures to maintain their integrity until the fire is extinguished by either stopping the fuel source or active fire protection methods.

The principal value of fireproofing is realized during the early stages of a fire when efforts are primarily directed at shutting down units, isolating fuel flow to the fire, actuating fixed suppression equipment, and setting up portable firefighting equipment. During this critical period, if nonfireproofed equipment and pipe supports fail due to fire-related heat exposure, they could collapse and cause gasket failures, line breaks, and equipment failures, resulting in expansion of the fire. Fireproofing may be applied to control or power wiring to allow operation of emergency isolation valves, vent vessels, or actuate water spray systems during a fire.

#### *7.3.2.1. Determining Fireproofing Needs*

Determining fireproofing requirements involves experience-based or risk-based evaluation as suggested in Chapter 6. An approach for selecting fireproofing includes the following steps:

- Conducting a hazard evaluation, including quantification of inventories of potential fuels.
- Developing fire scenarios, including potential release rates and determining the dimensions of fire-scenario envelopes.
- Determining fireproofing needs based on the probability of an incident considering industry experience, the potential impact of damage for each *fire-scenario envelope (see 7.3.2.2)*, and technical, economic, environmental, regulatory, and human risk factors.
- Choosing the level of protection (based on appropriate standard test procedures) that should be provided by fireproofing material for specific equipment, based on the needs analysis.

#### *7.3.2.2. Fire-Scenario Envelope*

The fire-scenario envelope is the three-dimensional space into which equipment can release flammable or combustible fluids capable of burning long enough and with enough intensity to cause substantial property damage (API, 1999). Determining the fire-scenario envelope, along with the nature and severity of potential fires within the envelope, becomes the basis for selecting the fire resistance rating of the fireproofing materials used.

An integral part of defining the fire-scenario envelope is determining the appropriate dimensions for use in planning fire protection. For liquid hydrocarbon fuels, a frequently used frame of reference is a fire-scenario envelope that extends 20–40 ft (6–12 m) horizontally, and 40 ft (12 m) vertically, from the source of liquid fuel. For pool or spill fires, the source is generally considered to be the extent of the fire as defined by containment such as dikes and curbs.

Where containment is not available, the extent of the fire can be modeled using the equations in Chapter 5.

For example, if LPG vessels are considered to be within a fire-scenario envelope, they require fireproofing unless protected by a fixed water spray system. API Standard 2510, *Design and Construction of Liquefied Petroleum Gas Installations* (API, 2001) recommends fireproofing pipe supports within 50 ft (15 m) of the LPG vessel, or within the spill containment area.

### 7.3.2.3. Fire Resistance Rating

The required duration of the fireproofing protection is commonly referred to as fire resistance rating. Ratings generally depend on the type and thickness of the material and range from 1 to 4 hours. Information on rating and application of fireproofing is contained in Chapter 8. The manufacturers of fireproofing materials provide design criteria for given fire resistance ratings based on fire tests. The following considerations help to determine the required fire resistance rating:

- The time required to provide isolation of fuel that may be released.
- The availability and capacity of an uninterrupted water supply.
- The time required to apply adequate, reliable cooling from fixed water spray systems or fixed monitors, including response time for personnel to operate them.
- Response time and capability of plant or other fire brigades to apply portable or mobile fire response resources (including foam for suppression).
- The time required for the area's drainage system to remove a hydrocarbon spill.

The fire resistance rating is a useful measure for comparing fireproofing systems. However, fire resistance ratings should be used with judgment, including some reasonable safety factor. The rating is not an absolute measure of how a particular system will perform in a fire. In different fires, the same system may last longer or shorter depending on the intensity of the fire.

In general, the number of hours of fire resistance selected would apply to most of the structural supports within the fire-scenario envelope. Increased fire resistance should be considered for supports on important equipment that could cause extensive damage if they collapse. Certain large, important vessels, such as reactors, regenerators, and vacuum towers may be mounted on high support structures. In these cases, fireproofing materials should be considered for the entire exposed support system, regardless of its height. In some other instances, particularly at higher elevations within the fire-scenario envelope, the fire resistance rating may be reduced.

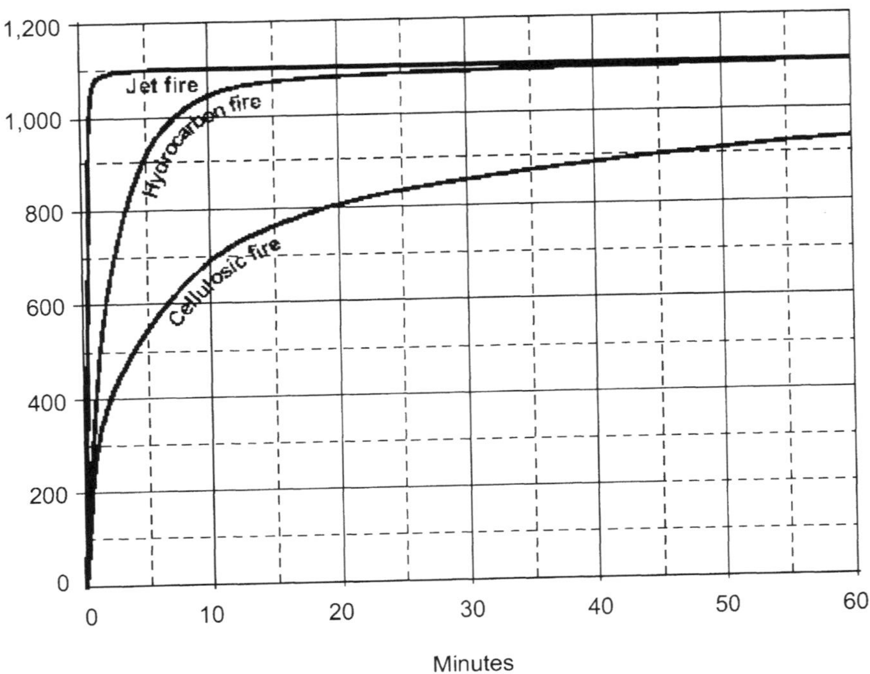


**Figure 7-2.** Time–Temperature Curve for Fire Tests

### *7.3.2.4. Testing for Fireproofing Systems*

There are several methods of testing fireproofing materials. Fireproofing systems should be tested and rated according to accepted test procedures that indicate how the material will perform when subjected to realistic fires. Two test procedures, UL 1709 and ASTM E 1529, have been developed to represent flammable liquids and pool fire conditions. Both test procedures reach 2,000°F (1,093°C) within 5 minutes and maintain that temperature for the remainder of the test. The primary difference is the UL 1709 test subjects the fireproofed assembly to a heat flux of 65,000 BTU/ft$^2$-hr vs. 50,000 BTU/ft$^2$-hr for ASTM E 1529. The 15,000 BTU/ft$^2$-hr difference translates to a 30% higher heat load on the test assembly. There are tests, such as ASTM E 119, that were developed to characterize fires involving ordinary combustibles (celluosic). Ratings based on ASTM E 119 should not be used in specifying the protection of chemical, petrochemical, and hydrocarbon processing facilities.

### *7.3.2.5. Fireproofing Materials*

Each type of fireproofing system uses a different combination of materials with various physical and chemical properties. Selection of fireproofing materials requires care to obtain the desired degree of protection during the service life. Besides providing a given extent of fire resistance, a variety of characteristics should be evaluated to ensure proper performance in the environment where it is installed. Principal characteristics governing selection of fireproofing materials are:

- Adhesion or bonding strength and durability
- Chemical resistance
- Coefficient of expansion
- Compatibility—stainless steel and aluminum are susceptible to corrosion from exposure to agents that can be in certain fireproofing material, especially chlorine.
- Density
- Ease of application
- Fire resistance rating (in hours)
- Friability
- Hardness or resistance to impact
- Porosity
- Reinforcing requirements
- Resistance to thermal shock
- Specific surface preparation (cleaning and priming, etc.) and support structures
- Tolerance to exposure from adjacent hot equipment
- Vibration resistance and flexural strength
- Weatherability
- Weight
- Toxicity—certain application materials contain toxic substances

### 7.3.2.6. *Types of Fireproofing Materials*

#### 7.3.2.6.1. DENSE CONCRETES

Concretes made with Portland cement have a specific weight of 140 to 150 $lb/ft^3$ (2,242 to 2,400 $kg/m^3$). Concrete absorbs the heat of a fire when chemically bound water is released from a crystalline structure and is reduced to lime. Dense concretes can be formed in place, or pneumatically sprayed to the required thickness using steel reinforcement. The corrosive effect of chlorides on the steel surface in moist saline environments (coastal or other chloride environments) dictates the use of protective primers and topcoat sealers.

Major advantages of dense concrete are:

- Easily maintained
- Durability, can withstand thermal shock and direct hose streams
- Can withstand direct flame impingement up to 2,000°F (1,100°C)
- Ability for most general contractors to satisfactorily apply (no specialty contractors required)
- Extensive proven performance for four or more hours of protection

Disadvantages of dense concrete include:

- Relatively high weight—in certain circumstances the strength of the supports for equipment, piping, or vessels has to be upgraded just to support the added weight of concrete
- Relatively high thermal conductivity
- Need for steel reinforcement
- The installation cost and time involved in forming in-place, especially when applied to existing facilities
- Susceptible to weathering and cracking

#### 7.3.2.6.2. LIGHTWEIGHT CONCRETE

Lightweight concrete uses very light aggregate, such as vermiculite or perlite (instead of gravel), with cements that are resistant to high temperatures. Dry densities range from 25 to 80 lb/ft$^3$ (400 to 1,300 kg/m$^3$).

Lightweight concrete is usually sprayed on, but may be troweled or formed-in-place using reinforcing mesh. Pneumatically applied material is about 20% heavier than poured-in-place lightweight concrete. As with all concretes, moisture creates a corrosive condition at the surface of the steel. Protective coating of the substrate surface is needed to protect against corrosion.

Advantages of lightweight concrete materials are:

- Better fire-resistant properties than dense concrete
- Fairly durable and have limited maintenance requirements
- Capable of withstanding direct flame impingement up to 2,000°F (1,110°C)
- Capable of withstanding thermal shock and high-pressure hose streams
- Satisfactorily applied by most contractors

Disadvantages of lightweight concrete materials include:

- Porosity, which can allow penetration by water or leaked hydrocarbons (corrosion)
- Moisture absorption can lead to cracking and spalling in freezing climates

- The need to maintain a top coating (and possible shielding or caulking) to prevent moisture or hydrocarbons from penetrating
- Susceptible to mechanical damage

#### 7.3.2.6.3. SUBLIMING, INTUMESCENT, AND ABLATIVE MASTICS

Mastics provide heat barriers through one or more of the following mechanisms:

- Subliming mastics absorb large amounts of heat as they change directly from a solid to a gaseous state
- Intumescent mastics expand to several times their volume when exposed to heat, and form a protective insulating ash or char at the surface that faces the fire
- Ablative mastics absorb heat as they lose mass through oxidative erosion

Mastics are sprayed on a substrate in one or more coats, depending on the desired degree of fire resistance. The final coat of all fireproofing mastics should be rolled or brushed to provide a smooth surface finish. The material should be applied with a sufficient number of coats to prevent running or slumping and sufficient drying time should be allowed between coats. Mastics may also be hand-troweled, if permitted in the manufacturer's specifications.

Reinforcing fabric or wire (which may be rigidly specified) is usually needed for fire resistance ratings of 1 hour or more. Substrate preparation is important to achieve adequate bonding in applying coatings; a specific primer may be required. After applying the mastic coating, some materials require a top-finish coating on the surface to prevent moisture penetration. The surface coating should be inspected and renewed according to the vendor's recommendations.

Advantages of mastics are:

- They can be quickly applied
- They are lightweight
- They are suitable for use on existing equipment supports that may not be able to handle additional weight

Disadvantages of mastics are described below:

- Because coat thickness and proper bonding to the substrate are important to satisfactory performance, application techniques specified by the manufacturer should be rigorously followed to ensure good long term performance. A specialist contractor is normally required.
- Some mastics tend to shrink while drying. Specifications should indicate the wet thickness that will yield the required dry thickness.

- Materials rated for protection with thin coats should be applied skillfully to maintain adequate thickness.
- Some mastics are less durable than more traditional concrete materials when subjected to mechanical impact and abrasion.

#### 7.3.2.6.4. INTUMESCENT EPOXY COATINGS

A wide range of intumescent epoxy coatings are available. These can be described as a mix of thermally reactive chemicals in a specific epoxy matrix formulated for fireproofing applications. Under fire conditions, they react to emit gases, which cool the surface while a low density carbonaceous char is formed. The char then serves as a thermal barrier.

Advantages of intumescent epoxy coatings are:

- Excellent bonding and corrosion protection
- Lightweight and durable under nonfire conditions
- Flexible and tolerates vibration
- Exceptional durability in severe jet-fire tests
- Because they are "organic-based" special characteristics can be designed into the coating
- Coatings are available that provide an attractive finish appearance

Disadvantages of intumescent epoxy coatings are:

- There is a possibility of damage to a char coating during a fire, if subjected to impingement by fire hose streams.
- They require expertise in application, and may require multiple coats or special equipment that can apply dual components simultaneously.
- Some manufacturers require factory-certified application personnel.
- Some concerns have been raised regarding potential toxicity of gases generated during fire conditions.

#### 7.3.2.6.5. PREFORMED INORGANIC PANELS

Preformed fire-resistant inorganic panels can be cast or compressed from lightweight aggregate and a cement binder, or from compressed inorganic insulating material, such as calcium silicate. The panels are attached to the substrate by mechanical fasteners designed to withstand fire exposure without appreciable loss of strength. When panels are used outdoors, an external weatherproofing system to prevent moisture penetration is typically required. All joints or penetrations through fireproofing (such as clips or attachments) should be rigorously caulked or sealed. The control of all these joints and seals is vital during the entire life of the facility. This leads to higher inspection and maintenance costs.

Advantages of preformed panels include:

- Clean application
- No curing time

Disadvantages of preformed panels are:

- Labor-intensive application when unit instruments and appurtenances are attached to steel members
- More susceptible to damage from impact than concrete

#### 7.3.2.6.6. MASONRY BLOCKS AND BRICKS

Masonry blocks of lightweight blast-furnace slag (used as coarse aggregate) are sometimes used. These units are laid-up with thin staggered joints not more than ⅓in (0.8 cm) thick. Joints should use only fire-resistant mortar.

Brick and block are no longer commonly used because of their high installation cost and fairly extensive maintenance requirements. Brick-and-block assemblies tend to crack and admit moisture, which can lead to serious corrosion and spalling.

#### 7.3.2.6.7. ENDOTHERMIC WRAP FIREPROOFING

Endothermic materials absorb heat chemically, generally with the concurrent release of water and physically through heat absorption by the released water. This flexible, tough, inorganic sheet material with a bonded aluminum foil outer layer is formed from an amount of inorganic, highly endothermic filler, and a minimum amount of organic binder and fiber. It can be wrapped around a wide variety of potentially exposed vulnerable equipment. In most applications, the wrap is held in place by stainless steel bands with foil tape and fireproof caulk on seams, gaps and termination points. For structural steel in new construction, surface preparation of the substrate should include fresh prime paint to provide corrosion protection.

Advantages of endothermic wrap fireproofing are:

- Easily reentered and repaired, allowing retrofitting over steel without dissembling wiring and other attached items.
- Does not catalyze corrosion (nor protect against corrosion).
- Can be applied directly over existing cement or block where additional protection is required.
- Can be applied directly over other fireproofing.

The major disadvantage of endothermic wrap fireproofing is when used outdoors, the system must be weatherproofed. Stainless steel jacketing or wrapping with a manufacturer's protective tape is necessary to provide the recommended level of protection.

#### 7.3.2.6.8. INSULATION

Insulation used for fireproofing should be rated for 1200°F (650°C) minimum service. Insulation should consist of one of the following types of materials:

- Calcium silicate, block or preformed
- Ceramic foam
- Expanded alumina silica fiber blanket with a minimum density of 6 lb/ft$^3$ (96 kg/m$^3$)
- Foam glass
- Mineral wool block with a minimum density of 12 lb/ft$^3$ (192 kg/m$^3$)
- Perlite, block or preformed

When insulation is used to fireproof steel that will be at or below ambient conditions, caution should be taken to prevent external corrosion of the steel caused by condensation when the steel is below the atmospheric dew point.

Before insulation is installed, carbon steel surfaces between 32°F and 200°F (0°C and 93°C) should be prepared and painted in accordance with manufacturer's recommendations. A vapor barrier should be provided over the outer layer of the insulation. Minimum thickness of insulation for fireproofing should be 2 in (5 cm). Moisture contamination to the insulation should be avoided to prevent damage to the insulation as a result of steam pressure caused by a fire.

Insulation used for fireproofing should be jacketed with 18Cr-8Ni stainless, vinyl-clad galvanized steel or uncoated galvanized steel. Uncoated galvanized steel should not be used below ambient temperature service where corrosion will be accelerated if the galvanized steel is kept wetted from condensation.

*Note*: Unless specified for fireproofing use, materials sold as pipe insulation might not survive the high temperatures generated in tests such as UL 1709 or ASTM E 1529. The user should ensure the fireproofing system components are fire-rated before they are specified.

#### 7.3.2.6.9. MAGNESIUM OXYCHLORIDE

The use of magnesium oxychloride plasters for fireproofing is not recommended. Field experience has indicated that corrosion of the substrate steel occurs as the topcoat (over the fireproofing) weathers and moisture combines with the chloride present in the plaster to form hydrochloric acid.

### *7.3.2.7. Considerations for Installing Fireproofing*

The most important aspect of fireproofing installation is using a contractor who is experienced in the type of fireproofing being applied. Dense concrete fireproofing can be applied satisfactorily by suitably trained facility personnel or general contractors. However, to apply lightweight concrete and mastics, the appliers must understand and have experience with the specific materials and their use. If improperly applied, the application may lose its bond, deteriorate, or fail to perform as expected during a fire. It is advisable to use contractors with a factory certification for installing these materials. Some manufacturers insist that their product performance guarantee is contingent upon using approved contractors.

All rated fireproofing systems should be carefully installed to specification and manufacturer's requirements. Substrate surfaces should be cleaned so they are free from oil, grease, liquid contaminants, rust, scale, and dust. If a primer is required, it should be compatible with the fireproofing. Specifications to be followed include the specified thickness or number of layers, adequate attachment, and proper caulking, sealing, or top-coating of the systems.

The following installation considerations apply to fireproofing coatings and wet cementitious materials:

- Shelf life should be determined and maintained.
- Materials should be stored onsite in accordance with the manufacturer's recommendations (some materials must remain upright in their containers for proper sealing; refer to manufacturer's specifications).
- Some materials are temperature sensitive and cannot tolerate extremes during storage and shipping.
- Fireproofing materials should be applied directly from their original sealed containers to avoid possible additions to, or changes in, their formulation.
- Some materials require a controlled curing period to develop full strength and prevent serious cracking in the future.
- Materials that contain free water require a drying period during above-freezing temperatures.
- Appliers should understand that the specified thickness is a dry thickness. Some mastic coatings shrink as much as 30% when cured.

Lightweight concretes or fireproofing cements should have a density of 35 to 75 lb/ft$^3$ (560 to 1200 kg/m$^3$), installed and dried. If a pneumatic gun is used, specified densities may be increased 20%. The density selected should be based on the fire resistance rating desired versus the strength required (i.e., low density enhances insulating properties; high density reduces insulation, but enhances strength). The finished lightweight concrete should be given two coats of a sealer for weather protection. The seal coating should be equivalent to the concrete manufacturer's recommendation for the possible extreme conditions that may be encountered in the geographical area of the structures or equipment.

Steel should be prepared by using a caustic solution and by scraping, wire brushing, or grit blasting.

For previously painted and fireproofed surfaces, the mastic manufacturer should be consulted to assure compatibility with the existing surface, proposed paint primer, and mastic. New galvanized surfaces should be mechanically abraded to ensure adherence of the prime coating.

For spray-on applications, consult the manufacturer on minimum application temperature. Some materials do not flow through the pump or sprayer below 40°F to 50°F (4°C to 10°C) and some materials do not cure well below 30°F to 40°F (–1°C to 4°C).

Intumescent materials, which are water-based systems, cure by evaporation. If applied during periods of high humidity, the areas of application may experience blistering when the humidity drops and the evaporative curing continues. In such environments, solvent-based materials are recommended.

When intumescent and subliming mastic coatings are to be installed in areas prone to extreme weather conditions, the manufacturer's recommendation for applying protective sealing should be followed. Also, note should be taken of these circumstances where a manufacturer advises against installing a material.

Satisfactory performance of the fireproofing material over its expected lifetime depends on the user's and the applier's knowledge of materials and application techniques and on continuing inspection by qualified personnel. Specifically, once a fireproofing system has been chosen, it is imperative that personnel involved in each phase of the project be familiar with the relevant aspects of the manufacturer's requirements and specifications.

The contractor should provide supervision to assure that the thickness and quality of the materials and workmanship provide the level of protection required.

### 7.3.3. *Containment and Drainage*

Local containment systems within process structures or areas are intended to retain spills released from process systems within the immediate area of their release. Containment systems are passive mitigation systems usually employed to prevent the spread of spills and releases of flammable or combustible liquids and other materials. Local containment systems are also used for controlling spills of powders or dusts and for molten materials that solidify at ambient temperatures. Containment systems are commonly used to:

- Contain potential fire damage to the zone of origin (which may be either expendable or provided with appropriate fire protection features).
- Prevent spread to areas not equipped with fire protection.
- Facilitate clean-up.

Containment systems are also used in conjunction with an automatic foam-water sprinkler/deluge system where rapid fire extinguishment is expected.

By design, local containment systems hold the spilled or released material within the process structure or area. Hence, the structure, equipment, vessels, and piping will be exposed to any fire resulting from the contained material and will require fire protection.

Local containment may be appropriate where the quantity of potential flammable or combustible liquid spills is small to moderate. Where larger quantities are involved, such as in continuous process operations, containment alone is usually not an appropriate design. See Chapter 8, Section 8.1.2, for more information on drainage and containment.

### *7.3.3.1. Local Containment Design Considerations*

Containment systems are typically designed with solid, impermeable floors with perimeter curbing, dikes, berms, or walls to contain the released material within the area of the release. The containment should be designed such that a flammable spill and fire water runoff flow away from and do not collect under process equipment. Concrete is commonly used, but other materials may also be appropriate. This containment zone may include the ground floor of a process structure or a portion (one or more bays) of a structure. Local containment may also be used on upper levels of a structure or in a building around specific equipment with identified potential for spills or releases.

Containment systems should be designed to contain a volume equal to the largest probable rapid process spill, release, or leak. Curb height should be calculated based on the depth of this maximum probable spill over the containment area plus a minimum freeboard allowance of 2 in (5 cm); the minimum recommended curb height is 4 in (10 cm). The containment volume should include an allowance for a probable quantity of firefighting water, assumed to be 10-20 minutes minimum. After this period, local containment will not be effective and an alternative means of protection should be provided.

Most containment will have some type of drainage to carry away rain water and liquid following an incident. Liquid containment should be designed so that the amount of liquid containment does not overflow the curb.

Access points into the containment area for personnel or wheeled-traffic may be provided with ramps over the curbing spanning the width of the access path.

### *7.3.3.2. Process Area Floor Trench Design Considerations*

A trench should be designed to collect and control spills within only one containment/drainage zone in order to prevent the trench from becoming a source of fire spread from one zone to another. Trenches can be designed into the perimeter of solid floors on elevated levels of a process structure. Trenches from each zone or floor level should have a liquid trap to isolate them from the piping system and from other zones feeding into the drainage piping.

Some design features of trenches should be noted. Commonly recommended trench dimensional proportions are a width of twice the depth. Under no circumstances should trench depth exceed the width to ensure adequate ventilation. Where heavy wheeled traffic, such as forklift trucks, may occur in a process structure or area with trenches, the trenches can be designed with extra-heavy duty grating that can support heavy equipment.

Trench covers, per NFPA 30, should be used where appreciable quantities of flammable liquids could be caught in the trench system and adjacent equipment could be exposed to direct flame impingement (NFPA 30, NFPA 15). Figure 7-3 shows the details of a wick-type cover where the center third of the grating is open and the other two-thirds are covered by either steel plates or

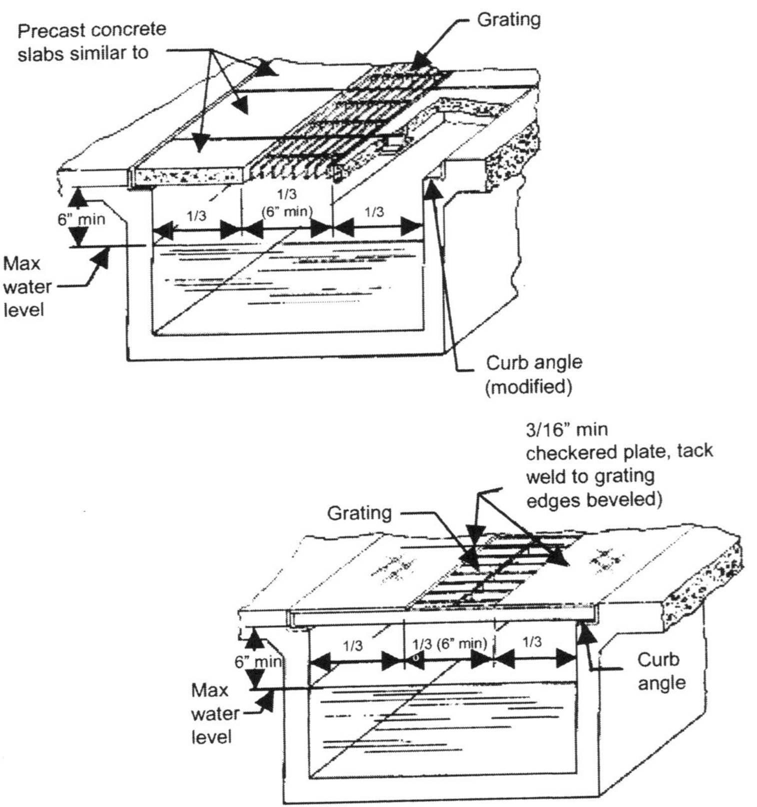


*Note: Grating type used must have a minimum of 50% open area to prevent flammable concentrations inside the trench and to provide venting.*

**Figure 7-3.** Drainage Trench with Steel Plate and Grating

concrete panels. The covers provide a flame dampening effect and significantly reduce flame height.

### 7.3.4. Electrical Area Classification

Flammable vapors, flammable gases, combustible dusts, and ignitable fibers are hazardous materials found in many manufacturing facilities. These materials are necessary for, created by, or are unavoidable byproducts of, the manufacturing process.

It has long been recognized that electrical equipment has the potential to act as an ignition source for such materials. Primarily the energy for ignition is generated in the form of a spark or due to the temperature generated by the operation of the device. The concept of the electrical classified area and the divisions within the area was created to provide a graded measure of the risk of an ignition event based upon the probability of there being a flammable mixture within the area. A hazardous (classified) location is a space containing any of the following:

- An atmosphere in which an ignitable concentration of flammable gas or vapor is present or might occasionally be present.
- An atmosphere in which combustible dust is, or could conceivably be, in suspension in sufficient quantity to cause an explosive or ignitable mixture.
- Electrical equipment on which combustible dust might accumulate, and interfere with heat dissipation from the equipment.
- Surfaces that contain easily ignitable fibers or flyings.

Most general purpose electrical equipment has an unacceptably high probability of igniting flammable concentrations of vapors, dusts, and fibers. In response to this situation, a range of electrical equipment evolved that is intended to minimize this probability by means of special design and construction features. This includes limiting the surface temperature of the equipment, minimizing the potential for sparking, and controlling vapor travel into or out of an electrical enclosure.

When selecting heat-producing, electrical equipment for a hazardous (classified) location, its hottest, external-surface operating temperature should be compared to the ignition temperature of the surrounding gas, vapor, or dust. Lowering of the ignition temperature for organic dusts that dehydrate or carbonize should be considered. Care should be taken to ensure that these special features of the equipment match the flammability and ignition characteristics of the materials to which it is likely to be exposed.

Appropriate precautions should be taken to maintain hazardous (classified) location equipment in a manner that does not jeopardize the integrity of protection. Guidance offered by the manufacturer, listing agency, or NFPA 70B should be followed.

Where practical, the possibility of electrical equipment igniting a combustible gas, vapor, dust, or flyings should be reduced or eliminated as follows:

- Eliminate the use of materials that are hazardous. Alternate processes or material substitutions may accomplish this.
- Limit hazardous (classified) areas by using ventilation, barriers, enclosures, or other suitable means.
- Locate electrical equipment outside of hazardous (classified) locations or replace electrical operators with manual or pneumatic operators.

#### *7.3.4.1. Traditional Electrical Classification System in the United States*

Hazardous (classified) locations have traditionally been designated by Class, Division, and Group. Equipment used in areas so designated is selected and systems are designed, based on requirements established for the classification. This approach is defined in Article 500 of the NEC and API 500.

Three distinct classes of hazardous (classified) locations have been established:

- *Class I hazardous atmospheres* are characterized as areas containing flammable vapors escaping from a flammable or heated combustible liquid and areas containing flammable gases.
- *Class II areas* contain combustible dust suspended in air or combustible dust accumulations that can interfere with heat dissipation from electrical equipment or can be ignited by that equipment.
- *Class III areas* contain accumulations of fibers or flyings.

For a Class I or Class II area, a Division 1 location is likely to contain the hazardous condition during normal operations or frequently because of maintenance and repair. A Division 2 location is likely to contain the hazardous condition only under abnormal circumstances, such as process upset or equipment failure. These two divisions, which are based on the likelihood of an atmosphere being hazardous, control or prescribe the design, construction, and operating features of equipment in that area. Engineering practice tolerates lower levels of protection where there is less likelihood of a hazardous material being present. Thus, Division 1 locations require equipment built to higher standards than equipment built for Division 2 locations.

For Class III areas, the division classification is based on whether the area is used for processing or storage. A manufacturing area is a Division 1 location: a warehouse is a Division 2 location.

Equipment protective features also depend on the degree or severity of a hazard to which equipment is exposed. For convenience, Class I hazardous materials are typically placed into one of four groups, depending on their physical properties and characteristics. Dusts, which are Class II materials, are similarly grouped by degree of hazard.

#### *7.3.4.2. Recent Expansion of Class I Designations in the United States*

Today, the NEC also provides an alternative method of classifying hazardous locations involving Class I atmospheres. This expansion brings the code more in line with standards used in other countries. Either the traditional method or the alternative method can be selected for a given hazardous (classified) location.

Table 7-3 provides a comparison between historical and alternative approaches for identifying electrical classification groups. Table 7-4 compares NEC Division and Zone Classification Systems.

This "new" approach has been in use for many years in most European countries. Originally, each country had their own national requirements. Then, based on the work done by the IEC (International Electrochemical Commission), standards were harmonized through the work of CENELEC (Comité Européen de Normalisation Electrotechnique). Harmonized standards, published as European Standards (EN), have to be adopted by the participating countries as national standards. Some of the countries which have already adopted these are: Belgium, France, Germany, and the United Kingdom.

NEC Article 501 defines electrical equipment suitable for use in Class I, Division 1 and 2 locations. Article 505 defines equipment suitable for use in Class I, Zone 0, 1 and 2 locations. Mixing classification methods in a single clas-

**Table 7-3**
*Comparison of Historical and Alternative NEC Class I Grouping Systems*

| Class I Content (Liquid, Gas or Vapor) | Historical Approach | Alternate Approach |
|---|---|---|
| Atmospheres containing acetylene. | Group A | Group IIC |
| Atmospheres containing hydrogen, fuel and combustible process gases containing more than 30% hydrogen by volume, or gases or vapors of equivalent hazard such as butadiene, ethylene oxide, propylene oxide and acrolein. | Group B | |
| Atmospheres containing ethyl ether, ethylene, or gases or vapors of equivalent hazards. | Group C | Group IIB |
| Atmospheres containing acetone, ammonia, benzene, butane, cyclopropane, ethanol, gasoline, hexane, methanol, methane, natural gas, naphtha, propane, or gases or vapors of equivalent hazard. | Group D | Group IIA |

**Table 7-4**
*Comparison of NEC Division and Zone Classification Systems*

| Historical Approach—NEC | Alternate Approach—Zone |
| --- | --- |
| ***Class I, Division 1****—Where ignitable concentrations of flammable gases (vapors) can exist:*<br>▪ Normally during operating periods.<br>▪ Frequently because of repair, maintenance, or leakage.<br>▪ Accidentally where the possibility exists of a simultaneous release of ignitable concentrations of gas (vapor) and failure of electrical equipment.<br>Fine Print Note No. 2 in the Code states where ignitable concentrations of flammable gases (vapors) can exist continually or for long periods of time, electrical equipment should be avoided altogether, or, intrinsically safe systems should be used. | ***Class I, Zone 0****—Where ignitable concentrations of flammable gases (vapors) can exist:*<br>▪ Continually.<br>▪ For long periods of time.<br><br>***Class I, Zone I****—Where ignitable concentrations of flammable gases (vapors) can exist:*<br>▪ Normally during operating periods.<br>▪ Frequently because of repair, maintenance, or leakage.<br>▪ Accidentally where the possibility exists of a simultaneous release of gas (vapor) and failure of electrical equipment that can cause ignition.<br>▪ Because the area is adjacent to a Zone 0 area, from which gas (vapor) can be communicated. |
| ***Class I, Division 2—***<br>▪ Areas containing volatile flammable liquids or gases, where these materials are normally confined to closed containers and systems from which the materials can escape only upon the rupture or breakdown of containment or upon abnormal operation of equipment or process.<br>▪ Where ignitable concentrations of gases (vapors) are normally prevented by positive mechanical ventilation, but which might become hazardous through failure or abnormal operation of ventilating equipment.<br>▪ Where adjacent to a Class I, Division I location from which ignitable concentrations of gases (vapors) might occasionally be communicated. | ***Class I, Zone 2—***<br>▪ Where ignitable concentrations of flammable gases (vapors) are not likely to occur during normal operations, and if they do occur, they will exist for only a short period of time.<br>▪ Areas containing volatile flammable liquids or gases, where these materials are normally confined to closed containers and systems from which the materials can escape only upon the rupture or breakdown of containment or upon abnormal operation of equipment or process.<br>▪ Where ignitable concentrations of gases (vapors) are normally prevented by positive mechanical ventilation, but which might become hazardous through failure or abnormal operation of ventilating equipment.<br>▪ Where adjacent to a Class I, Zone 1 location from which ignitable concentrations of gases (vapors) might occasionally be communicated. |

sified location is not permitted. Equipment requirements are established by the single classification method selected for that area. If the alternate classification system presented by Article 505 is used, area classification, wiring, and equipment selection must be under the supervision of a qualified licensed professional engineer. *(This is the only section of the NEC that requires such control.)*

European harmonization of the zone concept for Class II and Class III locations is still under development. Generally, the European community treats Class III locations the same as Class II locations.

### 7.3.4.3. *Equipment*

Once a hazardous location has been classified, appropriate electrical equipment must be chosen for that area. In general, equipment must be approved for use in that hazardous classified area. Testing labs such as UL test, label, list, or approve equipment suitable for installation in accordance with their legislated code.

Listed equipment for hazardous (classified) areas is marked to show the code-specified environments where it can be safely used. These markings often include the maximum surface temperature of the equipment under normal operating conditions.

The best-known type of hazardous location electrical equipment is explosion-proof equipment. This equipment is suitable for use in certain Class I, Division 1 locations and in Class I, Zone 1 locations when listed for use in those atmospheres. Explosion-proof equipment is not suitable for use in Class I, Division 1 locations where ignitable concentrations of gases or vapors can exist for long periods of time, or in Class I, Zone 0 locations. Explosion-proof equipment is designed to contain explosions without allowing the escape of enough energy to ignite the hazardous atmosphere in the area.

In recent years, electrical area classification has become more focused on risk of release and distance away from the release point than the risk of flammable vapor/air mixture. Therefore, equipment types are often mixed inside buildings or units instead of all being explosion-proof.

Comparable equipment suitable for use in Class II, Division 1 locations is called dust-ignition proof. Dust-tight equipment is designed for use in Class II, Division 2 locations. These terms should not be confused with equipment designated "dustproof." Dustproof equipment is constructed or protected so that dust will not interfere with its successful operation. This term does not imply the equipment is suitable for use in a hazardous (classified) area.

Other kinds of electrical equipment may also be used in hazardous locations. One kind is purged and pressurized electrical equipment. This equipment works by using air or nonflammable protective gas flow or pressure to prevent hazardous materials from entering the enclosure; NFPA 496 defines three types of purged and pressurized equipment as identified in Table 7-5.

**Table 7-5**
*Purged and Pressurized Electrical Equipment Enclosures*

| Type | Description |
|---|---|
| X | Reduces enclosures classification from Division 1 to nonhazardous |
| Y | Reduces enclosure classification from Division 1 to Division 2 |
| Z | Reduces enclosure classification from Division 2 to nonhazardous |

Another kind of electrical equipment suitable for use in hazardous locations is equipment whose maximum possible energy output is insufficient to ignite the hazardous material. The electrical input to this equipment must be controlled by a specially designed electrical barrier. Such electrical equipment must be compatible. ANSI/UL913 defines low energy intrinsically safe electrical equipment and associated apparatus permitted in Division 1 areas. Nonincendive electrical equipment is permitted in Division 2 locations. Table 7-6 describes intrinsically safe and nonincendive equipment and identifies permitted uses.

Table 7-7 identifies equipment protection for an alternate classification system.

**Table 7-6**
*Low Energy Electrical Equipment for Hazardous Locations*

| Type | Description | Where Used |
|---|---|---|
| Intrinsically Safe | Will not ignite the most ignitable concentration of the hazardous material at 1.5 times the highest energy possible under normal conditions, under 1.5 times the energy of the worst single fault, and under the energy of the worst combination of two faults. | Class I, Division 1<br>Class I, Zone 0<br>Class II, Division 1<br>Class III Locations |
| Nonincendive | Will not ignite the most ignitable concentration of the hazardous material under normal conditions. | Class I, Division 2<br>Class II, Division 2<br>Class III Locations |

**Table 7-7**
*Equipment Protection for an Alternate Classification System*

| Protection Type | Zone Application |
|---|---|
| Intrinsic safety (double fault) | 0 |
| Intrinsic safety (single fault) | 1 |
| Flameproof | 1 |
| Purged and pressurized | 1 |
| Increased safety | 1 |
| Encapsulation | 1 |
| Nonincendive | 2 |

Intrinsically safe equipment approved for use in the European community might not pass UL 913 tests for intrinsically safe designation in the U.S. The standards are similar, but not identical. Integration components intended for different codes or systems should be avoided, unless approved by an appropriately qualified electrical engineer.

Electrical equipment suitable for classified locations can be expensive and hard to maintain. Options to using this equipment are sometimes available. These options include eliminating the hazardous materials, separating the hazardous location from the electrical equipment, moving electrical equipment outside the hazardous location, or replacing electrical operations with manual or pneumatic operations. It is frequently possible to locate much of the equipment in less hazardous or in nonhazardous locations and, thus, to reduce the amount of special equipment required.

Special precautions are required to maintain equipment used in hazardous (classified) locations. Examples are identified in paragraph 4-7 of NFPA 70B. If maintenance work voids the listing applying to a device, the device should not be reenergized in the hazardous (classified) area. A replacement device should be obtained. Special attention should be given to replacement and proper tightening of enclosure bolts, covers, and other fastening devices following maintenance.

### 7.3.5. Ventilation/Exhaust

All enclosed spaces with the potential for flammable or toxic atmospheres should be ventilated, preferably at a rate of not less than six air changes per hour. Low level exhaust ventilation is important to minimize the potential accumulation of vapors. A rule of thumb for design is 1 cfm/ft$^2$ of floor area with exhaust points no higher than 12 in (30 cm) off the floor. The ventilation system should be designed in accordance with Chapter 5 of NFPA 30.

Where ventilation is installed to effect a reduction in the area electrical classification, the rate should be no less than 12 air changes per hour for an adequately ventilated area in accordance with API RP 500. Areas containing ignition sources, such as control and switchgear buildings, gas turbine acoustical enclosures, and power generators, should be pressurized in accordance with NFPA 496, if located in an electrically classified area.

### 7.3.6. Static Electricity, Lightning, and Stray Current Protection

The generation of electric charges, their accumulation on material, and the process of dissipating these accumulated charges causes static electric hazards (API RP 2003). Such charges (positive and negative) are generally generated as a result of friction. This does not present a hazard when conductive materials are involved, as the charge will quickly dissipate or "relax." The concern arises when the materials involved are poor conductors and the charges continue to accumulate until they are so strong that a discharge or sudden recombination of the generated positive and negative charges occurs. This process can produce sufficient energy to generate a spark capable of igniting a flammable mixture.

Static electricity is possible whenever materials are transferred. Some materials have more capability to generate static electricity than others. If the material is readily ignitable, such as flammable vapors or flammable/combustible dust, then steps are required to reduce the risk during design and operation. The primary risk control measure when flammable liquids and combustible dusts are flowing is to prevent accumulation of electrical charges. This is of particular concern in systems that are open to the air during material transfer, where a material/air mixture may exist. A medium size dust collector with a medium sized grid can become isolated from its own grounding system (see below) through the buildup of dust. The buildup may be as high as 50 mJ, which is sufficient to cause a flash fire upon discharge.

Loading and unloading operations of ships, barges, tank cars, and tank trucks or, in the case of solid material, hopper cars or trucks, are susceptible to static electricity generation. Filling operations should use down-comers or run down the side of the container to avoid splashing that causes static. Transferring from drums to small containers and some processing operations in open topped vessels can also be at risk.

### *7.3.6.1. Grounding and Bonding*

A static charge can be removed or allowed to dissipate. The predominant means to prevent accumulation of electrical charges is bonding, grounding, or a combination of both. Other means include prevention of free fall of liquids or solids through the air in the presence of a flammable vapor mixture or combustible dust mixture. Grounding and bonding are illustrated in Figures 7-4 and 7-5.

**Figure 7-4.** Grounding. Electrical conductors are attached to the tank and then connected to a grounding rod.

**Figure 7-5.** Bonding. To perform bonding, electrical conductors are connected to each side of a flange because the gasket is nonconductive.

Process structures and the equipment and vessels in them must be effectively grounded to ensure the dissipation of static, stray, and induced charges encountered in normal and abnormal operations and lightning strikes. Process equipment, vessels, and piping should be bonded together as well as grounded. Equipment and vessels in open ground level processing areas should have their own grounding systems. Normally, equipment and vessels in steel multilevel process structures are grounded to the process structure's steelwork tying them to the structure's grounding system (NFPA 780, 2000; API RP 2003, 1998).

The key issue is to design grounding and bonding systems so that the long term integrity of the grounding or bonding path is maintained. Static electrical charges can be neutralized to ground through a relatively high resistance of up to $1 \times 10^6$ ohms. The issue is more one of integrity of the path to ground over time and in a variety of conditions, than the electrical conductive capacity. Number 8 or Number 10 AWG wire is recommended for strength and is more than sufficient in conductive capacity.

Solid conductors are used for fixed connections to ground. Stranded or braided conductor is used where the wire must be frequently moved or connected and disconnected. Uninsulated conductors are recommended to facilitate inspection of the integrity of the conductor.

A well-administered static electricity prevention program depends on implementation of good procedures and adherence to them. Making the bonding connections, positioning loading arms to minimize free fall of material, and periodic inspection of grounding/bonding connectors are key procedural issues.

#### *7.3.6.2. Lightning Protection*

Lightning strikes have resulted in fires in processing facilities. They can also be the cause of electrical and computer control system malfunctions and result in process upsets.

Open structural steel process structures normally do not require specific lightning protection since the columns, beams, joists, and stringers are all metal, electrically continuous down through the structure, and are bonded to the building or structure's grounding system as required. Buildings of masonry construction or steel frame buildings with nonmetal side wall cladding or nonmetal roof or top decks usually require lightning protection.

Building structures that are nonconductive can be equipped with air terminal ("lightning rod") conductors and ground terminal systems to safely direct lightning strikes to ground. Buildings in the design stage should utilize conductive building supports or rebar in concrete walls and floors to provide conductive paths. The conductive path to ground for lightning charges should have less than 1 ohm resistance. NFPA 780 provides additional guidance on lightning protection.

Tanks, vessels, and equipment handling flammable or Class II combustible liquids and constructed of 3/16 in (4.8 mm) or thicker metal that may be exposed to direct lightning strikes are not normally required to have lightning protection. This presumes that the tank bottom is grounded and the ground conductors are periodically inspected, tested, and replaced if deteriorated. Even so, the highly charged condition of the tank wall can result in an arc if the pathway to ground is interrupted at any point. The resulting arc, if in a flammable vapor space, can readily cause an ignition. For example, vapor leaking from poorly maintained seals on floating roof tanks has been ignited by lightning. Sheet steel less than $\frac{3}{16}$ in (4.8 mm) in thickness might be punctured by a severe lightning strike (NFPA 780, Sections. 3.6 and 6.3.2, 2000). Lightning protection is not required on process structures since they are already grounded.

## 7.4. Active Protection Systems

### 7.4.1. *Water Supply*

The extent and capacity of the firefighting system and associated equipment for a facility should be based upon the assumption that only one major fire will occur at any one time.

#### *7.4.1.1. Lakes, Rivers, and Sea*

Lakes, rivers, and streams are water supply sources that can provide a virtually unlimited supply. However, these natural water supplies may be affected by drought, changing water levels, freezing, and debris which can result in the plugging of sprinkler piping with sediment or marine/biological organisms. Suction screens or approved strainers should be provided with a pump crib (area around the pump suction) to minimize collection of solids and other materials. The use of either traveling or double removal screens, cleanable with the pump in service, is preferred. One of the disadvantages of using untreated water from lakes and rivers is that the sprinkler systems may become plugged. Use of untreated water supplies requires that the distribution network and fixed system be tested and inspected at regular and frequent intervals.

Seawater can also provide an unlimited supply but can be corrosive to piping. The potential for corrosion can be reduced if seawater is accounted for in the design and provisions for flushing are included.

Quite often jetties are fitted with their own water pumps, taking suction into the sea or river. The jetty fire water grid is often connected to the main plant fire water system as a backup. Care should be taken to avoid seawater corrosion of the main site fire water system.

### 7.4.1.2. Tanks and Reservoirs

Limited capacity sources such as tanks and reservoirs can be provided as a source of water. The designs of tanks and reservoirs should be for the minimum judged necessary for fighting fire within the facility. This may be as little as a two hour supply for a relatively low risk plant, but a minimum of 4 hours is typical, based on the largest fire water demand. Within a facility, there may be certain units that require more than a four hour supply, i.e. due to a larger inventory of flammable materials. In such cases, additional sources of fire water will be required. This may mean temporary hook-up of a neighbor's system, use of cooling water, storm impoundment ponds, or reliance on municipal systems. A larger capacity may be warranted for larger and more complex facilities. Fire water pump suction tanks should be on ground level. Freeze protection should be provided as appropriate. Refer to NFPA 22 for more information on freeze protection for storage tanks. Reservoirs should be designed with additional capacity to account for icing. Storage tanks and reservoirs are shown in Figures 7-6 and 7-7.

Automatic refilling using an automatic float valve should be provided. A maximum refill time of 8-hours is recommended (NFPA 22). Low-level alarms to protect the pumping unit should be provided. Suction pits associated with reservoirs should be monitored by level alarms to detect suction screen plugging.

### 7.4.1.3. Municipal Water Supply

A municipal water system could be the source of an unlimited water supply. Reliability should be investigated, based on frequency of interruption and

**Figure 7-6.** Storage Tank

**Figure 7-7.** Reservoir with Fire Water Pumps

promptness of repairs, overall pumping and storage capacities, dependability of pumping stations, and provisions against the effect of drought and increased demand by other users in the area.

When a booster fire water pump takes suction from a public main, the design should be such that operation of the pump at 150% of rated capacity will not reduce the public main pressure below 20 psi (137.9 kPa). Provisions against contamination of the municipal system should be made by adding a backflow device is typically required. Local agencies generally have specific requirements. Some locations do not allow direct suction from public water mains.

#### *7.4.1.4. Wells*

The adequacy of continuous pumping of wells should be considered because wells are often of limited capacity. These systems are normally used as backup fire water or for refilling tanks and reservoirs. Also, most well pumps are only electrically driven, requiring special electrical feed requirements.

#### *7.4.1.5. Cooling Towers*

The use of cooling tower basins and process water pumps as fire water supply is not recommended. This water is usually treated with chemicals or may be contaminated with hydrocarbons that interfere with the use of foam extinguishing agents. At best, this supply could serve as a secondary system, should the primary supply be interrupted, using emergency connections.

### 7.4.2. Fire Water Demand

Fire water demand is the maximum rate of water that will be needed at any one time and applicable to a single fire. Thus, the requirements of the largest single fire contingency will determine the capacity and design of the fire water system. The fire hazard analysis should serve as a basis for determining fire water demand. Normally, this design will be based on the assumption that there will be a single unit involved in the fire. Where separation of units or hazardous equipment is less than 50 ft (15 m), the combined area should be considered as a single fire area. If the design water flow rate for the process unit requiring the largest flow is less than the requirement for the largest tank or group of tanks, the tank protection demand should become the design basis.

Design water rates should be based on the ground areas within each unit's battery limit when separated by at least the distance specified in the spacing table. When the spacing within units is such that the processes are separated by at least 50 ft (15 m), the area demand may be calculated separately by using Figure 7-8, but, at a minimum, should be 3,000 gpm (11,400 1/min).

Structures with equipment on two or more levels, not protected by water spray or deluge systems, should be counted as double the grade area beneath the multilevel structure plus the areas as noted above. Where fixed water spray systems are provided, their water requirements should be added to the fire water demand. Water demands for other areas within a plant may include:

- Utility plants — 1,500 gpm
- Loading racks — 1,500 gpm
- Office buildings, workshops, storehouses, etc. — 1,000 gpm, plus sprinkler requirements if specified
- Cooling tower — 2,500 gpm

Normally, fire water demands range between 2,000 and 10,000 gpm (7,600 to 38,000 lpm). The design capacity of the fire water system should be at a minimum four (4) hours of continuous operation of the largest fire water demand. The capacity is based on a number of factors, including:

- Sources of water available
- Reliability of make-up water supply
- Potential for escalation to other areas of the facility
- Isolation philosophy and the ability to depressure high pressure units

The reliability of the fire water supply should be such that the loss of any one source does not result in a loss of more than 50% of the flow requirements of the system. For example:

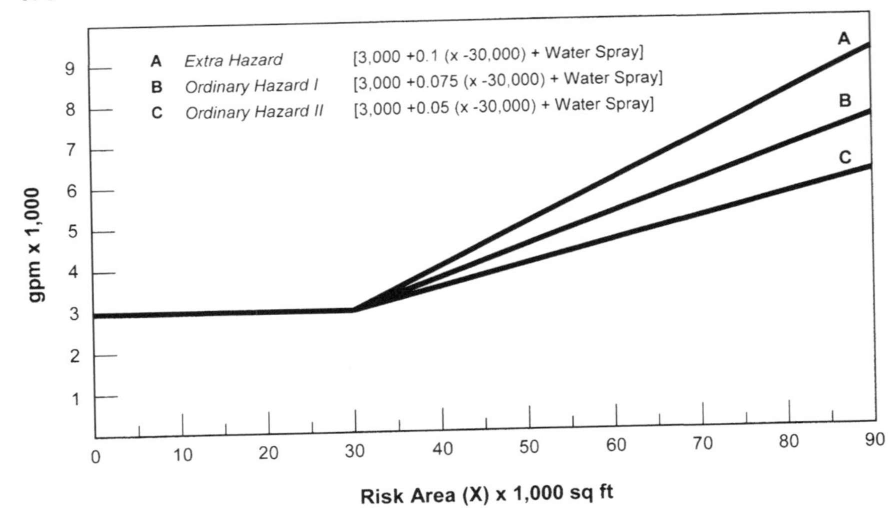


| | |
|---|---|
| *Extra Hazard* | Units with light material, vapor cloud explosion potential, liquid inventory over 10,000 gal |
| *Ordinary Hazard I* | Units with nonflashing material, inventory less than 10,000 gal |
| *Ordinary Hazard II* | Combustible materials |

**Figure 7-8**. Example Fire Water Demand Method

- A large facility connected to the city water supply should have two independent connections off different branches of the city underground piping.
- A facility drawing water from a stream or lake should either have independent locations from which to draw from or have a back-up supply from the city system or a private well. Care should be taken when using potable and nonpotable sources so that cross-contamination does not occur.

## *7.4.3. Water Distribution*

Water distribution within a facility involves fire mains, valves and fittings, and fire hydrants and monitors.

### *7.4.3.1. Fire Mains*

Fire water mains should be designed to handle the maximum pressures developed by fire water pumps. Systems operating at pressures over 150 psi (1,034 kPa) are discouraged, as this would exceed the normal design pressures for most fire protection assets such as monitors, hydrants, etc.

#### 7.4.3.1.1. HYDRAULIC DESIGN

A grid or looped piping distribution system should be used that is capable of supplying water on a single fire scenario to any part of the facility at the design rate determined for that specific area. The design should be based on water flow through all loops and mains of the grid system and the lines should be sized to provide 100 psig (690 kPa) minimum residual pressure at full design flow rate. Piping should be sized conservatively to account for fouling and increased roughness caused by aging. The residual pressure requirement or atmospheric storage tank farms may be reduced to 20 psi (138 kPa) where pumper trucks are immediately available and have the pumping capacity to meet the maximum fire water demand for the tank farm. If the pumpers are not available, then a residual pressure of 100 psi (690 kPa) should be available.

For calculation of flow in fire main systems, the design engineer should use the Hazen-Williams friction coefficient of $C = 100$ in the design of bare steel and concrete steel lined piping for fire water systems to allow for future deterioration as the system ages. A friction coefficient of $C = 140$ may be used for plastic pipe, such as fiber reinforced plastic (FRP) or polyvinyl chloride (PVC).

#### 7.4.3.1.2. PIPE DIAMETER

The pipe diameter for the main distribution lines should be hydraulically sized based on projected flow and pressure demand, but in no case less than 8 in (20 cm). Larger line sizes should be considered when there is a possibility of facility expansion. Typically the cost of installing pipe is mostly excavation cost. Increasing line size is insignificant when compared to the total cost of the project. Lines terminating at hydrants, monitors, and other protective systems should be at least 6 in (15 cm) in diameter.

#### 7.4.3.1.3. UNDERGROUND AND ABOVEGROUND INSTALLATIONS

Mains should be installed underground whenever practical and be located at the edge of roads or streets. Underground lines should not be routed under buildings, tanks, equipment, storage areas, or structural foundations. Steel pipe should be protected against corrosion by cathodic protection. In freezing climates, lines should be buried a minimum of 1 ft (0.30 m) below the frost line. Piping and connections above this level should be protected against freezing. Lines should generally be buried a minimum of 3 ft (0.91 m) to prevent damage from surface loads.

Thrust blocks or other approved methods or devices for anchoring mains and components should be provided where needed to prevent movement of underground lines and should not interfere with drainage for hydrants. Refer to NFPA 24 for additional guidance.

In areas not subject to freezing, fire, mechanical damage, or blast damage, aboveground piping is acceptable. Mains should be kept outside of areas

:re flammable and combustible liquids can accumulate. Protection from :hanical damage, due to vehicles etc., should be provided where neces- . Aboveground fire water mains to marine docks and other remote facilities are subject to freezing weather should be designed to prevent freezing to re usability of the mains. This can be achieved by provision of special valves gned to open at low temperature and release a small but constant flow to n.

.1.4. MATERIALS

erials acceptable for underground piping use include ductile iron, fiber- ;-reinforced epoxy plastic, polyethylene, polyvinyl chloride, reinforced :rete, and carbon steel. Plastic pipes are not acceptable in areas subject to ent exposure.

Plastic piping should only be used as specified by the manufacturers and d by UL or equivalent. New water lines should be subjected to a hydro- c test of 150% of design pressure for at least 2 hours (NFPA 24). Leaks ıld be repaired and tests repeated as necessary before lines are covered.

## 3.2. *Valves and Fittings*

'es used in a fire water system should be of a type such that their position, n or closed, can be readily determined, e.g., rising stem, post indicator. A -indicator valve is shown in Figure 7-9.

**Figure 7-9.** Post-Indicator Valve

Valves in underground lines should be operable from grade with the valve stem and packing protected against contact with soil and rock. Only valves that open counter-clockwise should be used. Valves should be clearly marked for easy recognition.

Sufficient sectional block valves should be provided so that sections of the underground system can be taken out of service for repairs without undue interruption of the fire water protection. For example, the sectional block valves should be located so that a combination of no more than five resources, such as hydrants, monitors, water spray systems, etc. would be out-of-service at any given time. Sectional block valves should be provided at no more than 800 ft (243 m) intervals for long runs of pipe.

Fire water lines within process unit areas feeding more than two monitors, hose reels, or hydrants, should be connected to two separate sections of the fire main and separated by a valve in the main. Piping to hose reels and monitors should contain an isolation valve.

If marine facilities are connected to a fire water system, an appropriate connection should be provided for the fireboat that could be used at the facility.

### 7.4.3.3. *Fire Hydrants and Monitors*

Hydrants and monitors should be spaced so that they can be safely utilized during a fire such that the total firefighting and equipment cooling water requirements for each fire area can be delivered. Hydrants are generally spaced a maximum of 200 ft (60 m) apart in process areas. Greater spacing is allowable in tank farms [up to 300 ft (91 m)] and low hazard areas, such as utility blocks. Hydrants and monitors should be located a minimum of 50 ft (15 m) from any equipment to be protected.

Hydrants should be located such that any piece of equipment can be reached by a hose stream comprising no more than 100 ft (30 m) of hose. Monitors should be located no more than 100 ft (30 m) from the equipment being protected. A hydrant is shown in Figure 7-10.

Hydrants should be located on the street or roadside of all pipelines or drainage ditches to allow access for the pumper truck. Where large pipelines or drainage ditches may hinder access, access across such obstructions should be provided. Hydrants and manually operated monitors should be located outside of tank dike areas. Monitors capable of remote operation from outside tank dike areas may be located within that area.

Hydrants equipped with monitors should have hose gate valves provided on the hose connections. Hydrants accessible to fire trucks should be equipped with compatible pumper connections as well as same type as public fire department. The pumper connection should face the roadway.

The 2½-in (5-cm) hose connections on the hydrant should be located so that the distance between the bottom of the connections and finished grade is

**Figure 7-10.** Hydrant

a minimum of 16 in (40 cm). Gate valves (outside screw and yoke type) and hose and pumper connections may be provided on the fire water lines in lieu of hydrants where aboveground dry pipe systems and pressure systems in non-freezing climates are used as shown in Figure 7-11.

Self-draining hydrants and monitors should be used on underground mains installed in freezing climates. Protection against mechanical damage should also be provided where necessary, usually by means of guard posts.

If a water source (pond or river) is located in the vicinity of the mains, a hydrant with a pumper connection should be placed near the source to allow water to be pumped into the system with the aid of a mobile pumper.

Monitors should be positioned to be accessible during a fire and located where they will provide effective water streams for the equipment to be protected. Typically, monitors are provided to cool equipment that, upon failure, would result in escalation of fire.

The exact spacing and location of monitors should be determined by considering the following:

- Elevation
- Solid decking
- Obstructions
- Stream trajectory

**Figure 7-11.** Nonconventional Hydrant

- Depth of structure
- Prevailing wind

Monitors range in size from 250–4,000 gpm (950–15,140 lpm), with 500 and 1,000 gpm (1,900–3,800 lpm) the most common sizes. In process areas that are not protected by water spray, monitors should be located so that each major piece of equipment can be covered by two monitors. Elevated monitors have application where it is necessary to deliver large volumes of water to areas that cannot be reached by ground level monitors or would be unsafe for manual firefighting. Elevated monitors can be fixed for remote use, self-oscillating or remote control from a safe location.

The last two applications, while convenient and initially appealing, require more maintenance and inspection than a simple, ground level monitor. Significant reactive forces can be generated and, therefore, the support structures must be properly engineered. Elevated monitors also require support and hydraulics to be protected by a water spray system if exposed to fires. An elevated monitor is shown in Figure 7-12.

### 7.4.4. Fire Water Pumps

Fire water pumping capacity (flow rate) should be sufficient to provide the required amount of water at required pressure to the fire areas having the greatest demand. At least 50% of the pumping capacity should be from diesel-driven pumps. Fire water pumps should have a minimum capacity of 1,500 gpm (5,700 lpm) and can range up to 5,000 gpm (18,930 lpm). Larger pumps

**Figure 7-12.** Elevated Monitor with Aerating Foam Nozzle

may be used in specially engineered applications; however, this larger pump should account for no more than one-half of the total demand. Figures 7-13 and 7-14 show examples of horizontal and vertical pumps.

It is common practice to provide pumping capacity so that when the largest fire water pump is out of service, the total fire water demand can still be met. In situations where the demand does not exceed 1,500 gpm (5,700 lpm), it may be acceptable to use a single pump.

**Figure 7-13.** Horizontal Pump

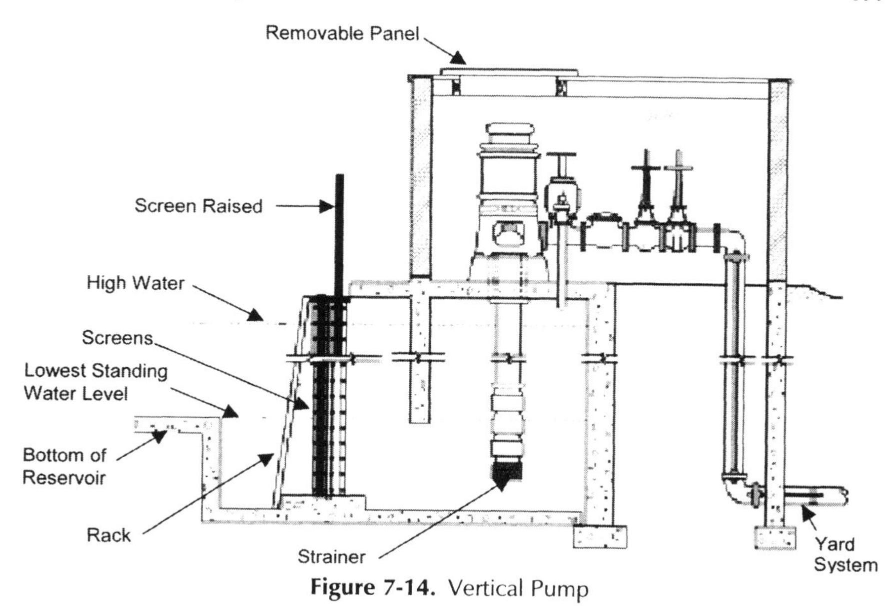


**Figure 7-14.** Vertical Pump

Listed or approved equipment should be specified for fire water pumps, drivers, controls, and associated equipment. The use of unlisted or unapproved equipment is acceptable if it is of equivalent construction and designed for fire services as specified in NFPA 20.

Fire water pumps should be dedicated solely to fire protection. They may be used to feed into a backup system for emergency process cooling, but not as the primary supply. Valving for this connection should be accessible for prompt shutdown when required under fire emergency conditions.

Provision should be made for testing of individual pump performance and should include a test header with either 2½ in (6.4 cm) outlets, or an approved flow meter. The latter is preferred because of its ease of testing and the water can be returned to the source tanks and reservoirs. If pulling from a river or waterway, there may be environmental requirements of returning water to the river or waterway.

Each pump should be subjected to a full acceptance test upon installation, as specified in NFPA 20. The manufacturer's shop test pump curves should be available for comparison during the acceptance test.

### *7.4.4.1. Construction and Location*

Pumps should be located close to the supply source and away from hazardous areas. If a pump house is required, it should be of noncombustible construction, free of combustibles, and sprinkler-protected if a diesel driver is used. Floor drains to handle leaks and condensation should be provided. Adequate air for combustion drivers should be provided through the use of self-operating louvers.

On large fire water systems, the location of pumps and storage tanks at various plant areas provides greater reliability of protection and results in less pressure drop between the pump and the area of demand. Net positive suction head (NPSH) requirements and friction loss in the piping should be considered in locating fire water pumps.

### *7.4.4.2. Pump Types and Ratings*

Horizontal centrifugal pumps should provide 150% of the rated capacity at 65% of the rated pressure, with a shutoff head of not more than 120% of the rated pressure. This pump should be used only when suction supply is under a positive head. Suction pipes should be designed to preclude the formation of air bubbles. A characteristic curve for a rated fire water pump is shown in Figure 7-15.

Vertical shaft, turbine type pumps have their impellers submerged, hence, they are ideal for conditions where a suction lift would be required for horizontal pumps. The second impeller from the bottom of the pump bowl assembly should be submerged below the lowest pumping water level. The submergence should be increased by 1 ft (0.30 m) for every 1,000 ft (305 m) of elevation above sea level. These pumps should provide 150% of the rated capacity

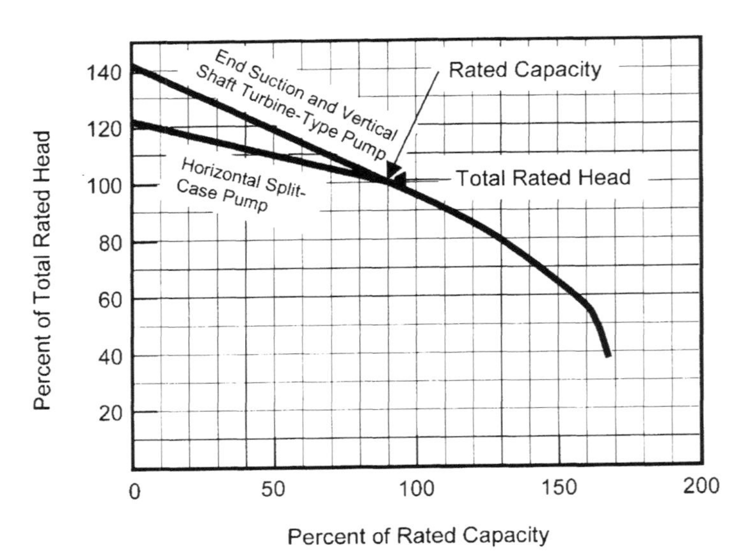


**Figure 7-15.** Rated Fire Water Pump Curve

at 65% of the rated pressure, with a shutoff head of no more than 140% of the rated pressure.

Pressure maintenance (jockey) pumps should be provided to maintain a predetermined pressure on the system and make-up normal leakage in the distribution system. Normally, the pressure maintenance pump will maintain 10–15 psi (69–103 kPa) above the starting pressure for the automatic starting of the main fire water pump. (See 7.4.4.3.9.)

When the fire water system will be routinely used for purposes other than firefighting (e.g., washdown), the pressure maintenance pump, typically 150–300 gpm (570–1,150 lpm), should have sufficient capacity for such use or a separate service pump should be provided. The service pump need not meet the requirements for fire water pumps.

#### *7.4.4.3. Fire Water Pump Driver and Miscellaneous Equipment*

The acceptable drivers for fire water pumps are electrical motor and diesel engine. If only one driver is used, it should be a diesel, unless the reliability of electrical power can be ensured by onsite diesel emergency power or there are two independent reliable power supplies.

##### 7.4.4.3.1. DIESEL ENGINE

The diesel engine is the most dependable engine for driving fire water pumps and should be the only type of engine acceptable for this service. The following minimum equipment should be provided for diesel engines:

- An adjustable governor capable of regulating engine speed within a range of 10% between shut-off and maximum load conditions; it should maintain pump rated speed at maximum pump load.
- An overspeed shutdown device with manual reset arranged to shutdown the engine at a speed of 20% above rated engine speed.
- A tachometer with totalizer provided to record total time of engine operation.
- An oil pressure gauge to indicate lubricating oil pressure and a temperature gauge for cooling water temperature indication.

The engine should have a reliable starting device, either from storage batteries or with an approved air starter. A reliable cooling system should be provided for the engine. This may be through a heat exchanger using water, a radiator, an engine-driven fan, or an ambient air cooling system (NFPA 20).

The fuel supply should have the capacity to operate the pump engine for not less than eight hours. If permitted by local regulations, the tank may be located inside the pump house, however, fill and vent lines should extend outdoors. Fuel lines should be steel. Similarly, the exhaust from the engines should be piped to a safe point outside, where it will not affect persons or endanger other structures, and arranged to exclude rain water.

The recommended location for the diesel fuel tank is outdoors, just outside of the building. Where location of the diesel fuel storage tank outside of the fire water pump building is not appropriate (cold weather), sprinkler protection of the fire-water pump house is recommended. Pump controllers should be sealed and splash resistant.

The temperature in the pump house should not be less than 40°F (4°C). For freezing environments, automatic heaters should be installed on the engine to maintain the jacket water temperature of liquid-cooled engines at or near operating temperature, but not less than 120°F (49°C).

#### 7.4.4.3.2. ELECTRIC MOTOR

The reliability of the power supply should be determined, taking into account the frequency of power outages and extent of interruption. Consideration should be given to connecting electrically driven fire water pumps to the emergency power system, where one exists.

The power cables to the motor should preferably be buried or otherwise routed away from or protected against fire or other potential source of damages. The feeder circuit should be independent so that plant power can be shut-off without interrupting the pump operation. Conventional feeder overcurrent protection is not permitted by NFPA 20.

#### 7.4.4.3.3. RELIEF VALVE

Pumps connected to variable speed drivers should be equipped with an approved relief valve. The discharge of the relief valve should be directed back to the source where possible or to drains. Piping of a relief valve back to pump suction piping should be avoided. Whenever the relief valve does not discharge to the atmosphere, a sightglass flow indicator should be installed. Constant speed pumps also require relief valves when the shut-in pump discharge pressure exceeds the rated pressure for the system components.

#### 7.4.4.3.4. CIRCULATION RELIEF VALVE

Each pump should be provided with an automatic circulation relief valve set below the shut-off pressure at minimum expected suction pressure (see NFPA 20). Pumps up to 2,500 gpm (9,500 lpm) capacity require a ¾-in valve; 3,000 to 4,000 gpm (11,000 to 15,000 lpm) capacities require a 1-in (2.5-cm) valve to circulate sufficient water and prevent overheating of the pump when operated with no discharge. Provisions should be made for discharge of the valve to a drain. Diesel-driven pumps, for which engine cooling water is taken from the pump discharge, do not require circulation relief valves.

7.4.4.3.5. PRESSURE GAUGES

Discharge and suction gauges should be provided at each pump. The exception is that suction gauges are not required for vertical shaft, turbine-type pumps taking suction from a well or open wet pit. Gauge dials should be at least 3½ in (9 cm) in diameter and be provided with ½-in (1.2-cm) gauge valves.

7.4.4.3.6. SUCTION AND DISCHARGE PIPING

Sizes of suction and discharge piping should be based on pump capacity. When two or more pump suctions or discharges are manifolded together, a minimum of a 16 in (40 cm) pipe should be used.

7.4.4.3.7. CONTROLLERS

A controller should be provided for each fire water pump. The controller should be designed and installed in accordance with NFPA 20 specifications. An example fire water pump controller is shown in Figure 7-16.

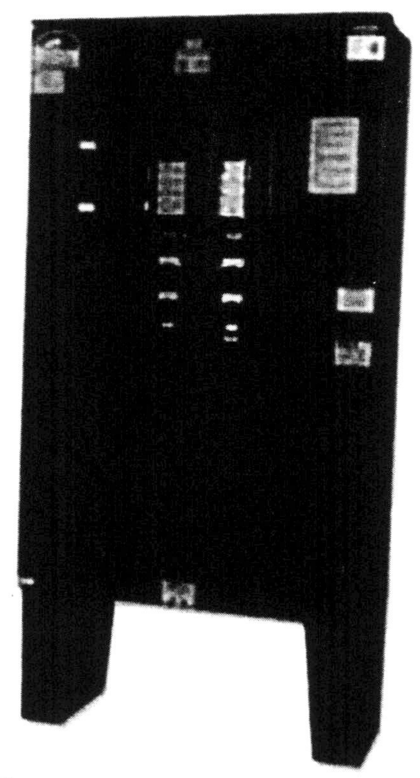

**Figure 7-16.** Fire Water Pump Controller

#### 7.4.4.3.8. OPERATION

The basic methods of starting fire water pumps are automatic, remote manual, and local. All pumps should be arranged for local manual shutdown only at the pump.

#### 7.4.4.3.9. AUTOMATIC STARTING

Automatic starting of a fire water pump should be initiated by a fire alarm signal, by low pressure in the fire water system, or both. When systems contain combinations of electric- and diesel- driven pumps, an electric pump should be the one designated as the first automatic start because there is less wear on an electric pump from starting and stopping. The remaining pumps can be brought on manually or by including these pumps in the automatic start sequence, so that they will be started if the low pressure signal persists.

#### 7.4.4.3.10. REMOTE MANUAL STARTING

Remote manual starting is initiated from a remote point rather than from the fire alarm system or a pressure switch. The remote point should be at a constantly attended area. Pump running and trouble indicators should be provided at the remote start point.

Remote controllers for fire water pumps or remote operational status indicator panels for sprinkler systems and fire water pumps will typically be protected according to the protection protocol of the room or building in which they are housed.

#### 7.4.4.3.11. MANUAL LOCAL STARTING

Manual starting usually consists of a push button at the pump.

#### 7.4.4.3.12. PUMP SHUTDOWN

The only acceptable method for shutting down a fire water pump is by means of a local "stop" push button.

### 7.4.5. Detection and Alarm

A system for reporting fires and alerting the plant fire brigade and the municipal fire department should be provided. This system should be as simple as possible to minimize the potential for confusion in emergencies. The preferred design is a multiplex system that alarms in the control room or some other 24-hour constantly attended location and activates visual devices, such as strobes or beacons, and audible notification devices, such as a steam whistle, air horn, or tone generator. Complete systems are available from recognized vendors.

A number of factors must be considered in the design of a reliable fire detection and alarm system including:

- Data on the nature and arrangement of power sources
- Coverage provided by the system
- Alarm function on loss of system operability
- Suitability of detection devices for the risk involved
- Testing and maintenance procedures to ensure a reliable system
- Consequence associated with false alarms

Where a detection system is part of an automatic, fixed fire extinguishing system, complete compatibility between the systems is essential.

It is important to select detection devices that are appropriate for the type of fire most likely to occur. Failure to do so will result in either a very slow response or the possibility of a large number of spurious alarms. The latter should be particularly avoided where the detection system is used to activate a fixed fire extinguishing system.

### 7.4.5.1. *Facility Emergency Alarm System*

#### 7.4.5.1.1. GENERAL

This section includes guidelines for the central control station equipment, emergency alarm stations, supervisory devices, and visual and audible alarm services. These systems can be used for all types of in-house emergencies, such as fires, explosions, vapor releases, liquid spills, and injuries.

The type of fire alarm system should be chosen based on personnel resources available at the facility. For continuously staffed facilities, proprietary supervised systems are preferred. For facilities staffed less than continuously, remote supervised station or central station fire alarms system are usually considered. In these systems, alarms are monitored by an outside firm responsible for alerting appropriate personnel or by the local fire department.

Process facilities may be divided into primary operational divisions called zones. These zones are divided into areas. A coded signaling system can be used to indicate the zones and area numbers. A third number may be used to indicate a station within the area boundaries. The plant emergency alarm system should:

- Supervise all signal receiving and sending circuits as well as power supply, open circuits, and grounds
- Provide notification of fire alarm, supervisory, and trouble conditions
- Alert facility occupants
- Summon aid
- Control fire safety functions
- Sound an "All Clear" signal when the emergency is over

Emergency alarm system design and installation should be in accordance with NFPA 72. Electrical aspects of the fire alarm systems should be designed and installed in accordance with NFPA 70. When devices are located in hazardous areas, they should meet the electrical requirements suitable for that hazardous area.

In addition to the abbreviated LCD display on the fire alarm panel, an annunciator should be provided to graphically display facility conditions in all areas. Often this annunciator is an additional panel provided with the fire alarm system, which contains a graphic display or representation of the plant or area being monitored as well as illumination capabilities which identify the status of all monitored plant areas.

More and more commonly, fire alarm panel data is transferred to a safety instrumented system (SIS) for graphic annunciation though the SIS human-machine interface (HMI).

Fire alarm bells or horns should be located in all buildings and on all levels of process structures and other open areas in sufficient number to ensure that all parts of the facility are alerted. Areas with high noise levels should have alarms with a sufficient sound level to overcome ambient noise; otherwise, revolving beacons or strobe lights should be used in high noise areas.

Where releases of flammable or toxic chemicals are possible, a flammable or toxic gas alarm system is often established as part of the plant emergency alarm system. Best practices require different types of alarms to be annunciated differently, both audibly and visually. For example, the toxic alarm stations may be provided with a blue light to distinguish them from fire alarm stations that are red. A consistent color system for lights should be adopted.

Fixed installations, such as water spray systems, halon systems, sprinkler systems, carbon dioxide extinguishing systems, explosion suppression systems, and other fire protection installations are often provided with flow and trouble detection switches connected to transmitters. A signal indicating the condition of the system should be sent to the attended location(s).

Where buildings are provided with an individual fire, gas, or smoke detection system, that system commonly actuates a relay upon alarm and transmits a fire alarm, gas alarm, or trouble alarm signal to the attended location(s).

Other supervisory signals may come from fire protection system components such as supervised control valves, system air and supervisory air pressure transmitters, water tank level and temperature transmitters, valve house and fire water pump building temperature transmitters, and fire water pumps.

Emergency alarm systems should always have a manual means for initiating alarms. Manual alarm stations are normally located on the periphery of the processing or storage areas near an expected path of exit travel or at control rooms. The alarm boxes should not be located in areas that are electrically clas-

**Figure 7-17.** Manual alarm station

sified as Division 1. Boxes used in Division 2 areas should be approved for that service. An example manual alarm station is shown in Figure 7-17.

#### 7.4.5.1.2. ALARM AND DETECTION CONTROL PANEL

The fire alarm and detection control panel is the central point of the fire detection system. This panel should be capable of:

- Powering all field detectors
- Supervising all field detectors
- Powering all output functions such as alarms, extinguishing system releases, etc.
- Providing all fire alarm and detection logic

Power for the control panel should be provided with a suitable uninterrupted power supply (UPS). The panel will provide a DC current to field detectors. This power will enable the panel to monitor all input circuits, output circuits, and trouble signals within the detectors, such as shorts, ground faults, and detector disconnects. It will also provide an AC powering signal to field output devices. All output circuits should be similarly supervised for trouble. An example alarm and detection control panel is shown in Figure 7-18.

In general, the fire detection system should include the necessary logic and actuators to initiate applicable fire systems. The input from fire detection systems may be used, where appropriate, to initiate the shutdown of ventilation systems. In some circumstances it may be justified to also use these inputs to initiate process related shutdown systems. In these latter situations consideration has to be given to balancing the advantages associated with a rapid shutdown during an emergency with the consequences associated with the possibility of spurious shutdowns during normal operations. The fire alarm system should be connected to the overall facility alarm system.

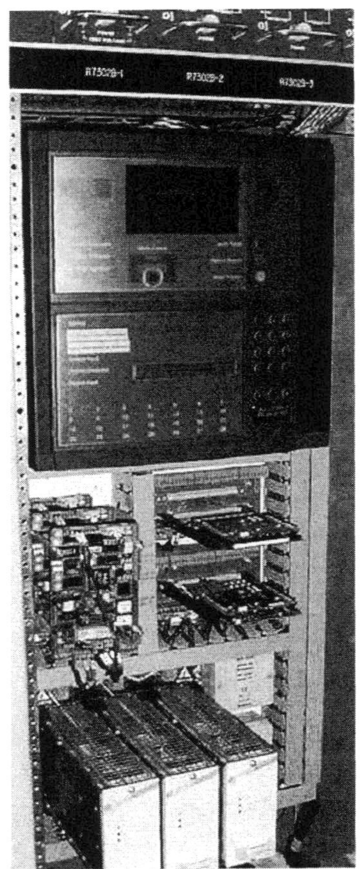

**Figure 7-18.** Alarm and Detection Control Panel

Fire detection system circuits should be completely supervised as described below:

- Circuit malfunctions should result in separate and distinct audible and visual trouble signals on the panel designating a circuit fault.
- The malfunction audible signal should be capable of being silenced after operation of an acknowledgment "trouble alarm silence" switch. The visual trouble signal should remain illuminated.
- Correction of the circuit malfunction should result in sounding the distinct audible trouble signal until the alarm is restored to the normal position.
- Each detection circuit should be shown in trouble condition by a separate visual signal. One audible trouble alarm signal may be used for more than one circuit.

- A normal power-on signal lamp should be included in the panel. Failure of normal power should actuate an audible and visual signal.
- End-of-line resistors installed for system circuit supervision should be mounted in covered electrical junction boxes. The boxes should be labeled with plastic or metal signs and mounted in an accessible location.

A main control and annunciator panel should be installed when the fire alarm system requires more than a single alarm zone. The panel should be installed in the control room or other continuously staffed location. Separate detection zones should be provided for each distinct fire area and identified by a permanent label. A detailed map of the area should also be provided at the annunciator that identifies which zone relates to which annunciator lamp. Systems with more than ten separate zones should be provided with an electric or electronic zone "mimic" panel showing the location of all alarms on the graphic display of the platform. Basic arrangements of equipment and system design should be in accordance with NFPA 72. A locked main fire panel and control cabinet should be provided.

Electric power should be available from two highly reliable sources. The usual arrangement is an alternating current (AC) power supply, with trickle charger supplying an emergency battery system. Batteries should be sized for loss of primary power for a period of no less than 8 hours, and for at least 12 hours if the supply is not reasonably reliable. An exception is power for alarm bells or horns, which require only 1 hour of emergency power. The power supply should be monitored by a power-on lamp on the control panel and a main power failure alarm.

#### *7.4.5.2. Detectors*

Fire detectors generally fall within one of three categories: heat, smoke, and flame. Heat detectors work by sensing the heat from a fire. Smoke detectors sense the combustion products from the fire. Flame detectors identify flame by sensing the IR or UV light it emits.

Each detector type has advantages and limitations that are discussed in the following sections. Table 7-8 shows the different type of detectors by category.

To function effectively, detection devices, particularly the spot, fixed-temperature, heat type, must be properly positioned. Detailed requirements for spacing can be found in NFPA 72. These spacing requirements are based on limitations contained in UL listings and FM Global approval lists and industry experience.

Where detectors manufactured outside the United States are used, the location and spacing recommendations of the AHJ should be followed. The manufacturer's recommended location and spacing distance should also be taken into account.

**Table 7-8**
*Types of Fire Detectors*

| Heat Detectors | Smoke Detectors | Flame Detectors |
|---|---|---|
| Fixed Temperature: | Photoelectric | Infrared |
| ▪ Fusible plugs or sprinkler heads with pressurized tubing or pipe | Ionization | UV |
| ▪ Plastic pneumatic tubing | Resistance bridge | IR/UV combination |
| ▪ Fusible links and quartzoid bulbs | High sensitivity (air sampling) | CCTV flame |
| ▪ Thermostats—bimetallic strip, snap-action disk, and thermostatic cable | Oil mist | $IR^3$ |
| ▪ Continuous discrete conductors | | |
| ▪ Thermistor sensors | | |
| Rate of temperature rise: | | |
| ▪ Copper tubing heat-actuated devices (HADs) | | |
| ▪ Thermoelectric | | |
| ▪ Combined rate of rise/fixed temperature | | |
| ▪ Rate compensation | | |

7.4.5.2.1. HEAT DETECTORS

There are two common types of heat detectors. Fixed temperature detectors operate when the detection element is heated to a predetermined set temperature. Rate-of-rise detectors respond when the temperature rises at a rate exceeding a predetermined amount.

Where process equipment is provided with fixed-temperature detectors, these should be located as near as possible to the potential fire source; for example, above flammable liquid pump seals, immediately over a solvent draw-off point, or mounted above a crude tank mixer stuffing box. As a general rule, fixed-temperature detectors directed at a potential hazard should be considered only for process equipment where specific fire problems are anticipated.

Fixed temperature detectors are preferred because they require less calibration and maintenance. Heat detectors are normally more reliable than other types of detectors because of the simple nature of their operation and ease of maintenance. These factors tend to lead to fewer false alarms. The main disadvantage of heat detectors is that they are unlikely to detect fires in the incipient stage, where little heat is generated, but much smoke is likely. .Since heat detectors are inherently slower in operation than other types of detectors, they should be considered for installation in areas where high speed detection is not required.

**The most widely used heat detector is the automatic sprinkler head.** These are fixed temperature devices that operate when the fusible link or element on the head is melted by heat from the fire. The sprinkler head is normally used to distribute water, but it can also be used to actuate water spray systems. When used in a detection and release system, the sprinkler head is called a "pilot head."

One effect that a flaming fire has on the surrounding area is to rapidly increase air temperature in the space above the fire. Fixed temperature heat detectors will not initiate an alarm until the air temperature near the device exceeds the design operating point. The rate-of-rise detector, however, will function when the rate of temperature increase exceeds a predetermined value, typically around 12 to 15°F (7 to 8°C) per minute. Rate-of-rise detectors are designed to compensate for the normal changes in ambient temperature [less than 12°F (6.7°C) per minute] which are expected under nonfire conditions.

There are several technologies of linear heat detectors, most designed to monitor air temperature and a few that detect radiant heat. This type of detection is wire or plastic tubing and can be used where other types of detectors are difficult to install. Generally, they are used to supplement other forms of detection in difficult areas, such as heavily congested areas, rim area of floating roof tanks, on pumps, etc. Linear types can be pneumatic, electrical, or optical. Electrical linear heat detectors come in three types that:

- Respond to an average temperature along their length.
- Respond to the highest temperature at a point along their length.
- Detect both a localized hot spot and low level temperature increases along their entire length.

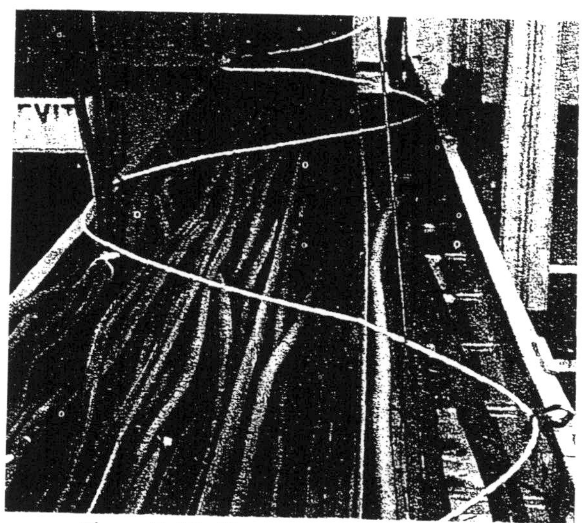

**Figure 7-19.** Linear Heat Detection Wire

Heat detectors should be spaced to ensure prompt detection of heat given off by incipient fire conditions. Spacing considerations should include the degree of hazard, type of detector used, geometry of the protected area, effects of air handling equipment if it is indoors, or environmental factors (wind, temperature, etc.) if it is outdoors. See NFPA 72 for more information on spacing and installation requirements.

#### 7.4.5.2.2. SMOKE DETECTORS

Smoke detectors are primarily used where smoldering fires can be expected and where electrical equipment is located indoors. Examples of their use are in offices and sleeping quarters, computer rooms, control rooms, electrical switchgear rooms, etc. Their response is typically faster than that of heat detection devices. Smoke detectors are more susceptible to false alarms and usually multiple detectors are required to be in alarm before an extinguishing system is activated.

Smoke detectors should be located and spaced to ensure prompt detection under incipient fire conditions. Spacing requirements should consider the type of detector used, manufacturer recommendations, geometry of the protected area, and effects of air handling equipment. See NFPA 72 for more information on spacing.

There are several types of smoke detectors. The photoelectric type detects smoke particles which interrupt a light beam resulting in the generation of an electrical signal. Photoelectric detectors are used where very slow evolving, smoky fires are expected.

Ionization smoke detectors contain a small radioactive source (Americium 241) which ionizes air in a small chamber. The ions flow to a charged plate giving a measurable current. Products of combustion in the chamber are not easily ionized and absorb the radiation and reduce the current. The low current trips the alarm circuit. The size and composition of the particles are crucial to successful detection so that some types of smoke or vapor are detected at very low (invisible) levels.

Ionization smoke detectors can be used in general accommodation areas such as corridors and public restrooms, to monitor the escape routes. They may be used in some types of equipment spaces, depending on ventilation.

High sensitivity smoke detection (also known as air sampling) is used to protect critical and high value equipment that generally requires better performance from detectors than is provided by the standard point type. These detectors are normally used with a multi-point sampling system to monitor extracted air from rooms containing high criticality electrical or electronic equipment. Their sensitivity is adjustable and at maximum is far higher than generally required. They are more than capable, when used with appropriate alarms and procedures, of detecting an overheated component in a multi-cubicle equipment room. The efficiency of the system is dependent on the design of the sampling system and must be checked on commissioning. Thereafter, testing, calibration, and maintenance are all performed in-situ.

Oil/mist smoke detectors are particularly useful in engine and turbine enclosures for indicating the release of high-pressure fuel or lubrication oil, which can lead to particularly aggressive and damaging fires. If the oil mist ignites, the instrument doubles as an optical smoke detector. The detectors are obscuration instruments. An infrared light source is transmitted across the area to be monitored, returned via a retro-reflector, and the intensity of the returned light monitored. Software signal filtering prevents unwanted alarms. These detectors are also useful in large open areas requiring smoke detection when numerous point detectors would otherwise be required.

Many manufacturers are developing and marketing new detectors that have infrared, photo-electric, and heat detection built into one unit. Many of these detectors also include a computer chip with an algorithm that monitors the response of each sensor and calls for an alarm when certain parameters are met. One advantage to this type of detector is that they are less susceptible to unwanted alarms.

#### 7.4.5.2.3. FLAME DETECTORS

The available types of flame sensing detectors are ultraviolet (UV), infrared detection (IR), combination of these, and monitored closed circuit television (CCTV). These devices operate on the detection of certain wavelengths of light emitted by flames. They are used when there is a potential for fires that rapidly produce flame such as flash fire.

Flame sensing detectors generally respond the fastest to fires and may be used in a number of ways, including activating extinguishing systems or emergency shutdown systems. It should be noted that some detectors are sensitive to sunlight, arc welding, lightning, and other flash type flame situations. Because of this, UV detectors should only be arranged to operate on a cross-zone system. The cone of vision for a UV detector is 60 to 90° with a maximum practical pick-up distance of about 35 ft (11 m).

IR responds to energy radiated from a flame in two adjacent bands of the mid-infrared spectrum. Two spectral sensing channels, each looking at a different wavelength of the flame spectrum, permit this detector to discriminate between flames containing hydrocarbon based material and spurious stimuli from ultraviolet and other spectral radiation sources, such as welding areas, sunlight, and flashing lamps. This type of detector does not require cross-zoning.

The $IR^3$ detector utilizes a combination of three IR sensors of extremely narrow band response. One covers the typical $CO_2$ emission spectral band, and the two other sensors cover different adjacent specially selected spectral bands. While the $CO_2$ emission band sensor is responsible for the detection of the flame radiation, the other sensors are sensitive to all other nonfire radiation sources.

The $IR^3$ employs dedicated algorithms processed by a microprocessor analyzing radiation, ratios, correlations, threshold values, and flickering signals obtained from the three sensors.

The high sensitivity of the $IR^3$ is achieved by extracting extremely low signals deeply buried in noise by adopting digital correlation techniques. This counts for the high sensitivity and long detection range while the immunity of the $IR^3$ detector to false alarms is maintained.

The IR/UV flame detectors are used to sense fires. Flame detectors that use only IR or UV can experience false alarms. The IR/UV detector is designed to recognize a different type of flame signature from the detectors while rejecting common false sources. When the conditions of any one of the several fire conditions are met the detector indicates a fire. IR/UV flame detectors generally have a cone of vision from 60 to 120-degree solid cone field of view.

CCTV technology is used to analyze digital images and identify the characteristics of a fire. CCTV can provide both a control action and display the picture in a staffed control room. The latter has a particular benefit as it allows very rapid incident assessment and also reduces the need for personnel to enter the area to investigate the fire conditions. In general terms, the CCTV flame detection system is comprised of three basic components: the camera/detection element, the control panel, and the display computers.

The camera/detection unit contains, in a small enclosure approved for a Class 1, Div. 1 hazardous area, the camera and computing facilities to determine whether or not the unit is "seeing" a fire. The device is programmed with

**Figure 7-20.** CCTV Flame Detectors

a range of algorithms to determine whether the changes within its field of view are a fire. If there is a fire, the view from the camera is automatically displayed at the staffed control room. This enables an operator to see in real time what is happening, take appropriate action and, if necessary, guide firefighting teams.

Such systems can generally recognize fires of 34,000 BTU/hr (10kW) or greater at 33 ft (10 m) within a 90°cone of vision, and are immune to common false alarms such as hot $CO_2$ emissions, reflections from flare radiation, black body radiation, and hot work. An added benefit is that the video image can be recorded for later analysis.

### 7.4.6. Gas Sensing Detectors

The objective of an area gas detection system is to detect gas clouds of sufficient size that, if ignited, could cause damage as a result of a flash fire or explosion overpressure. The size of a gas cloud requiring detection is based on the volume of the area, the level of building confinement, and equipment congestion. In assessing the need and value of installing combustible gas detection, the following should be considered: the nature of the hazard, consequence of ignition, gas composition, confinement of the area, equipment congestion, and required response time. There are several types of gas detectors available:

- *Catalytic bead sensors* (Figure 7-21) are one of the most common types of gas detector found in industry. The sensor consists of two alumina beads, each surrounding a platinum wire operating at approximately 450°C. One bead is passive; it will not react with combustible gas. The other bead is catalyzed to promote a reaction with combustible gas. Effects of changes in ambient temperature and relative humidity are nullified by placing the two beads in separate legs of a wheatstone bridge circuit. When the catalyzed bead reacts with combustible gas it heats up, which increases its resistance and, in turn, increases the output of the wheatstone bridge circuit used as a sensor signal. This type

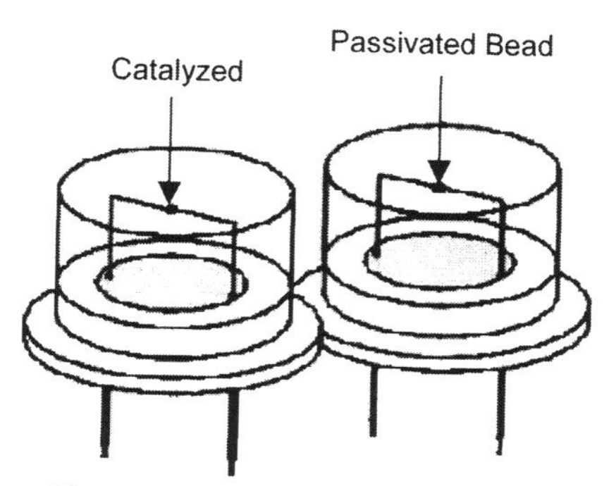


**Figure 7-21.** Catalytic Bead Sensor

of detector is prone to rapid poisoning or contamination by large releases, water, and other chemicals found in facilities. This poisoning effect can result in the detectors not functioning properly or at all.

- *Electrochemical sensors* (Figure 7-22) are like a fuel cell device consisting of an anode, cathode, and electrolyte. The components of the cell are selected so a subject gas, allowed to diffuse into the cell, will cause a chemical reaction and generate a current. The cells are diffusion limited, so the rate the gas enters the cell is solely dependent on the gas concentration. The current generated is proportional to the rate of consumption of the subject gas in the cell. Some detectors provide improved reliability by allowing the gas to diffuse into the sensor through a capillary port, rather than diffusing through membranes. The result is an extremely stable sensor with very low temperature and pressure coefficients and the capability to monitor gas as a percent by volume (oxygen) and ppm (toxics).
- The principle of an *infrared gas detector* (IR) is based upon the adsorption of the infrared light at a specific wavelength as it passes through the gas. Most combustible gases absorb infrared light energy at defined wavelengths, providing an adsorption signature for that gas. The more of the adsorbing gas that is present, the more light is adsorbed. The sensor compares the energy emitted by the light source to that received by the detector, resulting in the amount of light adsorbed by the gas.
- *Open Path Gas Detectors* (OPGDs) (Figure 7-23) produce an IR beam that is directed across the area to be monitored. The received light is analyzed at two or more frequencies, some of which is absorbed by the target gas or gases; the reference frequency is not. Given the initial and final intensities, the average concentration of gas in the path is calculated and transmitted. Some instruments operate with separate trans-

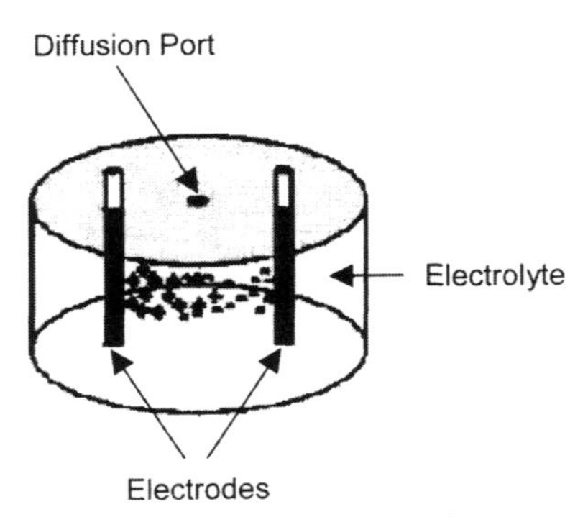


**Figure 7-22.** Electrochemical Sensors

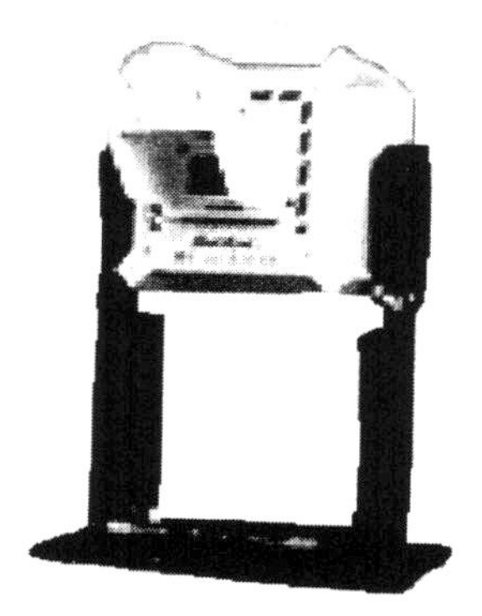

**Figure 7-23.** Open Path Gas Detectors

mitter and receiver units, one at each side of the area to be monitored. The other arrangement combines the two units and bounces the beam back from a retro-reflector. The separate transmitter and receiver units' arrangement is generally preferred since it improves the signal to noise characteristics of the device, improving its immunity to environmental conditions (e.g. fog, snow, sunlight). This results in a smaller diameter beam, reducing the potential for false gas readings from partial blockage and allowing easier installation in a congested facility. The separate transmitter and receiver layout can be used in process areas and HVAC ducts.

- *Point Combustible Gas Detectors* (IR) are used to indicate the presence of gas at a particular location (e.g., in a congested area of the plant or in small ducts.) IR technology has proven to be more reliable than catalytic bead detectors. The point detector functions in the same manner as the open path detector, by comparing absorbed and reference frequencies of IR light. The main difference between these and open path type is that the path length of the point type is short (3 inches) and is kept within the confines of the instrument.

#### 7.4.6.1. *Installation*

Gas detection location is important in detecting releases, however, obvious placement such as high up for lighter-than-air gases and down low for heavier-than-air gases may not always work.

For instance, if a high pressure, lighter-than-air gas is released in a downward direction, it will continue down until the jet-mixing phase is completed and it has been diluted by at least one order of magnitude. At that time, the buoyancy of the gas has little influence. Similarly, a heavier-than-air gas directed upward by pressure will continue traveling upward. The concept of buoyancy is only relevant to weeps and very small leaks in determining electrical area classifications.

Temperature effects are most noticeable in liquefied releases. These releases include those gases that are stored as a liquid as a result of being liquefied by pressure upon release. The temperature of the resulting gas/air mixture reduces, due to evaporation of the liquids, and therefore the density increases significantly. It can take a long time for such mixtures to reach ambient temperatures and, hence, achieve neutral buoyancy. The opposite effect can be seen when hot gas is released. This is usually less marked, as there is generally less energy available to maintain the temperature differential to ambient.

Air currents may strongly affect the behavior of gas clouds once the initial velocity of the release is dissipated in turbulent mixing. In many cases, the air currents will overcome any buoyancy effects and dominate the behavior of the gas cloud.

Some important gas and vapor detection system features to consider when selecting a system are: self-checking, auto-calibrating or self-diagnosing capabilities with sensor out-of-service, or malfunction alarming.

When a gas and vapor detection system is used either for fire protection or for personnel exposure protection, the system should be managed, inspected and tested on a scheduled basis, and maintained similar to any other major element of the fire protection system or any process safety interlock or safety instrumented system.

### *7.4.7. Sprinklers*

The success of fixed water suppression systems depends on the following:

- Adequate and reliable water supply
- Adequate water density
- Automatic actuation of the systems
- Effective water spray patterns
- Alarms that indicate the operation or malfunction of the system
- Effective maintenance and testing

#### *7.4.7.1. Types of Sprinkler Systems*

There are four basic types of sprinkler systems: wet pipes, dry pipes, pre-action, and deluge. These are described in the following sections.

##### 7.4.7.1.1. WET PIPE

The most common and least complicated of all sprinkler systems is the wet pipe system, where all the pipes are normally full of water. When the fusible link fails in one or more sprinkler heads, water is immediately discharged through those

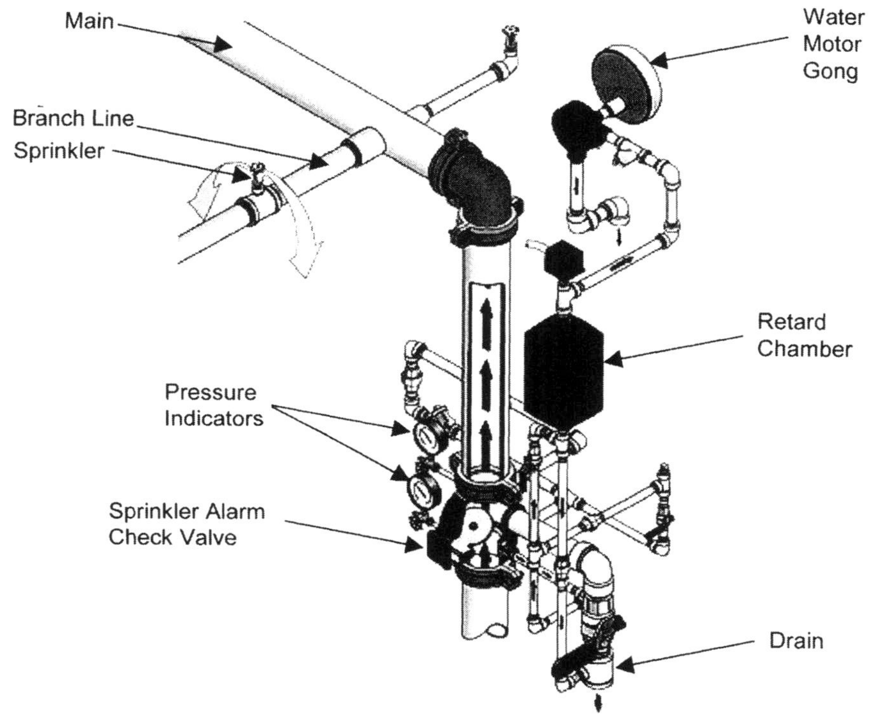


**Figure 7-24.** Wet Pipe Sprinkler System

sprinkler heads onto the fire in its early stages of development. An alarm is activated by the flow of water through the sprinkler valve into the pipe. This system is highly reliable and requires very little maintenance. The major disadvantage of this system is that it requires freeze protection in cold environments. An example of a wet pipe sprinkler system is shown in Figure 7-24.

#### 7.4.7.1.2. DRY PIPE

Dry pipe sprinkler systems were developed to provide sprinkler protection for unheated buildings in cold climates. Also, a dry pipe sprinkler system can be used to minimize corrosion. A differential dry pipe valve prevents water from entering the overhead piping system, which is filled with compressed air. Due to the mechanical advantage designed into the valve, one pound of air pressure can hold back approximately 5 psi (34.5 kPa) of water pressure. Normally closed sprinkler heads are provided on the overhead piping system, as in the wet piping system. When the sprinkler fusible element operates, the air trapped in the piping is released. As the air pressure decreases, the differential dry pipe valve operates and water enters the overhead piping to the open

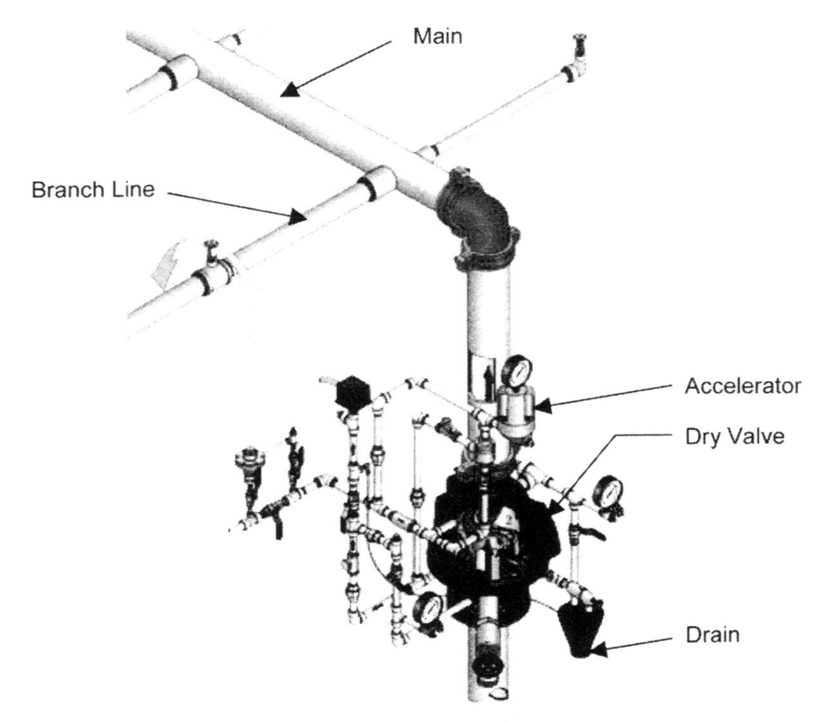


**Figure 7-25.** Dry Pipe Sprinkler System

sprinkler head and discharges as in the wet pipe system. The dry pipe valve must be located in a heated area or enclosure. An example of a dry pipe sprinkler system is shown in Figure 7-25.

### 7.4.7.1.3. PRE-ACTION

The principal difference between a pre-action system and a standard dry pipe system is that in the pre-action system, the water supply valve is actuated independently of the opening of the sprinklers. The water supply valve is opened by the operation of an automatic fire detection system, which immediately allows water to enter the overhead piping. Then, when the fire generates enough heat to fuse the sprinkler head, water is immediately discharged on the fire as with a wet pipe system. This system reduces the inherent time lag of a dry pipe system, but can still be used in unheated areas. An example of a pre-action system is shown in Figure 7-26.

Any suitable fire detection system can be utilized to actuate the control valve. Since any other fire detection device is likely to operate faster than a sprinkler head, it is common for water to have reached the sprinkler by the time

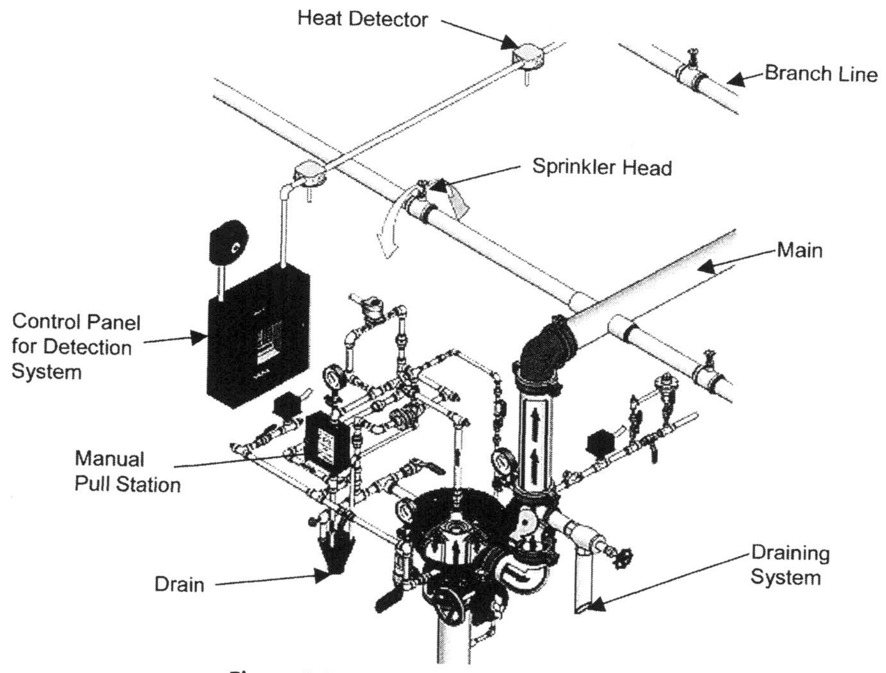


**Figure 2-26.** Pre-Action Sprinkler System

the head opens. Thus, pre-action systems usually have an operating time similar to that of a wet pipe system. In addition, pre-action systems reduce the risk of inadvertent water discharge since two events must occur for this to happen; detector actuation and sprinkler operation. Pre-action systems are generally used for MCCs, computer rooms, and areas where water can freeze.

System integrity can be monitored (it is required when more than 20 sprinkler heads are involved) by providing low air pressure, approximately 5 psi (34.5 kPa), in the piping. A common mistake is made by using high pressure air, which results in delayed delivery of water.

The availability of pre-action systems depends heavily upon the availability of the associated independent detection system. To compensate for this, systems should be designed so that the valve can be opened manually, both locally and at a constantly attended location. In addition, because of the complex mechanical valve arrangement, pre-action systems are less reliable than wet and dry pipe sprinkler systems.

The limitations and the higher cost of pre-action systems compared with wet or dry pipe systems limit the selection of a pre-action system to situations involving:

- An unacceptable risk associated with accidental system operation.
- Areas subject to freezing where an operating time faster than possible with dry pipe systems is required.

7.4.7.1.4. DELUGE

A deluge sprinkler system is a sprinkler system designed to NFPA 13 with open sprinkler heads. A water spray system is hydraulically designed with open spray heads to protect a specific hazard. Water spray systems are discussed in Section 7.4.8.

### *7.4.7.2. Design*

Sprinklers are the most common fire protection system used today. Most local building code officials and fire marshals have adopted NFPA 13 as law in their areas of jurisdiction. Many local authorities have added minor modifications unique to their areas.

Sprinklers are available in various temperature ratings. Sprinkler frames are color coded to identify the temperature rating as described in Table 7-9.

Further detail concerning proper temperature rating can be found in NFPA 13.

The design criteria of the sprinkler system should be established by a fire protection engineer, whereas the design of sprinkler systems is generally performed by licensed sprinkler contractors. Design calculations and plan drawings should be reviewed by a competent individual. Some important highlights and exceptions to NFPA 13 are:

**Table 7-9**
*Sprinkler Color Coding and Temperature Rating*

| Rating | Max. Temp. at Sprinkler Level °F (°C) | Rated Temp. of Sprinkler °F (°C) | Frame Color | Glass Bulb Color |
|---|---|---|---|---|
| Ordinary | 100 (38) | 135–170 (57–77) | Uncolored | Orange/Red |
| Intermediate | 150 (66) | 175–225 (79–107) | White | Yellow/Green |
| High | 225 (107) | 250–300 (121–149) | Blue | Blue |
| Extra High | 300 (149) | 325–375 (163–191) | Red | Purple |
| Very Extra High | 375 (191) | 400–475 (204–246) | Green | Black |
| Ultra High | 475 (246) | 500–575 (260–302) | Orange | Black |
| Ultra High | 625 (329) | 650 (343) | Orange | Black |

- The contractor should submit drawings and calculations for approval before ordering materials. The contractor may be required to submit drawings to insurance underwriters, inspection bureaus, or other regulatory agencies.
- The contractor should furnish copies of the "Contractor's Material and Test Certificate."
- Most sprinkler systems (except deluge systems) are not designed for simultaneous actuation of all sprinkler heads. For this reason, sprinkler systems are designed to protect an area based upon the fire anticipated for a given type of occupancy. Through a combination of testing and experience, the size of an anticipated fire and the amount of water required to control it should be calculated. The system is then designed to provide that amount of water over the anticipated area of fire spread at the most remote area, which would give the worst case hydraulically for designing the system. For the purpose of calculation, the remote area should be at least 3,000 $ft^2$ (278 $m^2$).
- A dry pipe system will take longer to operate than a wet pipe system, thus, a fire will have more time to develop and spread over a larger area. This means that a larger water supply and larger operating area must be considered when using a dry pipe system. The remote area for a dry pipe system should be increased by 30% over the same area required for a wet pipe system.
- Dry pipe system components containing water should be protected from freezing.
- Small wet pipe systems can be protected by an antifreeze solution. The extent of the system should be limited to 20 sprinkler heads or less. For systems over 20 sprinkler heads, a dry pipe system should be used.
- Sprinkler systems should be subjected to a hydrostatic test of 200 psi (1,379 kPa) for 2 hours during the acceptance test.
- A fire department connection with check valve should be provided on all sprinkler systems in buildings excluding deluge systems. A hydrant should be located within 50 ft (15 m) of the fire department connection.
- All systems should be hydraulically designed.
- Corrosion resistant or wax-coated sprinklers have a slower operating time than uncoated sprinklers. Their use should be carefully evaluated.
- Sprinkler heads that are exposed to potential mechanical damage should be protected with guards.
- A strainer should be provided for all systems that are not supplied from a clean water source.

- Corrosion resistant piping should be provided where corrosive atmospheres exist, such as marine environment, chlorine plants, etc.

### 7.4.8. Water Spray Systems

The term "water spray" refers to the use of water discharged from nozzles having a predetermined pattern, droplet size, velocity, and density. While deluge systems are for the overall protection of a given area in accordance with NFPA 13, *Installation of Sprinkler Systems*, water spray systems can be installed for the protection of a given area or specific equipment/hazards. Design guidance for water spray systems can be found in NFPA 15, *Water Spray Fixed Systems*. Figure 7-27 shows examples of water spray systems.

#### *7.4.8.1. Area versus Specific Application*

Deluge protection can be applied over the entire hazard area with open sprinkler heads located at various floor levels or fixed water spray systems can be applied specifically on the equipment to be protected. Specific water spray application is preferred because it provides better equipment cooling and reduces the water wasted due to wind and fire draft.

It can be difficult to install fixed water spray protection for equipment located in multilevel process structures, in congested areas, or in areas containing numerous pieces of small equipment. In such situations, the use of deluge protection of an area is an acceptable and cost-effective alternative to the specific protection afforded by fixed water spray.

#### *7.4.8.2. Applications*

Water spray systems are used for protection against hazards involving gaseous and liquid flammable materials and combustible solid materials. These systems are used to:

**Figure 7-27.** Water Spray

- *Cool Metal*—by applying water directly onto the protected surface to prevent distortion or rupture from flame exposure or radiant heat.
- *Control Fire Intensity*—by spraying water on equipment or areas, where a fire is likely to originate from leakage or where firefighting would be unusually difficult, in order to control the rate of burning and thereby limit the heat release from a fire until the fuel can be eliminated or extinguishment effected. Some examples are pumps, regulator manifolds, and piping at high-temperature pressure vessels.
- *Prevent Ignition*—by keeping materials cooled below their ignition temperatures prior to ignition and by quenching and cooling ignition sources.
- *Protect Vital Instruments Runs*—by cooling outside trays or multiple conduits.
- *Prevent Formation of Flammable Vapor Clouds*—by dispersing leaks and spills and inducing air into vapor releases to reduce the vapor-air concentration below the lower flammable limit and to prevent vapor travel.
- *Control Toxic Vapor Clouds*—by creating turbulence and mixing of air with the toxic vapor cloud.

Even though extinguishing fires may not be the prime purpose of a water spray system, extinguishment may result when a water spray system is used for cooling or fire intensity control. It is good practice to allow fires fueled by pressurized gases to continue to burn until the fuel source is isolated. To extinguish it incurs the greater risk of a flammable cloud and a subsequent vapor cloud explosion. Where such a scenario is a credible event, water spray systems should be designed only for control of fire. Water spray systems have the following advantages:

- Exposure protection is accomplished by application of water spray directly to the exposed structures or equipment to remove or reduce the heat transferred from the fire. Water spray curtains are less effective than direct spray.
- Prevention of fire spread or explosion is sometimes possible by the use of water spray to dissolve, dilute, disperse, or cool flammable or combustible materials.

The water demand for enclosed process structures may depend on the degree of compartmentalization by solid floors and the number of separate systems used to protect the structure. At a minimum, ground floor systems should be assumed to operate in a spill fire scenario. But, if water spray systems are used, it should be assumed that all systems will operate in order to estimate maximum water demand.

For open structures protected by water spray systems, it should be assumed that all systems within the structure will operate. The water demand for each system is summed to get the total demand for the unit plus manual firefighting.

Fixed water spray systems are most commonly used to protect flammable liquid and gas vessels, piping and equipment, process structures and equipment, electrical equipment such as transformers, oil switches, rotating electrical machinery, and openings through which conveyors pass. The type of water spray will depend on the nature of the hazard and the purpose for which the protection is provided.

The design densities to be used in designing water spray systems are given in Chapter 8.

### *7.4.8.3. Design*

The practical location of the piping and nozzles, with respect to the surface to which the spray is to be applied or to the zone in which the spray is to be effective, is determined largely by the physical arrangement and protection needs of the installation requiring protection. Once these criteria are established, the size (rate of discharge) of nozzles to be used, the angle of the nozzle discharge cone, and the water pressure needed can be determined.

The design of water spray systems should be in accordance with NFPA 15. Highlights and exceptions to NFPA 15 are:

- The size of the system should be limited to avoid overtaxing the fire water drainage systems. For locations with multiple systems it is common to activate 3 or 4 systems to ensure adequate protection. Many designers use 2,000 gpm (7,571 lpm) and 65 psi (207 kPa) as a reasonable limit on size to avoid the issues discussed.
- A minimum orifice size, should be used to avoid pluggage. The size is related to the hydraulic pressure available in the system. For example, ½-in (1.3-cm) nozzles should have a minimum of 20 psig (138 kPa); ⅜-in (1.0-cm) nozzles should have a minimum of 30 psig (207 kPa) to avoid pluggage.
- Self-cleaning strainers should be provided on all systems. The strainers should have stainless steel baskets with openings of 0.13 in (0.33 cm) diameter and be provided with a blow-off valve.
- Corrosion-resistant piping should be provided where corrosive atmospheres exist, such as marine environment, chlorine plants, etc.

#### 7.4.8.3.1. DETECTION AND ACTUATION

For plant process areas, water spray or deluge systems should be designed to actuate by all of the following means:

- *Fire detectors*—dry-pilot head and pneumatic rate-of-rise heat detectors are the most frequent devices used. A dry pilot head detection systems uses 165°F (74°C) ½-in sprinkler head with air pressure maintained in the piping. When the sprinkler opens the air is released resulting in the sprinkler valve opening. Other types of detectors include ultraviolet, infrared flame detectors, or thermostatic cable heat detection.
- *Combustible gas detectors*—where a flammable vapor or gas hazard exists, gas detectors are typically installed. Gas detectors should be arranged to a low set point to sound an alarm and the upper set point to activate the water spray or deluge system.
- *Manual stations*—Local and remote manual trip stations should be readily accessible and conspicuously labeled. The number of trip stations will depend on the hazards present, size of the system, and personnel escape routes. All operating personnel should be trained on the location and use of these manual trip stations. Some facilities have only remote valves that can be manually opened without remote actuators; not all systems have deluge valves.

There are many advantages to actuating the water spray or deluge system as soon as a flammable vapor or gas is detected:

- Sprayed water acts as an air mover, drawing in air that dilutes the cloud and also helps to drive the vapors from under structures and away from equipment. If vapors were to remain trapped, the cloud would be partly confined, increasing the chance of a vapor cloud explosion.
- Sprayed water will increase the relative air humidity therefore reducing the probability of static discharge igniting the cloud.
- Sprayed water will wet down critical and highly vulnerable electric wiring and instrumentation leads, thus protecting them from damage should the cloud ignite in a flash fire.
- Sprayed water wets equipment and could reduce the probability of a mechanical spark developing.
- Some gases and vapors (such as ammonia) may be water soluble and spraying with water will dilute the cloud.

#### 7.4.3.8.2. EXPLOSION PROTECTION

Water spray and deluge systems are subject to damage due to an explosion. To limit damage to such systems and therefore reduce the potential for impairment, the following design guidance should be considered:

- Deluge valves should be located remotely [at least 50 ft (15 m)] for the area to be protected or located in blast resistant buildings.

- Piping should run underground when possible. Risers should rise above ground behind a protecting steel column or other structural member providing shielding from overpressure and flying debris.
- The number of deluge valves manifolded together should be strictly limited to no more than three.

The piping and fittings used can greatly influence the explosion resistance. The following extra material and design specification should be considered:

- All piping 2½ in (63 mm) or larger should be of the welded flanged type.
- All welded piping and flanges should be hot dipped after fabrication.
- All piping from ½ to 4 in (15 to 100 mm) should be galvanized steel and Schedule 40.
- Screwed fittings should be Class 150 screwed galvanized malleable iron.
- Welded fittings should be standard weight seamless steel, ASTM A-234 grade WPB.
- Flanges should be Class 150.

#### 7.4.3.8.3. WATER SPRAY NOZZLES

The selection of the type of spray nozzles should be made with proper consideration given to such factors as discharge characteristics, physical characteristics of the hazard involved, water quality and potential for pluggage, ambient conditions, material likely to be burning, and the design objectives of the system. Also, some spray nozzles are more subject to pluggage than other ones.

#### 7.4.3.8.4. POSITION

Water spray nozzles should be permitted to be placed in any position necessary to obtain proper coverage of the protected area. The positioning should consider the following areas:

- The shape and size of the area to be protected.
- The nozzle design and characteristics of the water spray pattern to be produced.
- The effect wind and fire draft may have on spray with drops of very small diameter or with drops of large diameter, but low initial velocity.
- The potential to miss the target surface.
- The effects of the nozzle orientation on coverage characteristics.
- The position of the nozzle to be self draining.
- The potential for mechanical damage.

Water spray systems should be provided with a water flow alarm installed in accordance with NFPA 13, *Installation of Sprinkler Systems*. Electrically operated alarm attachments should be installed in accordance with NFPA 72, *National Fire Alarm Code*.

At least one remote manual actuation device independent of the manual actuation device located at the system actuation valve should be installed for all automatic water spray systems.

#### 7.4.3.8.5. GAUGES

Pressure gauges should be installed as follows:

- Below the system actuation valve
- Above and below alarm check valves
- On the air or water supply to pilot lines

Pressure gauges should be installed to permit easy removal and should be located where they will not be subject to freezing.

Provisions should be made for test gauges at or near the highest or most remote nozzle on each major separate section of the system. At least one gauge connection should be provided at or near the nozzle calculated as having the least pressure under normal flow conditions.

#### 7.4.3.8.6. STRAINERS

Main pipeline strainers should be provided for all systems utilizing nozzles with waterways less than 3/8 in (1 cm) and for any system where the water is likely to contain obstructive material. Mainline pipeline strainers should be installed accessible location for flushing or cleaning.

### 7.4.9. Water Mist Systems

Water mist systems are intended for rapid suppression of fires using water discharged into completely enclosed limited volume spaces. Water mist systems are desirable for spaces where the amount of water that can be stored or that can be discharged is limited. In addition, their application and effectiveness for flammable liquid storage facilities and electrical equipment spaces continues to be investigated with optimistic results. Water mist systems are also used for gas turbine enclosure protection.

A water mist system is a proprietary fire protection system using very fine water sprays. The very small water droplets allow the water mist to control or extinguish fires by cooling of the flame and fire plume, oxygen displacement by water vapor, and radiant heat attenuation. These systems are single shot systems. A water mist system and nozzles are shown in Figure 7-28.

Properly designed water mist systems can be effective on both liquid fuel (Class B) and solid fuel (Class A) fires. Research indicates that fine (i.e., smaller than 400 microns) droplets are essential for extinguishing Class B fires, although

**Figure 7-28.** Water Mist System and Nozzles

larger drop sizes are effective for Class A combustibles due to the wetting of the fuel. Other potential applications for these systems include:

- Flammable and combustible liquids
- Electrical hazards, such as transformers, switches, circuit breakers, and rotating equipment
- Electronic equipment

A significant challenge with using mist systems is determining whether the conditions of a particular manufacturer's limitation are representative of the actual conditions in which the system will be used. An engineering analysis should be performed to evaluate the significance of any variations between the actual fire hazard and the known performance capabilities of the water mist system. It is important that the test protocol contain these factors:

- Fuel package
- Volume
- Height

- Ventilation
- Obstructions
- Duration of protection

### 7.4.10. Foam Systems

Foam is primarily used for extinguishment of two-dimensional surface fires involving liquids that are lighter than water. Foams may be used to insulate and protect against exposure to radiant heat. They also act to prevent ignition of flammable liquids that are inadvertently exposed to the air (typically due to a spill), by separating them from air by spreading foam completely over the exposed surface. However, progressive foam breakdown can render the protective foam coating useless; thus, frequent reapplication may be necessary.

Because of water content, foams may be used to extinguish surface fires in ordinary combustible materials such as wood, paper, rags, etc. Foams are arbitrarily subdivided into three ranges of expansion roughly corresponding to certain types of usage:

- *Low-expansion foam*—expansion up to 20 times foam to solution volume
- *Medium-expansion foam*—expansion from 20 to 200 times foam to solution volume
- *High-expansion foam*—expansion from 200 to 1,000 times foam to solution volume

Foam systems have more components requiring higher maintenance and have somewhat lower reliability than water-only systems. Low expansion foam systems do offer the advantage over high expansion foam of an effective "water only" discharge after the foam agent has been expended. "Water only" discharge can be detrimental in most situations that require foam application. Limitations on the use of foam include:

- Foams are not suitable extinguishing agents for fires involving pressurized gases and liquefied petroleum gases (LPG), such as butane, butadiene, propane, etc.
- Foam breaks down and vaporizes its water content when exposed to heat and flame. Therefore, it must be applied to a burning surface at a sufficient rate to compensate for this loss and to ensure a residual foam layer over the liquid.
- Three-dimensional liquid fires, such as those involving elevated equipment, flowing fires, or pressure leaks, are not readily extinguishable with foam.

- Foam should not be used to fight fires in materials such as metallic sodium and metallic potassium, etc., which react violently with water.
- Foam is a conductor and should not be used on electrical equipment fires.

Judgment must be used in applying foams to hot oils, burning asphalts, or burning liquids, where the bulk temperature of the liquid is above the boiling point of water. Although the comparatively low water content of foams can cool such fuels at a low rate, it can also cause violent frothing and "slop over" of the contents of a tank.

Certain wetting agents and some dry chemical powders may be incompatible with some foams. If they are used simultaneously, an instantaneous breakdown of the foam blanket may occur. Precautions must be taken to ensure that such agents are fully compatible with the foams being used.

Some types of foam are not suitable for water miscible or polar solvent liquids. Special foams designed for fighting fires involving these materials are available. For more information on low-expansion foam systems refer to NFPA 11, Low-Expansion Foam Systems. For additional information on medium- and high-expansion foam systems, refer to NFPA 11A, *Medium- and High-Expansion Foam Systems.*

### *7.4.10.1. Application Methods*

Foam can be applied with fixed, semi-fixed, or portable systems.

#### 7.4.10.1.1. FIXED

Fixed systems are complete installations piped from a central foam station to tanks or equipment, discharging through fixed delivery outlets. These outlets take the form of foam chambers, monitor nozzles, foam–water spray nozzles, etc. Any required pumps are permanently installed.

#### 7.4.10.1.2. SEMI-FIXED

Semi-fixed systems employ fixed discharge outlets connected to permanent piping that terminates at a safe distance from the tanks or equipment to be protected. The fixed piping may or may not include a foam maker. Necessary foam producing materials and equipment are transported to the scene after the fire starts and are connected to the permanent piping. Perhaps the most common example of a situation requiring the use of a semi-fixed system is a cone-roof or floating roof tanks. The surface of its contents, when burning, cannot be approached directly; however, the fire must be brought under control quickly to prevent shell collapse. This can be accomplished by having a fixed foam system on the tanks fed by portable equipment from outside the tank dike. This system has several advantages over the portable system described below. For instance, the presence of ground fires in the vicinity of the tank dike area would not delay the establishment of foam flow to the tank, as there is no need to enter the dike area to establish foam flow.

The system can be operated with fewer emergency responders and foam loss is minimized. Foam chambers cause the foam to flow down the tank shell and onto the liquid surface; therefore, thermal updrafts have little effect on foam application. A similar effect is seen with the use of foam chambers to protect the seals on floating roof tanks.

#### 7.4.10.1.3. PORTABLE

Portable systems contain no fixed foam piping. Rather, the foam making apparatus, foam producing materials, hose, and other delivery equipment are transported to the scene after the fire starts. Foam is delivered to the spill or burning surface by hose streams or portable monitors. Experience has shown that, in general, when manpower is available to handle the equipment, it is preferable that as much equipment as possible be portable or mobile. The amount of equipment required is less than that required under a fixed system, and the cost of installation and maintenance is reduced. In addition, a portable system permits the greatest concentration of equipment at a given fire. On the other hand, for portable systems to be effective, the firefighting effort must be more organized and the fire team better trained. Also, it should be understood that more time is required to put portable equipment into operation.

### *7.4.10.2. Foam Proportioning Methods*

In order that a predetermined volume of liquid foam concentrate may be taken from its source and placed into a water stream to form a foam solution of fixed concentration, the following two general method classifications are made:

- Methods that utilize external pumps or pressure to inject concentrate into the water stream at a fixed ratio (balanced pressure systems).
- Methods that utilize the energy of the water stream by venturi-action and orifices to induce concentrate (typically, such devices impose a 35% pressure drop on the water stream).

#### 7.4.10.2.1. BALANCED PRESSURE

The use of a separate pump for transporting and pressurizing foam concentrate to be delivered in the correct proportions to a flowing water stream offers the greatest advantage for reliable and accurate operation of a foam concentrate proportioning system which must function at varying rates of volume or pressure during its use.

The balanced pressure proportioning system is ideal for large systems, concentrate requirements exceeding 1,000 gal (3,800 l), and for most mobile foam apparatus.

#### 7.4.10.2.2. PRESSURE (BLADDER) TANK

The principle of operation of this device is simple. A small amount of the flowing water volumetrically displaces foam concentrate from the tank into the main water stream. The working pressure of the vessel must, of course, be above the maximum static water pressure encountered in the system. This type of proportioner may consist of one tank or pressure container with a watertight divider so that it operates as two tanks, two tanks separately connected to the water and foam solution lines, or the tanks in the system may each be fitted with flexible diaphragms or bladders to separate the "driving" water from foam concentrate or they may rely simply on differences in density of the two liquids to retard mixing during operation.

Water is allowed to enter the foam tank from the main stream with as little friction loss as possible, while pressure in the main stream is dropped about 10% through use of an orifice. Liquid in the tank is metered into the low pressure area by a second orifice. The pressure proportioning system offers the advantages of low pressure drop, automatic proportioning over a range of flows and pressures, freedom from external power and the absence of moving parts. Its disadvantages are that the concentrate cannot be resupplied while the system is in operation, the bladders tend to leak and there is an economic maximum limit on size.

#### 7.4.10.2.3. AROUND THE PUMP

Around the pump proportioners consist of an eductor installed in a bypass line between the discharge and suction of the water pump. A small amount of water passes through the eductor drawing the required quantity of foam concentrate into the system. These systems have limited uses and are not recommended for industrial purposes.

### *7.4.10.3. Application Rates*

Successful use of foam depends on the rate at which it is applied. Application rates are described in terms of the amount (in gallons) of foam solution reaching the fuel surface (in terms of total square footage) every minute. Increasing the foam application rate over the minimum recommended rate will generally reduce the time required for extinguishment. However, little is gained if application rates are increased more than three times the minimum recommended because it becomes ineffective and results in foam waste. If application rates are less than the minimum recommended, extinguishment will be prolonged or may not be accomplished at all.

If application rates are so low that the rate of foam loss equals or exceeds the rate at which foam is being applied, the fire will not be controlled or extinguished. The critical application rate is the lowest rate that foam will extinguish a given fire under a particular set of conditions. The minimum recommended

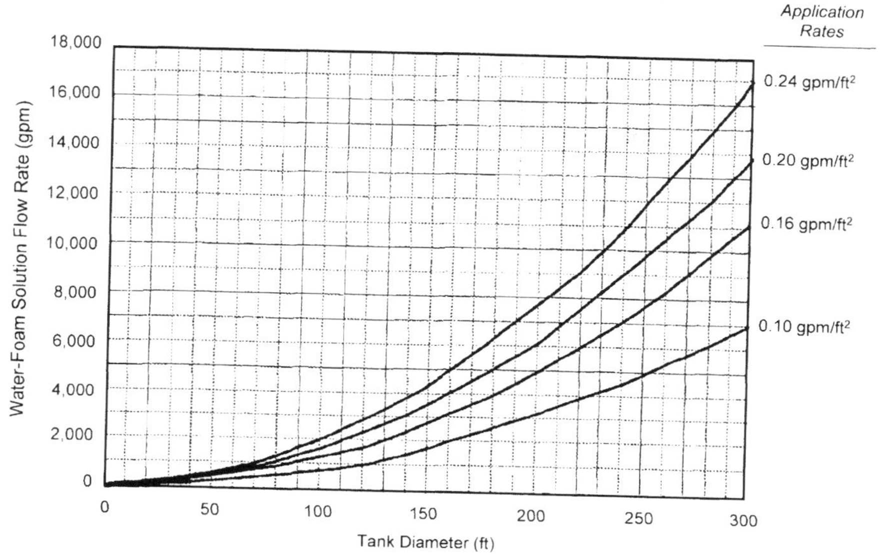


**Figure 7-29.** Flow Requirement for Full Surface Fire

application rate is the rate determined by tests to be the most practical in terms of speed of control and amount of agent required. Figure 7-29 provides a quick method to estimate water foam solution flow rate based on tank diameter and application rate.

In general, foams will be more stable when they are generated with water at ambient or lower temperature. Preferred water temperatures range from 35 to 80°F (2 to 27°C). Either fresh or seawater may be used. Water containing known foam contaminants, such as detergents, oil residues, or certain corrosion inhibitors, may adversely affect foam quality and appropriate steps should be taken to ensure an adequate supply of suitable quality water. See NFPA 11 for more information on application rates.

#### 7.4.10.3.1. GENERAL DESIGN

In addition to the requirements of NFPA 11, the following, where applicable, should be included in the specification prepared for purchasing foam equipment and systems:

- Physical details of the asset to be protected including the location, arrangement, and hazardous materials involved
- Type and percentage of foam concentrate
- Required solution application rate and duration
- Available water supply

- Required amount of concentrate
- Type of proportioning system
- Identification and capacity of all equipment devices to be protected
- Location of piping, detection devices, operating devices, generators, discharge outlets and auxiliary equipment
- Explanation of any special features
- Plan approval, acceptance test
- Contractor's guarantee

The contractor should submit drawings and calculations for approval before ordering materials. The contractor may be required to submit drawings to insurance underwriters, inspection bureaus, or other regulatory agencies.

### *7.4.10.4. Application of Foam for Suppression of Tank Fires*

Properly applied firefighting foam can be an effective fire suppressant for most flammable or combustible liquids stored in vertical cylindrical tanks. Foam may be applied to the:

- Surface of the liquid from above by use of "multiple chambers" located high on the tank's interior side wall
- Secondary containment area
- Subsurface, allowing the foam to float to the liquid surface

Foam may be applied to containment areas around a storage tank by either fixed, semi-fixed, or manual systems. Typical applications include:

- Areas where dike height exceeds 6 ft (2 m)
- Manual application is restricted by physical obstacles

#### 7.4.10.4.1. FLOATING ROOF TANKS—MULTIPLE-CHAMBER METHOD

Fluoroprotein and alcohol resistant type AFFF concentrates are suitable for seal protection of open top floating roof tanks. Foam may be applied from fixed chambers in an "over-the-top" fashion or by discharge from "foam chambers" located at the top of the tank shell allowing the foam to run down the inside wall of the tank shell onto the seal area. Subsurface foam on floating tanks is generally not recommended as a sunken roof would block the foam distribution. Typical protection is shown in Table 7-10.

The circumference of the tank will determine the number of points needed for foam application. The maximum spacing between chambers should be 40 ft (12 m) measured around a tank's circumference using a 12-in (30-cm) high dam and 80 ft (24 m) measured around the tank circumference using a 24-in (61-cm) high dam.

**Table 7-10**
*Protection for Open-Top Floating Roof Tanks*

| Protection | Open-Top Floating Roof Tank Diameter [ft (m)] | | |
|---|---|---|---|
| | **<120 (36.6)** | **120–150 (36.5–45.7)** | **>50 (45.7)** |
| Foam dam on roof | Yes | Yes | Yes |
| Foam chamber | 1 | 2 | 4 |
| Fixed foam system | Not required | Yes | Yes |
| Seal Access for firefighting | From top of tank stairs | Yes, provide hand rails on wind girder | Yes, provide hand rails on wind girder |

This method requires a foam dam to retain the foam over the primary and secondary seals or weather shield. This dam is normally 12–24 in (30–61 cm) high. Complete construction details of the foam dam may be found in NFPA 11. Figure 7-30 illustrates a typical foam chamber and dam for floating roof tank protection.

7.4.10.4.2. CONE ROOF VERTICAL CYLINDRICAL TANKS

*Chamber Method.* The foam chamber method for cone roof vertical cylindrical tanks consists of one or more foam chambers installed on the shell of the tank just below the roof joint. A foam solution pipe is extended from the proportioning source, outside the dike wall, to the foam maker located inside the chamber. A deflector is located inside the tank shell to deflect the discharge

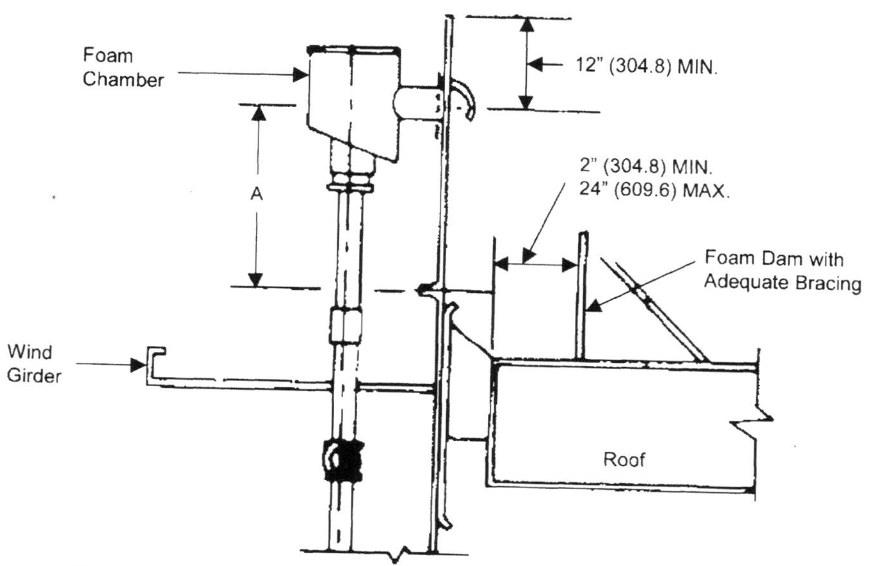


**Figure 7-30.** Typical Foam Chamber and Dam for a Floating Roof Tank

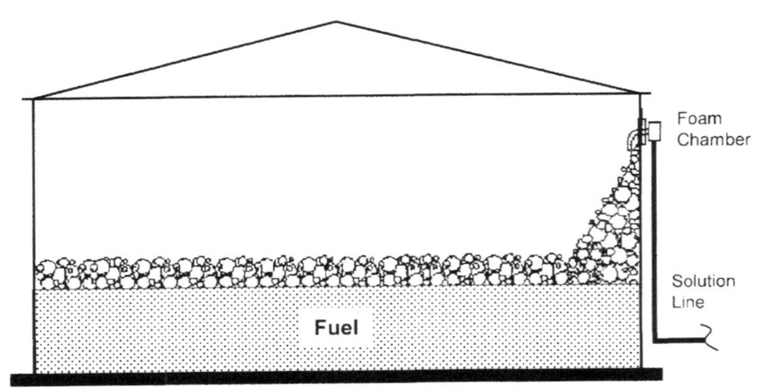


**Figure 7-31.** Foam Chambers in a Cone Roof Tank Application

against the shell. Protein, fluoroprotein, or AFFF foam concentrates are suitable for use with foam chambers. Figure 7-31 illustrates foam chambers in a cone roof tank application.

The minimum foam solution application rate for extinguishment of liquid hydrocarbons is 0.1 gpm/ft$^2$ (0.3 lpm/m$^2$) of product surface areas. Foam liquid supplies must be sufficient to operate the system for a minimum period of time as required in Table 7-11 (modified from NFPA 11).

The number of foam chambers required is determined by the tank diameter. Where two or more chambers are required, they should be designed to deliver foam at approximately the same rate. Table 7-12 (Modified from NFPA 11) indicates the number of chambers required for various diameter cone-roof tanks.

*Subsurface Method*. The subsurface method requires a high back-pressure foam maker and forces this foam through a pipe into the bottom of the tank. This pipe may be the existing product line or a line installed specifically for

**Table 7-11**
*Foam Application Time—Chambers and Subsurface Methods*

| Properties | Typical examples | Minimum Operating Time (min) |
|---|---|---|
| Liquids with flash points above 200°F (93°C) and viscous liquids | Lubricating oils; dry viscous residuum [more than 50 sec Saybolt Furol (1.068 × 10 sq m/s)] at 122°F (50°C); dry fuel oils, etc. | 25 |
| Liquids with flash points from 100°F to 200°F (37°C to 93°C) | Kerosene, light furnace oils, diesel fuels, etc. | 30 |
| Liquids with flash points below 100°F (37°C) | Gasoline, naphtha, benzol, and similar liquids including crude petroleum | 55 |

**Table 7-12**
*Number of Foam Discharge Outlets*

| Tank Diameter ft (m) | Minimum Number of Foam Chambers | Subsurface: Classes IB, IC, II Liquids | Subsurface: Class III Liquids |
|---|---|---|---|
| Up to 80 (24 m) | 1 | 1 | 1 |
| Over 80–120 (24–36 m) | 2 | 2 | 1 |
| Over 120–140 (36–42 m) | 3 | 3 | 2 |
| Over 140–160 (42–48 m) | 4 | 4 | 2 |
| Over 160–180 (48–54 m) | 5 | 5 | 2 |
| Over 180–200 (54–60 m) | 5 | 6 | 3 |
| Over 200 add one nozzle for each additional | 5,000 ft$^2$ | 5,000 ft$^2$ | 7,500 ft$^2$ |

foam application. The foam travels up through the product to form a vapor resistant blanket on the surface. Subsurface foam injection is illustrated in Figure 7-32.

The foam concentrate used with subsurface systems should be a fluoroprotein type for best results, although some AFFF foams are listed for substitutable application (because of their "fuel shedding" properties). The minimum foam solution rate should be 0.3 gpm/ft$^2$ (12 lpm/m$^2$). The supply of foam liquid should be adequate to operate the system for 20 minutes. The foam injection point must be above the level of any residual water in the bottom of the tank. Subsurface foam application is not recommended for open or covered floating roof tanks or cone roof tanks with internal floating covers.

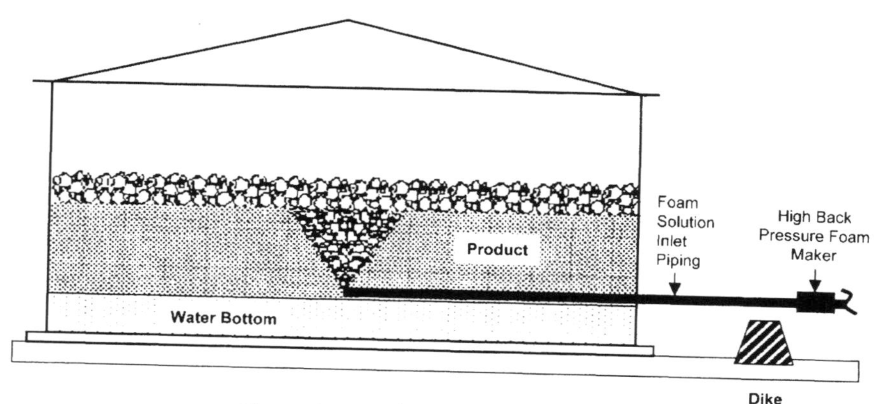


**Figure 7-32.** Subsurface Foam Injection

The most viscous fuel which has been extinguished by subsurface injection when stored at ambient temperatures of 60°F (15°C) is a material with a viscosity of 25 Saybolt Furol (SSF) at 122°F (50°C).

### *7.4.10.5. Foam Concentrate Stocks*

The base quantity of foam liquid concentrate that should be stocked is the greatest amount calculated to be needed for any fire in a fire risk area. Normally, this involves either the largest cone roof tank or the seal of the largest floating roof tank and includes hose streams for ground fires around the tank. In addition, a supplementary supply of foam concentrate equal to 100% of the base supply should be readily available within 24 hours.

In selecting the type of foam to be used in the system and also the amount of foam liquid concentrate to be maintained as a supplementary supply, consideration should be given as to what back-up compatible foam is available from nearby sources, such as other facilities, municipalities, the military, etc. This is especially desirable in areas where foam liquid concentrate may not be readily available from foam supply companies. If a facility is depending on back-up foam from other sources (other plants, military, etc.), it should be made certain that compatible foam is available.

The minimum stored supply should permit foam application at recommended rates for the periods of time indicated in earlier sections for tanks and portable applications.

The formula for determining total foam concentrate stock requirements (excluding the supplementary supply) is based on criteria presented and is shown below.

| | | |
|---|---|---|
| *Area* ($ft^2$) × *Application rate* ($gpm/ft^2$) × % Foam concentration × Operating time (minutes) | = | *A* |
| *Number of hose streams* × *gpm/stream* × % Foam concentration × Operating time (minutes) | = | *B* |
| *Total base quantity of foam concentrate (gallons)* | = | *A* + *B* |

**Example**

Assume the largest cone roof tank is 100 ft in diameter, contains gasoline, and is equipped with foam chambers. What would be a reasonable quantity of 3% foam concentrate to stock?

| | | |
|---|---|---|
| 7,854 $ft^2$ × 0.1 gpm/ $ft^2$ × 3% foam concentration × 55 minutes | = | 1,296 |
| 3 hose streams × 500 gpm/stream × 3% foam concentrate × 30 minutes | = | 135 |
| *Total base quantity of foam concentrate (gallons)* | = | 1,431 |

### 7.4.11. Foam–Water Deluge and Water Spray Systems

The foam–water sprinkler system is basically the same as a sprinkler system except foam concentrate is proportioned into the water causing foam to be discharged. These systems can flow either water or foam effectively.

#### 7.4.11.1. Applications

Foam–water deluge systems are especially applicable to the extinguishment, prevention and control of most flammable liquid hazards. They may be used for any, or a combination, of the following purposes.

- The primary purpose of such systems is the extinguishment of fire in the protected hazard area. For this purpose, suitable foam-solution discharge densities should be provided by system design and by provision for adequate supplies of air/water at suitable pressures.
- Controlled burning of combustible material, where extinguishment is not practical, and for exposure protection by reducing heat transfer may be accomplished by water and foam spray. The degree of success is related largely to the discharge densities provided by the system design.

#### 7.4.11.2. Design

The foam system should be designed in accordance with NFPA 16.

The foam discharge rate for a foam sprinkler system should be at least 0.16 gpm/ft$^2$ (0.6 lpm/m$^2$). The foam discharge should be continuous for a minimum of 10 minutes. However, a density of 0.25 gpm/ft$^2$ is recommended in order to provide adequate cooling even when the foam supply is gone.

With the exception of high expansion foams, all other foams can be used in foam–water sprinkler system. However, AFFF is the most versatile agent because of its spreading capability and it can be applied through ordinary sprinkler heads.

Existing compatibility with fixed water systems can be modified to allow for the injection of foam from a mobile foam unit by providing hose connection downstream of the system's alarm check valve. A check valve should be provided in the foam injection line as part of this modification.

### 7.4.12. Clean Agents

Clean agents are electrically, nonconductive, volatile, or gaseous fire extinguishing agents that do not leave a residue. Cleans agents fall within two categories: halocarbons and inert gases. Typical halocarbons include hydrofluorocarbons, (HFCs), hydrochlorofluorocarbons (HCFCs), perfluorocarbons (PFCs or FCs), and fluoroiodocarbons (FICs). Typical inert gases include argon, nitrogen, carbon dioxide, or combinations of these agents.

Clean agent fire extinguishing systems are used primarily to protect enclosures. Clean agents can be used to protect enclosures containing:

- Electrical and electronic equipment
- Subfloors and other concealed spaces
- Flammable and combustible liquids
- Telecommunication equipment

An example of a clean agent system is shown in Figure 7-33.

Clean agent systems can also be used for explosion prevention and suppression where flammable materials are confined. Clean agents should not be used for fires involving:

- Materials that are capable of rapid oxidation such as cellulose nitrate and gunpowder
- Reactive metals such as lithium, sodium, potassium, magnesium, titanium, zirconium, uranium, and plutonium
- Metal hydrides
- Chemicals capable of decomposition such as organic peroxides and hydrazine

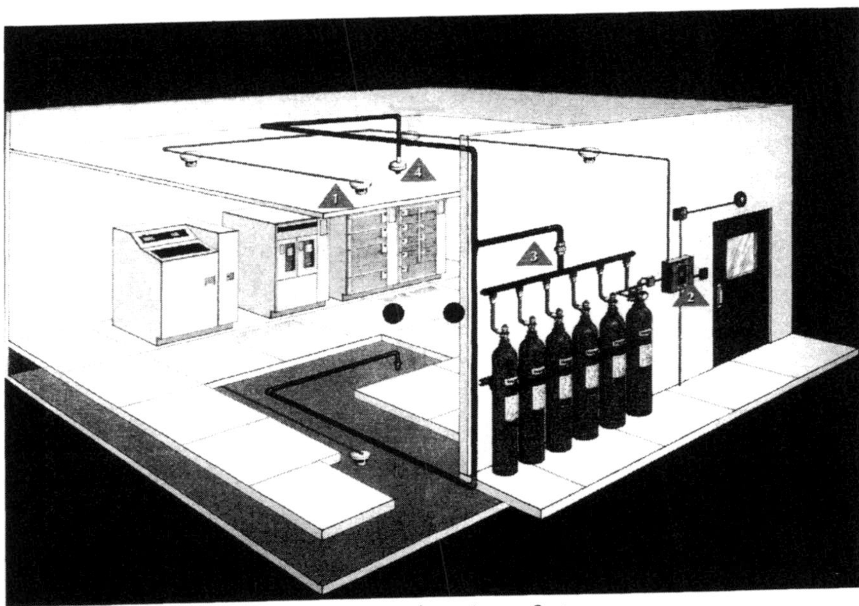


**Figure 7-33.** Clean Agent System

The design of clean agent fire protection systems should follow NFPA 2001. Key design features include:

- The fire extinguishing or inerting concentrations should be determined based on the hazard present. In many cases, the clean agent supplier can provide guidance on the concentration. Each clean agent has its own limitations. It is important to fully understand the clean agent test conditions and limitations prior to using it for a specific application.
- Where the clean agent is used for inerting and extinguishment, the higher of the recommended concentrations should be used. Concentration safety factors should be used as follows for:
  - Class B fires: 1.3× extinguishing concentration
  - Class A and C fires: 1.2× extinguishing concentration
  - Inerting of flammable liquids and gases: 1.1× inerting concentration
  - Manually activated system: 1.3× extinguishing concentration
- The design concentration must be maintained for a certain duration to allow effective emergency actions. In general, the duration should be a minimum of 10 minutes. Along with the duration, the time to discharge the agents is equally important. For halocarbons, the time to reach 95% concentration is 10 seconds. For inert gases, the time to reach 95% concentration is 60 seconds.
- It is important that a performance test is carried out to ensure the design concentrations can be met. Performing the testing using the clean agent is an expensive method of confirming the tightness of the room. A door fan test, as described in NFPA 2001, *Clean Agent Fire Extinguishing Systems*, is the preferred alternate method for testing the enclosure without releasing the clean agent.
- Doors, windows, or ventilation louvers should be closed automatically prior to discharge of clean agents.

### 7.4.13. Carbon Dioxide Systems

Carbon dioxide systems should be designed and installed and tested in accordance with NFPA 12. Fixed $CO_2$ systems may be total flooding or local application systems as described in the following sections. An example of a carbon dioxide system is shown in Figure 7-34.

#### *7.4.13.1. Total Flooding Carbon Dioxide Systems*

This type of system may be used where there is a permanent enclosure around the area or equipment to be protected that is adequate to enable the required concentration to be built up and to be maintained for the required period of

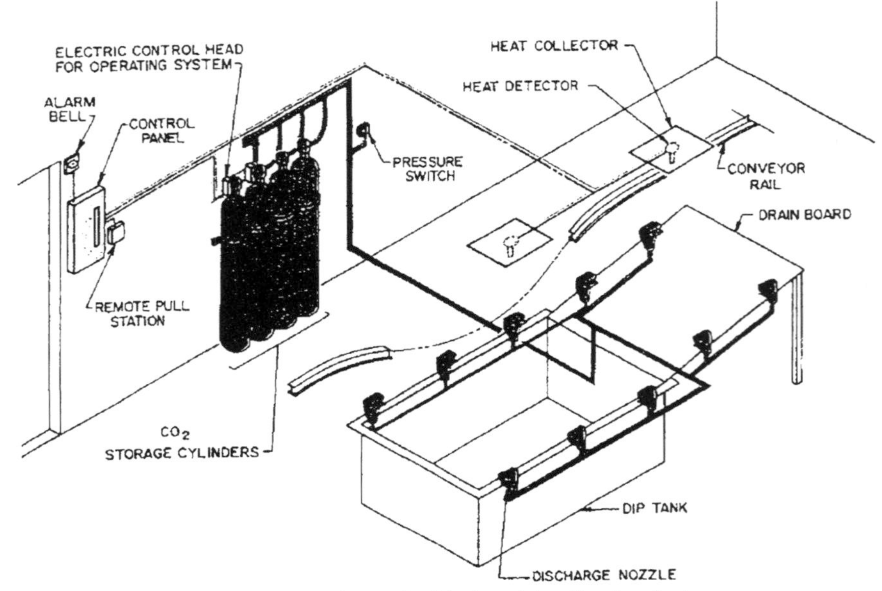


**Figure 7-34.** Carbon Dioxide Local Application System

time to ensure complete and permanent fire extinguishment. The minimum inert gas concentrations for the suppression of flammability in air are provided in Table 7-13 (Modified from NFPA 12).

Examples of areas or equipment that may be successfully protected by total flooding $CO_2$ systems are rooms, vaults, and enclosed machines spaces. An example of a total flooding carbon dioxide system is shown in Figure 7-35.

Doors, windows, and air handling systems should be arranged to automatically close or shut down prior to $CO_2$ discharge. The concentrations of $CO_2$

**Table 7-13**
*Minimum $CO_2$ Gas Content for Suppression of Flammability in Air*

| Compound | Carbon Dioxide (%v/v) |
|---|---|
| Methane | 24 |
| Ethane | 33 |
| Ethylene | 41 |
| Propane | 30 |
| *n*-Butane | 28 |
| *n*-Pentane | 29 |
| Benzene | 31 |

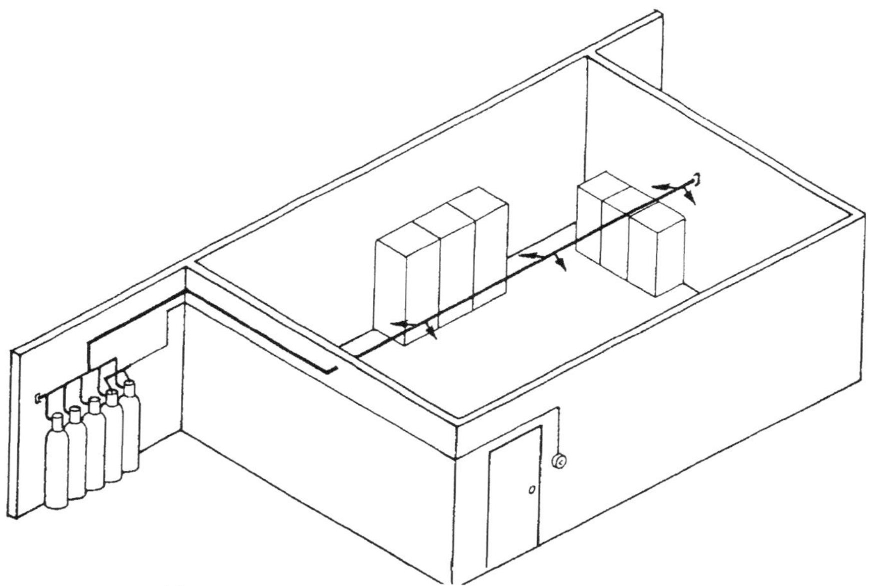

**Figure 7-35.** Total Flooding Carbon Dioxide System

necessary to achieve extinguishment pose a serious risk of asphyxiation to any human in the enclosures, thus, the design incorporates a suitable time delay (30 seconds) to accommodate emergency egress or activation of the abort.

In large facilities, semi-fixed $CO_2$ systems can be found. The $CO_2$ supply is on fire trucks and the required $CO_2$ amount is injected into the room through a quick-connection between the fire truck and the fixed piping $CO_2$ distribution system in the room.

### *7.4.13.2. Local Application Carbon Dioxide Systems*

This type of system may be used for the extinguishment of surface fires in flammable liquids, gases, and solids where the hazard is not enclosed. Examples of hazards that may be successfully protected by local application systems include dip tanks, quench tanks, and spray booths. Since local application systems do not utilize enclosures to maintain design concentrations, two methods for determining the quantity of carbon dioxide required for extinguishment are used. These are:

- *The rate by area method*—used for flat surface fires or low level objects associated with horizontal surfaces.
- *The rate by volume method*—used where the fire hazard consists of three dimensional, irregular objects that cannot be easily reduced to equivalent surface areas.

### *7.4.14. Dry Chemical*

Application of these chemicals in portable or wheeled extinguishers, hand-held hose line systems or fixed nozzle systems provides for quick fire knockdown and extinguishment.

Dry chemicals are recognized for their unusual efficiency in extinguishing two-dimensional fires involving flammable liquids. Fast extinguishing action is achieved provided the agent engulfs the fire without interruption of the application. The finely divided powder acts with a chain-breaking reaction by inhibiting the oxidation process within the flame itself.

These agents are effective on small spill fires. If there is risk of re-ignition from embers or hot surfaces, these ignition sources should be quenched or cooled with water and secured with foam, or the source of fuel should be shut off before attempting extinguishment. An example of a dry chemical system is shown in Figure 7-36.

Dry chemicals agents are nontoxic and are suitable for extinguishing clothing fires on individuals. The agent should not, however, be directed toward a person's face. The discharge of large amounts of dry chemicals may reduce visibility and make breathing difficult. When personnel may be exposed to sudden releases of large amounts of dry chemicals, warning signs, pre-discharge alarms, air-supplied breathing equipment and training should be provided. Disadvantages of dry chemical systems are:

- Potential damage to noninvolved instrumentation or lab equipment
- Difficulty in subsequent clean-up
- Glazing of hot electrical connections or other sliding contacts

**Figure 7-36.** Dry Chemical System

For these applications it is preferable to use a clean agent as the extinguishing agent.

Different types of dry chemical should not be mixed in extinguishers or containers. Under no circumstance should the type of dry chemical used in a system be changed unless the change is recommended or approved by the equipment manufacturer. Such a change may void the equipment's listing or approval.

#### *7.4.14.1. Hand Hose Line Systems*

These systems employ large central storage tanks containing up to 3,000 lb (1,410 kg) of agent. Nitrogen gas is generally used as an expelling agent. Their use is generally limited to areas where large application of dry chemicals is required and where access is generally limited. They are also recommended for areas where quick knockdown is desirable prior to the introduction of water and foam, such as in fueling operations or loading racks where there is a shortage of water for foam-making systems. The compatibility of the particular dry chemical with a particular foam should be checked with the foam manufacturer prior to specifying and purchase.

#### *7.4.14.2. Fixed Nozzle Systems*

Although these systems can be provided for both total flooding and local applications, suggested usage is in the prepackaged or local application mode. Kitchen hoods, dip tanks, etc. are areas for which these systems can be provided.

#### *7.4.14.3. Twin Agents*

Hose reels equipped with twin hoses and twin nozzles or turrets for simultaneous or alternate application or aqueous film forming foam (AFFF) and dry chemical have been developed to provide greater firefighting capability. Usually, AFFF and potassium bicarbonate are the two agents used, although multipurpose dry chemicals and water with a wetting agent are available. The AFFF concentrate usually is premixed in a solution with water and stored in a container that can be pressurized with nitrogen for instantaneous use, similar to a dry chemical extinguisher.

This design permits effective firefighting against three dimensional pressure leaks and large spill fires. The AFFF secures against reflash, as the dry chemical provides for fast knockdown and extinguishment. See NFPA 13 for more information on combined agent systems.

### 7.4.15. Steam Snuffing

Steam snuffing is a method to assist in controlling a fire in a confined space. The snuffing occurs when steam is discharged into the confined space. The confined space can be a firebox for a furnace, boiler, or other heating mechanism. Snuffing is used when the application of fire water into the confined space cannot be done safely or the application of water could cause the loss of the mechanical integrity of the equipment or piping. Special care must be taken to ensure that the chemicals in the space will not react with the steam or steam condensate and cause additional damage to the equipment or create a safety hazards for firefighting personnel.

### 7.4.16. Portable Fire Suppression Equipment

Based on the physical layout of the site, the hazards of the process and the fixed fire protection systems, additional fire suppression equipment may be required to effectively manage a fire emergency. Portable fire suppression equipment assists in providing fire protection for the equipment involved in an emergency. The use of portable equipment provides protection, where the cost of fixed system may not be acceptable or the fire water supply may be limited. Examples of portable equipment include:

- Fire water monitors
- Foam monitors
- Foam eductors
- Foam tanks
- Water tanks
- Special high flow nozzles
- Trailer mounted fire water pumps
- Water/foam pumper
- Aerial and ladder fire trucks

Portable fire water monitors and foam monitors give the fire suppression team the flexibility to set up the equipment in a location that provides maximum coverage of the water or foam spray with limited exposure to the fire personnel. Portable monitors can be set in place, with the stream on the equipment being protected, and fire personnel do not have to constantly monitor the operation.

Portable foam eductors provide flexibility in providing foam suppression capability. By attaching foam eductors to suitable fixed or portable monitors, the foam stream can be directed on the equipment.

Supplies of foam firefighting agents can be stored in portable trailers and totes taken to the site where they will be used. Trailers come in a variety of sizes. Smaller wheeled carts can be pushed by employees, while others need to be transported by a vehicle.

Portable water tanks can be useful where water supplies are limited. Some water tanks are on fire water pumper trucks or in storage tanks on flatbed trucks.

Numerous vendors build complete lines of fire trucks. The configuration of the truck is designed to meet the needs of the site. The types of trucks vary from water pumper trucks, foam pumper trucks, combination water and foam trucks, aerial trucks, and ladder trucks. Sizing of the fire water pumps on the trucks varies from 500 to 4,000 gpm (1,900 to 15,000 lpm). Foam storage on the trucks varies from 250 to 2,000 gal (950 to 9,500 l).

Portable equipment comes in a variety of sizes, shapes, and usage. Selection of portable equipment should be based on an assessment of the site's fixed fire suppression equipment, availability of personnel, and access to mutual aid. Portable equipment assists in being able reach the problem area with fire suppression equipment in a timely and efficient manner.

#### *7.4.16.1. Fire Extinguishers*

This section covers recommendations for the selection, location, and installation of portable fire extinguishers.

Portable fire extinguishers purchased in the United States should be listed by Underwriters Laboratories (UL). Extinguishers for marine use should bear the label of the U.S. Coast Guard or other Authority Having Jurisdiction (AHJ). Extinguishers and agents purchased outside the United States should be approved by the AHJ, such as the governmental authority.

To simplify the training of personnel and the stocking of spare parts at each facility, extinguishers should be made by the same manufacturer. Recommended extinguisher types and sizes and their minimum UL classifications are given in Table 7-14, along with the five most common extinguisher types and sizes.

For ease in handling, fully charged hand extinguishers should not weigh more than 55 lb (25 kg). All hand extinguishers should be furnished with the manufacturer's standard wall hook or bracket for mounting. If the extinguisher is to be mounted on a vehicle, it should be purchased with a bracket suitable for this purpose.

Hand, wheeled, and skid-mounted extinguishers located outdoors should be protected from the weather by mounting them in protective cabinets or protected with covers of heavy red vinyl (preferably iridescent) or another suitable material.

## 7.4.16.2. Fire Extinguisher Types

### 7.4.16.2.1. CARBON DIOXIDE EXTINGUISHERS

Carbon dioxide extinguishers should be used with care in closed spaces where oxygen may be displaced by the $CO_2$, reducing the oxygen concentration. $CO_2$ extinguishers are pressurized and should conform to the appropriate DOT shipping container specification or, if purchased overseas, should conform to the requirements of the AHJ. They should be designed for a minimum pressure of 1,800 psig (kPa) and should be provided with a suitable pressure relief. For more information on $CO_2$ extinguishers refer to UL 154.

### 7.4.16.2.2. DRY CHEMICAL EXTINGUISHERS

Potassium bicarbonate (Purple K) dry chemical agent in cartridge-type or pressurized extinguishers is compatible with foam. Sodium bicarbonate agent may be used where potassium bicarbonate is not available or approved for general

**Table 7-14**
*Portable Fire Extinguishers*

| Type | Nominal Size | | UL Classification (minimum) |
|---|---|---|---|
| | lb | kg | |
| *Wheeled and Skid-Mounted* | | | |
| ▪ Potassium bicarbonate dry chemical | 350<br>150 | 150<br>95 | 640-B:C<br>320-B:C |
| ▪ Potassium bicarbonate and urea dry chemical | 175 | 75 | 480-B:C |
| ▪ Premixed AFFF foam | 33 gal | 125 l | 20-A:160B |
| *Hand* | | | |
| ▪ Potassium bicarbonate dry chemical | 30<br>22<br>5 | 12<br>9<br>3 | 120-B:C<br>80-B:C<br>20-B:C |
| ▪ Potassium bicarbonate and urea dry chemical | 30<br>20<br>7.5<br>5 | 12<br>10<br>3.5<br>2.5 | 160-B:C<br>120-B:C<br>80-B:C<br>- |
| ▪ Multipurpose dry chemical | 30<br>20<br>7 | 12<br>9<br>3 | 20-A:80-B:C<br>10-A:60-B:C<br>2-A:10-B:C |
| ▪ Carbon dioxide | 20<br>15<br>5 | 9<br>7<br>2 | 10-B:C<br>10-B:C<br>5-B:C |
| ▪ Pressurized water | 2.5 gal | 10 l | 2-A |

use. When sodium bicarbonate is used, it should be UL listed. Dry chemical powder should be foam compatible.

It is preferable that extinguishers not have to be returned to the manufacturer for servicing. Thus, the extinguisher should be designed so that it is suitable to be refilled and recharged by the user. The cartridge-operated type meets this requirement better than the pressurized type.

Dry chemical extinguishers subject to extremely low temperatures, −20°F (−29°C) or below, should use nitrogen rather than $CO_2$ as the expellant gas ($CO_2$ will liquefy at these temperatures). Special low-temperature hose, nozzle, and seals are also required. For more information on dry chemical extinguishers, refer to UL 299.

##### 7.4.16.2.3. PRESSURIZED WATER EXTINGUISHERS

Pressurized water extinguishers should be hydrostatically tested to at least 200 psig (1,379 kPa). The design should provide automatic pressure release for safety upon disassembly. A listed fire extinguisher containing antifreeze agent suitable for the minimum expected temperature should be supplied when necessary.

##### 7.4.16.2.4. WHEELED EXTINGUISHERS

In order to provide optimum flexibility for a one-person operation, dry chemical wheeled extinguishers should have 100 ft (30 m) of hose. The extinguisher tank should be of welded steel construction designed for a minimum working pressure of 200 psig (1,379 kPa). The tank should normally be at atmospheric pressure and should be pressurized prior to use by a nitrogen cylinder with a quick-opening valve. The nitrogen cylinder should be of steel construction.

Wheeled fire extinguishers containing a 3% AFFF foam–water solution are intended for applying foam on small flammable-liquid spill fires where manpower is limited. They can be moved from one area to another and can be operated by one person. AFFF solution should be applied by means of a 1 in (2.5 cm) hard rubber hose, 15 ft (4.5 m) long, connected to the tank.

#### *7.4.16.3. Fire Extinguisher Location*

Potential fire hazard areas should be considered when locating fire extinguishers. Portable fire extinguishers should be distributed within the areas, in conformance with NFPA 10. Guidance on the location of portable fire extinguishers necessary to protect property is outlined below.

##### 7.4.16.3.1. GENERAL OCCUPANCIES AND LABORATORIES

Multipurpose dry chemical extinguishers (2-A:10B-C) should be provided for office buildings, auditoriums, field offices, change rooms and other low hazard occupancies. Travel distance from any hazard to an extinguisher should not exceed 75 ft (23 m). Pressurized water extinguishers may be used in lieu of multipurpose extinguishers.

Multipurpose dry chemical extinguishers (20-A:80B-C) should be provided in warehouses and shop buildings where wood, paper, plastics, and other flammable solids, as well as flammable liquids, may be stored. In grease manufacturing and compounding or similar operations where little or no wood, paper, and similar combustibles are stored, but where flammable liquids may be prevalent, potassium bicarbonate dry chemical extinguishers should be used. Extinguishers should be located not more than 75 ft (23 m) away from any potential Class A fire hazard and not more than 50 ft (15 m) from Class B fire hazards.

Laboratories should be provided with at least one dry chemical extinguisher (80-B:C) in each lab room. For Class B fires, at least one potassium bicarbonate dry chemical extinguisher (120-B:C) should be provided on each floor. Additional dry chemical extinguishers should be located in areas where hydrocarbon samples are stored in quantity or where large Class B spill fires may occur. Multipurpose dry chemical extinguishers (20-A:80B-C) should be provided near storage areas where Class A fires may occur. Electrical rooms should be provided with $CO_2$ or clean agent extinguishers.

#### 7.4.16.3.2. PROCESS AREAS

Within process unit battery limits of processing facilities, potassium bicarbonate dry chemical extinguishers should be provided along main access ways. On elevated structures, the extinguisher locations should be near stairway landings.

Dry chemical extinguishers should be located on elevated main platforms with stairway access and on air cooler platforms. When determining the proper location for extinguishers, consideration should be given to hydrocarbon pump and compressor areas, hot-oil areas, or similar potential hazards, as well as access to the extinguisher from control rooms or battery limits. Extinguishers should be located so that one can be reached without traveling more than 50 ft (15 m) to any hazard.

- At least one potassium bicarbonate dry chemical extinguisher (120-B:C) should be provided at each fired process heater handling liquid fuel or a liquid process stream. They should be installed on opposite sides, or ends, and adjacent to fire aisles.
- At least one multipurpose extinguisher (20-A:80-B:C) and one 9 lb (4 kg)clean agent extinguisher should be strategically located inside each control room adjacent to the main door.
- At least one $CO_2$ extinguisher (10-B:C) or equivalent other clean agent extinguisher should be located outside the entrance door to the unit electrical substation for Class C fires.

- At main electrical substations containing large oil-immersed transformers, at least one potassium bicarbonate dry chemical extinguisher (120-B:C) should be located outside the fence enclosure gate for Class B fires. Extinguishers should be mounted outside and adjacent to each substation entrance. $CO_2$ or equivalent other clean agent extinguisher should be placed inside larger substations.
- At least one potassium bicarbonate dry chemical extinguisher (120-B:C) should be located in pump and compressor areas.
- Cooling towers should be provided with one multipurpose dry chemical extinguisher (20-A:80-B:C) at the top of each stairway landing.
- Steam-generating plants, air compressor plants, and similar plants should be provided with potassium bicarbonate dry chemical extinguishers (120-B:C) for Class B or Class C fires in the areas containing hydrocarbons or other flammable liquids. For fires in electrical equipment, at least one extinguisher (10-B:C) should be provided. If Class A fire hazards exist, multipurpose dry chemical extinguishers (20-A:80-B:C) or water extinguishers should be provided.
- Tank car loading racks should be provided with one potassium bicarbonate dry chemical extinguisher (120-B:C) at the base of the platform stairway. One (160-B:C) unit should also be provided on the platform at every fourth tank car station. In addition, at least one potassium bicarbonate wheeled extinguisher (640-B:C) should be located near the rack.
- One dry chemical extinguisher (3-A:10-B-C) should be provided on a combustion engine driven vehicle, compressor, or welding machine.

### *7.4.16.4. Fire Extinguisher Positioning*

Portable fire extinguishers should be positioned for easy access and should be protected from moving equipment and mechanical or environmental abuse. Consideration should be given to providing the clearance necessary for removal of process equipment and piping during maintenance periods. Locations subject to steam condensate or to product drips or drains should be avoided. To avoid corrosion on the bottom of the unit, hand extinguishers should be hung in position rather than allowed to rest on grade. Locations within the manufacturer's recommended ambient temperature range should be chosen.

Hand fire extinguishers should be installed in the following manner:

- 120:B-C and 10:B-C dry chemical units, 10-B:C $CO_2$ units, and pressurized water units on wall hooks so that the top of the extinguisher is not more than 3 ft (1 m) above the floor.

- 2-A:10B-C multipurpose, 80B-C potassium bicarbonate, 5-B:C $CO_2$ units, and 5-B:C $CO_2$ units on wall brackets so that the top of the extinguisher is not more than 5 ft (1.5 m) above the floor.

When an extinguisher is installed on a column, the column should be painted on all sides with a band of "fire engine red" paint. The band width should exceed the overall height of the extinguisher by 12 in (30 cm) at top and bottom. When such banding will be obscured by storage, it should be continued to a point 3 ft (1 m) from the ceiling.

When the extinguisher is mounted on a wall, the area behind the extinguisher should be painted "fire engine red." This area should be the same height as the extinguisher and 3 ft (1 m) wide and the mounting should be centered within it.

The area behind extinguishers need not be painted "fire engine red" for light-hazard occupancies (such as offices, office hallways, and cafeterias), where decor is important and the extinguishers are usually enclosed in a cabinet. Where practical, a clearly visible sign should indicate the extinguisher location.

Dry chemical wheeled extinguishers, when outdoors, should be located on a concrete pad to facilitate mobility.

Portable fire extinguishers should be shown in their proposed locations on process unit or area plot plans and on building floor plans. A list of the extinguishers and their locations should be made for inspection and maintenance purposes.

# 8

# SPECIFIC DESIGN GUIDANCE

While Chapters 7 and 8 could be written as a single chapter, its length would make it very difficult for practical use. Chapter 7 describes a fire protection system in general terms, while Chapter 8 concentrates on where it may be applied. Some overlap is provided where necessary for clarity. Chapter 8 provides:

- Guidance on where fire protection may be necessary
- Suggested design criteria for general fire protection applications. For industrial fire protection, the hazards are generally very specific and the guidance provided here should be reviewed to determine if it is appropriate for the hazard being protected.

Chapter 7 provides a basic understanding of fire protection systems (both passive and active), general design information on fire protection commonly used in process industries, and the advantages and disadvantages of different fire protection systems.

The recommended design criteria presented in this chapter are based on generally accepted codes and standards and individual company standards. **Where values are given for application densities, etc., these should be regarded as typical rather than definitive. The optimum amount of fire protection is affected by a number of factors and greater or lesser values may be acceptable or warranted.** Historically, fire protection strategies have varied from company to company and between different branches of the process industries (e.g. chemical, petrochemical, and refining). Whether to provide passive fire protection, manual systems, or automatic systems was traditionally part of a company's fire protection strategy or culture. As fire protection decisions become more performance-based, a company's fire protection strategy will incorporate different fire protection options.

Fires in processing facilities may include vessel and equipment fires (internal or external), ground level pool fires, multilevel and three-dimensional fires resulting from spills or releases at elevated levels, liquid or gas jet fires from leaks, gas fires from vaporizing liquefied gas releases, or combinations of these.

The selection of appropriate fire protection for a specific type of facility or item of equipment should be based on the lifecycle stage of the facility and the results of a fire hazard analysis as described in Chapter 5. Typically, the protection features available will include one or more of the following:

- Elimination of the hazard and its resulting scenarios
- Prevention (reduction of probability) of its occurrence
- Detection and control
- Mitigation of its consequences
- Emergency response (see Chapter 11)

The elimination of a fire hazard may be the ideal solution, but it is often not possible. In general, the optimum level of fire protection is achieved by selecting from the other appropriate prevention and mitigation options. The higher the performance availability (or lower the probability of failure-on-demand) of each selected fire protection feature, the more effective the overall fire protection system. The generally preferred approach to improve effectiveness is to select a combination of *passive* and *active* fire protection features.

All fire protection features (passive and active) require periodic inspection and maintenance.

## 8.1. Process

A *process* may be viewed as the equipment used to produce or process a specific product or group of products, its directly related support utilities, and its physical location. In this context, process equipment includes vessels, tanks, piping, instruments, and controls, as well as specific items of equipment that handle solids, liquids, or gases. This section provides guidance on suggested fire protection for selected items of process equipment, support utilities, and the facility itself.

### *8.1.1. Process Structures and Areas*

Processing facilities may be constructed at grade in open air, in elevated multilevel open-air process structures, or in enclosed process structures or buildings.

A *process structure* is an industrial occupancy designed for, and suitable only for, a particular type of processing operation. It is characterized by a relatively low density of employee population with much of its volume occupied by equipment, vessels, piping, and machinery that is typically located at various elevations within the structure and with platforms or floors at multiple levels for personnel access. Process structures may be further described as:

- *Open process structures* have no exterior or interior walls to impede air circulation. They may have roofs or solid floors and decks associated with specific equipment. The natural air circulation associated with open process structures assists in the dilution and dissipation of vapors from spills or releases of flammable and other hazardous materials. Open process structures may also permit access for more effective manual firefighting from outside of the structure.
- *Enclosed process structures* have partial or complete walls and roofs or solid floors or decks enclosing part or all of the structure, usually for the purpose of weather protection. The handling of hazardous materials in an enclosed process structure presents significant additional concerns regarding fires and explosions. Given this description of process structures and the variety of processing operations they may contain, it should be expected that there is a wide range of possible fire protection options.

Separation distance provides the basic passive fire protection feature for process structures. The separation distance between different process structures and from storage areas, utility operations, and important buildings or facilities should be based on the hazards and risks involved. Refer to *Guidelines for Facility Siting and Layout* (CCPS, 2003b) for further spacing guidance.

#### 8.1.1.1. Open versus Enclosed Process Structures

Outdoor facilities are preferable for processing significant quantities of flammables. Relatively minor releases in an enclosed process structure or building could create flammable concentrations, increasing the risk of ignition and fire or explosion. Similar releases in an outdoor location or open process structure are more readily dispersed. Greater ease of personnel egress is typical of open structures. Firefighting in open process structures is typically easier because access to the fire can often be obtained from outside the structure with monitor nozzles or hose streams. Runoff from firefighting is more readily accommodated. Enclosed buildings and structures typically require entry by specially trained interior structural firefighters.

In colder climates, enclosures may be necessary to mitigate freeze-related hazards. In enclosed process buildings, additional fire protection features should be added to compensate for reduced ventilation and dissipation of flammable vapors, limited access for firefighting, and handling of runoff from spills.

Enclosed process structures or enclosed areas within process structures require careful consideration of the following:

- Forced ventilation to limit hazardous gas concentrations in the event of a spill or release (the electrical area classification of the area may be contingent on the use of forced ventilation)
- Fixed fire protection systems
- Safe personnel emergency escape routes
- Solid versus grating floors within the enclosed space process structures

Within a process structure, the design of the individual levels and flooring can have a significant impact on fire prevention and protection. Table 8-1 provides a summary of process structure design impacts.

Explosion effects are significantly influenced by the confinement and congestion in the area of a vapor cloud. Confinement is created by solid floors, walls, or densely packed equipment and increases or prolongs the existence of flammable mixtures following a release. Congestion is a combination of the amount and proximity of obstacles (vessels, pumps, or piping) that are included

**Table 8-1**
*Summary of Process Structure Design Impacts*

| Process Structure Type | Floor Design | Involved Process Materials | Impact | Comments |
|---|---|---|---|---|
| Open | Grating | Gas/Vapor | + | More rapid dissipation of a release and reduced probability of an ignited vapor cloud that creates overpressure |
| | | Liquid | – | Could result in a three-dimensional fire |
| | | Solid | – | Could result in housekeeping problems |
| | Solid | Gas/Vapor | – | May slow dissipation of a release and increase overpressure in the event of a VCE |
| | | Liquid | + | Facilitates fire containment and control |
| | | Solid | + | Localizes spills for easier housekeeping |
| Enclosed Fully or Partially | Grating | Gas/Vapor | + | May be easier to ventilate the structure |
| | | Liquid | – | Could allow a three-dimensional fire |
| | | Solid | – | Could result in housekeeping problems |
| | Solid | Gas/Vapor | – | May be more difficult to ventilate the structure |
| | | | | Decreases volume of confinement used in confined vapor explosion calculations |
| | | Liquid | + | Facilitates fire containment and control |
| | | Solid | + | Localizes spills for easier housekeeping |

in an area. These obstacles increase the turbulence (and thus the overpressure) associated with a vapor cloud explosion. Minimizing confinement and congestion minimizes potential explosion effects. Where solid decking is necessary, floor to ceiling distance should be increased at least three times the height of the expected vapor cloud that might accumulate. As a rule of thumb, a height of 45 ft (14 m) between solid floors or decks is sufficiently high for most applications.

*Open grating floors* allow spilled liquids or solids to fall through onto lower levels, possibly resulting in housekeeping problems and personnel exposure to spilled materials. In addition, if cascading liquid is ignited, a three-dimensional fire can result. These three-dimensional fires can be difficult to control by manual firefighting or even with fixed fire protection system.

*Solid floors* in multilevel process structures can provide a passive means of containing any spilled liquids or solids and preventing materials from falling onto lower levels. To maximize the effectiveness of solid floors, the floor design should include appropriately located drainage for spills and fire water runoff. Fire protection systems can be designed to effectively manage liquid pool fires.

#### *8.1.1.2. Small Enclosures for Process Equipment*

Process equipment and ancillary support or utility equipment are often placed in either partial or complete enclosures for:

- Weather protection
- Noise abatement
- Control of fugitive emissions (dust or vapor)
- Fire resistant protection for exposure from adjacent equipment

Some common enclosures are small buildings or enclosures for utility air or process gas compressors, gas turbines, refrigeration compressors, process pumps, filters, additive systems, dust collectors, field instruments, etc.

Enclosures, even partial enclosures, containing equipment handling flammable, combustible, or toxic materials may permit the accumulation of hazardous concentrations of these materials within the enclosure, potentially resulting in fire, explosion, or personnel exposure. Where the possibility of a flammable spill or release within an enclosure exists, the enclosure design should include a relevant selection from the following features: noncombustible construction, adequate ventilation, drainage, appropriate electrical classification, flammable vapor detection, isolation and alarm, and internal automatic sprinkler or water spray protection.

Enclosures, such as analyzer houses, providing a protective and temperature controlled atmosphere for the enclosed equipment or instruments may still warrant internal fire protection if the housed devices contain flammables or combustibles.

Where there is exposure from surrounding flammable or combustible process equipment, external fire protection of the enclosure may be needed, based on the importance or value of the protected equipment or instruments. In addition, the interior of the enclosure may require electrical classification, unless the enclosure is purged or pressurized as described in NFPA 496.

### *8.1.2. Drainage and Containment for Process Structures and Areas*

The design of process structures or areas should consider the management of liquid spills and releases, rain water, and fire water. The provision of appropriately selected and designed drainage and containment is particularly important for those areas handling flammable and combustible liquids. The options are:

- Local containment within process structure/area
- Drainage out of process structure/area into an external containment area. The drainage may be achieved either by surface swales or berms, ditches or trenches, or via an underground piping system
- Drainage out of the process structure/area onto surrounding ground

Drainage onto surrounding ground with grade slope away from the process structure or area is the simplest form of drainage. This type of drainage has been used in the past, but environmental impacts need to be considered. Additional information on the design of drainage systems is contained in Section 7.3.3.

#### *8.1.2.1. Local Containment*

Local containment is a passive mitigation system intended to retain liquid spills or releases from process systems within the immediate area of the release. Local containment may be appropriate where the quantity of potential flammable or combustible liquid spills is well defined and small to moderate. Where larger quantities of liquids are involved, e.g. continuous process operations, local containment alone is usually not an appropriate design. Containment systems are commonly used for the following purposes:

- Contain potential fire damage to the zone of origin (which may be either expendable or provided with appropriate fire protection features).
- Contain spills of powders and dusts.
- Contain molten materials that solidify at ambient temperatures.
- Used in conjunction with an automatic foam-water sprinkler/deluge system where rapid fire extinguishment is expected.
- Prevent spread to adjacent areas.
- Facilitate clean-up.

Local containment systems are designed to retain the spilled or released material within the process structure or area. Hence, the structure, equipment, vessels, and piping in that area will be exposed to any fire resulting from the contained material. Usually the exposed portion of the structure will require appropriate fire protection, unless it is considered of low value and expendable.

#### *8.1.2.2. Drainage Systems*

Drainage systems are passive mitigation systems intended to collect, control, and route spilled process liquids, rainfall, and fire water out of the process structure or area. Drainage systems provide gravity drainage of these liquids from the process structure or area in order to:

- Reduce potential fire hazards to operating personnel and to emergency response personnel.
- Limit the quantity of flammable or combustible liquids that can remain in the structure or area available to fuel a fire and, thus, limit potential damage to the process equipment.
- Prevent spread of spill or fire to adjacent areas.
- Facilitate clean-up by allowing the controlled wash-down of spills out of the process structure.

Drainage systems should be provided wherever liquid spills of flammables or combustibles can occur that, if allowed to accumulate, will result in spreading and increasing the potential fire hazard. In addition, accumulations of liquid will interfere with operations and may dangerously impair emergency response and the mobility of firefighters during a fire. Drainage systems are typically provided at grade or ground levels, and may be provided in elevated process structures and buildings on all floor levels or selected parts of those levels where solid floors may allow a liquid spill or pool fire to occur.

#### *8.1.2.3. Drainage System General Design Considerations*

Within practical limits of plant and individual unit layout, the high point of grade should pass through process buildings and areas, large equipment pads and centerlines of roadways and pipeways so that drainage will be away from these facilities and equipment.

Drainage from one unit should not pass through any other unit. It is a good practice to isolate individual plant units by installing peripheral roadways that are 6 in (15 cm) higher than the high-point grade of adjacent units. This will prevent spills, fire water, and rain water from flowing across the road from one unit to adjacent units of the plant.

A drainage system includes a containment system using solid, impermeable floors with perimeter features (curbing, floor slope, or trenches) to limit and control the spread of liquids and a collection system (floor drains or trenches, catch basins, sumps, piping, and manholes) to direct the liquids into a gravity drain system that routes them to an appropriate location out of the unit.

The decision process to determine when to provide drainage for a process unit or structure is illustrated in Figure 8-1, *Drainage Logic Diagram*.

### *8.1.2.4. Chemical, Process, or Oily Water Sewers*

In addition to a system for disposing of rain, fire, and wash-down water, many process units require special dedicated sewer systems (i.e., chemical and oily water sewers) for routine nonemergency drainage of process waste due to environmental, waste disposal, cross-contamination, or reactivity reasons. Chemical, process, or oily water sewers are usually not appropriate in capacity or purpose for use in drainage of large uncontrolled process spills, rain water, or fire water.

Chemical, process, or oily water sewers are typically designed to receive drainage directly from tanks, vessels, pumps, equipment, and piping. Their

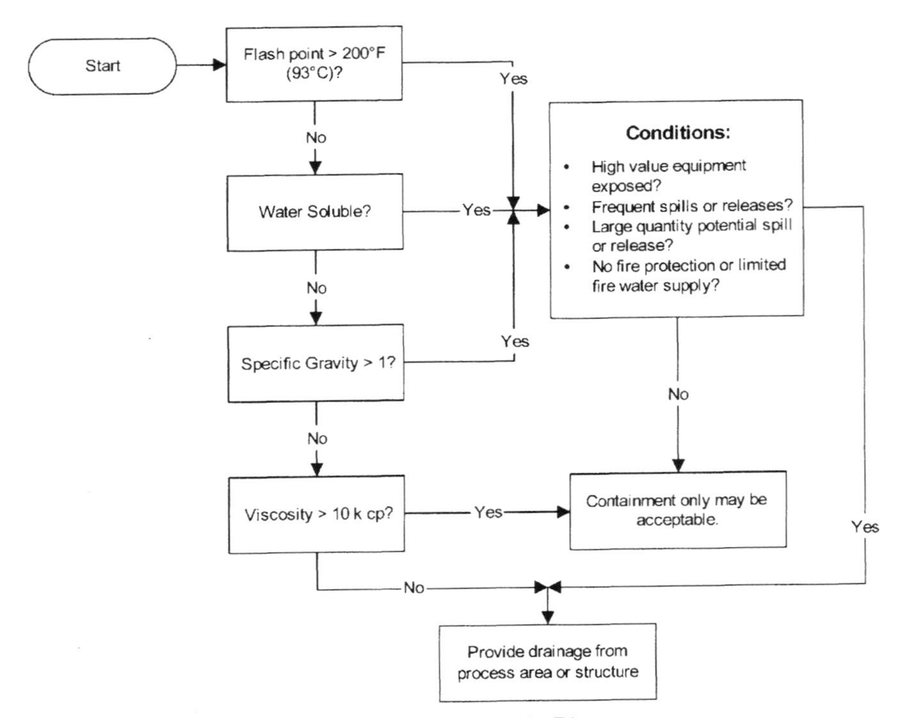


**Figure 8-1.** Drainage Logic Diagram

drain inlets (sometimes called "hubs") may be elevated 6–24 in (15–24 cm) above the surrounding floor grade to ensure that rain, fire, or wash-down water cannot flow into these dedicated sewer systems.

### *8.1.2.5. Drainage System Design and Features*

In most cases, for a process unit, the fire water flow rate establishes the individual unit's drainage system design basis. Drainage system capacity should be adequate to carry away the unit's maximum fire water application rate. When increasing the fire water protection for an existing unit due note should be taken of the potential impact of this water load on the drainage system. Where the total applied fire water (from fixed systems and manual streams) can exceed 125% of the drainage system flow rate capacity, then the drainage system capacity should be increased. Where it is subsequently found that the drainage capacity has not been upgraded where necessary then the pre-fire plans should limit the water application rate until appropriate changes can be made. When sizing a drainage system it is not necessary to combine fire water and rain water flow rates, the greater flow rate of the two is sufficient. The system for any individual process units is usually based on fire water demand while the capacity of the overall facility is normally determined by the 100 year storm rainfall.

Although individual drainage areas will vary in size and shape, all should have the following characteristics:

- All areas under process equipment should be paved and sloped to assure that liquids will drain away from equipment.
- Drainage patterns should minimize pooling of runoff near equipment and prevent drainage from unpaved areas from entering floor drain inlets in paved areas.
- Drains and trenches should be centrally located and as far away from equipment and overhead piping and pipe racks as possible. A minimum spacing of 10 ft (3 m) from major equipment is preferred.
- Outer edges of paved areas and ridge (high point) lines for each drainage zone area should be at a constant elevation.
- The preferred slope should be ¼ in per ft (2%) for paved areas including around pumps and other areas where leaks are anticipated. For example, the preferred differential elevation from any ridge line to a drain inlet should be 6 in (15 cm) for a 50 $ft^2$ (4.6 $m^2$) area and 9 in (23 cm) for a 75 $ft^2$ (7 $m^2$) area.
- The maximum slope should be no more than ½ in per ft (4%).

### *8.1.2.6. Curbs, Trenches, Drains, Catch Basins, and Oil Separators*

Either curbs or trenches may be used at the perimeter of the containment zone. Curbs or high grade can be used to separate drainage areas. Curb heights for drainage systems should be calculated, similar to containment systems, based on the maximum probable flow into the containment/drainage area plus a minimum freeboard allowance of 2 in (5 cm); the minimum recommended curb height is 4 in (10 cm). The floors themselves should be sloped at ¼ in per ft (2%) toward the trenches or floor drains. Figure 8-2 illustrates a typical floor drain and trench layout (FM Date Sheet 7-83).

Conventional floor drains are grating-covered flush floor boxes with a 4–6-in diameter (10–15-cm) outlet pipe connection with the cover grating area usually twice that of the outlet pipe. When using floor drains, a safety factor of 25% additional floor drains should be added to allow for flow restriction due to debris accumulation plugging the cover grating (a well designed floor box will contain a sump that will allow it to collect occasional debris without impairing the drain path). Maximum flow through a typical conventional floor drain requires a liquid head of approximately 3–4 in (7–10 cm). This is a significant pool depth when the objective is to drain the material out of the process area or structure. This pool depth may not be acceptable in all situations.

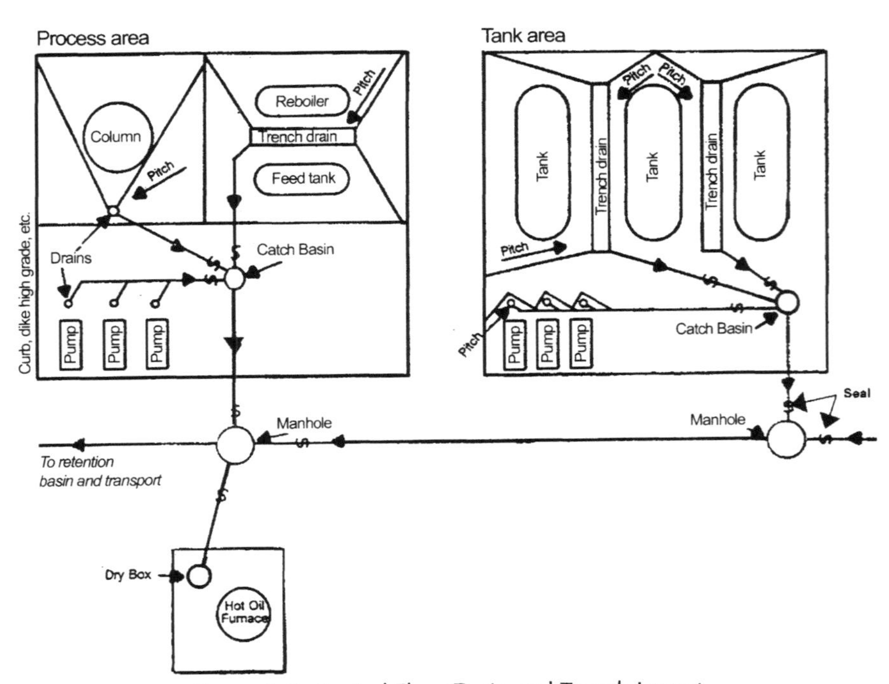


**Figure 8-2.** Typical Floor Drain and Trench Layout

Trenches formed into the floor can provide a higher drainage flow rate than conventional floor drain boxes, resulting in minimal depth of the liquid runoff within the area served. Illustrations of drainage trenches are shown in Chapter 7, Figure 7-3.

Isolation and containment of flammable or toxic vapors and gases within a drainage system is important to prevent their spread through the system to other locations well away from where the release or spill occurred. Isolation and containment of vapors and gases can prevent fires and explosions from propagating back to process areas or between process areas. The primary means of isolation and containment of flammable or toxic vapors and gases in drainage systems are:

- Liquid seals that block vapors and gases from being conveyed through the system to other areas.
- Vent pipes directed to an appropriate location for the safe disposal of any gases trapped in, or released by, the drained liquids.

A *liquid seal* uses a liquid (usually water) trap to isolate vapors or gases to one part of the drainage system and to block them from being conveyed to other areas served by the drainage system and the flashback of any subsequent ignition.

For fire protection purposes, drainage systems that may contain flammable vapors should be sealed and vented both inside the battery limit and outside the battery limit. In the absence of these provisions, flammable or toxic vapors and gases may migrate into an area well away from where the release occurred, thus spreading the hazard.

Process, chemical, or oily water sewer system branch and lateral lines should enter main lines through vapor-sealed and vented manholes. Branches and laterals in clean or storm water drainage systems may enter main lines without vapor seals if liquid-sealed catch basins are used on the inlets to these branches and laterals.

Catch basins should be provided for each fire zone of the process area to collect the flow from the floor drains and trenches. All catch basins within the process area battery limits should be sealed for vapor control. Catch basin seals should be of types that are easily cleaned out. P-traps should not be used. Catch basins require a periodic check to verify that a liquid seal is in place and clean-out plugs are installed.

Piping with appropriate traps is the preferred method for routing drained liquids away from the process structure or area to a containment system. Drainage piping systems must be sized and configured to ensure adequate flow capacity to minimize the retention of spilled liquids and fire water in the process structure or area. Common practice is 4 in (10 cm) diameter minimum drain piping size to avoid plugging by debris. Liquid traps should be provided at the inlets to the drainage piping from each zone containing flammable or com-

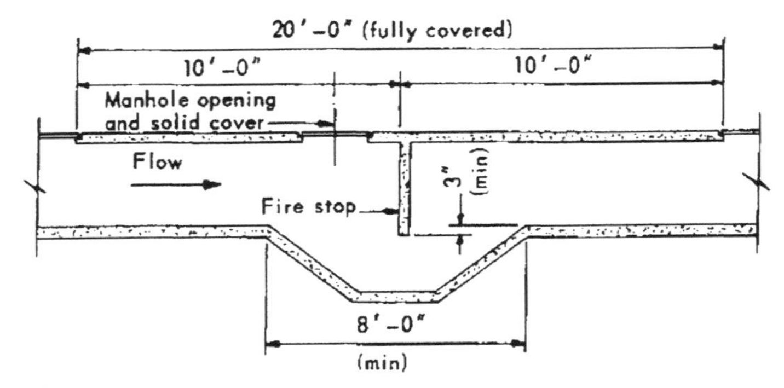


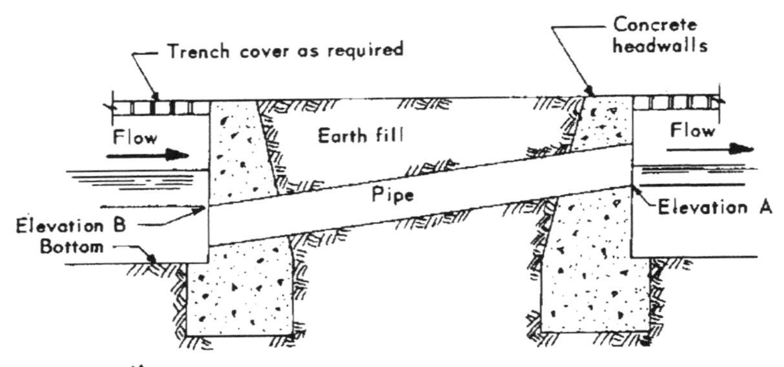


**Figure 8-3.** Drainage Trench Flame Trap Designs

bustible liquids (see Figure 8-3). Drainage piping materials should be compatible with the materials likely to be handled. Compatible plastic piping may be considered when it is buried underground or encased in concrete. Metal drainage piping should be used where it may be subjected to heat from a fire, such as drain lines from an upper level of a process structure.

Oil–water separators are often used to separate nonmiscible process liquids from water. Oil–water separators should be liquid-sealed at both their inlet and outlet lines to prevent fires and explosions from propagating back to process areas or between process areas.

Manholes inside process area battery limits should be vented at least 18 in (46 cm) above the highest pipelines or equipment within a 10-ft (3-m) radius or at least 12 ft (3.7 m) above grade, walkway or work platforms within a 10-ft (3-m) radius. Vent pipes of 2–4 in (5–10 cm) diameter should be used. The vents should be located at least 50 ft (15 m) horizontally from furnaces and fired process heaters. The vent pipe opening should be pointed straight up to disperse vapor.

Manholes outside of process area battery limits should be vented at least 12 ft (3.7 m) above grade; 2 in (5 cm) diameter vent pipes are normally adequate. The vent pipes should be located well away from roads and ignition sources and pointed straight up to disperse vapor.

Flame arrestors are not considered necessary on the discharge of vent pipes from drainage systems designed primarily for fire water and rain water drainage. There are a number of reasons for this:

- Practice indicates that the vapors emanating from these vents are rarely within the flammable region, as the nominal atmospheric pressure of the drainage system is not adequate to allow vapor or gas to pass through the 3–4 in (7.5–10 cm) deep catch basin water seals.
- Even if these vapors are in the flammable range, they are rarely ignited if the vents terminate as described above.
- Flame arrestors are typically constructed of a close mesh material and, in this type of service, have a tendency to plug due to atmospheric moisture and dirt and, thus, require frequent inspection and cleaning.
- Blockage of flame arrestors can result in the gases/vapors from a spill not being vented and the system being overpressured. The result is that other liquid seals may be "blown," allowing these gases to vent in noninvolved portions of the facility; thus, spreading the hazards

#### *8.1.2.7. Maintenance*

Drainage systems and special dedicated sewer systems must be clean and maintained in good repair. Volatile liquids or hot liquid condensate in the systems can generate pressure allowing vapor to escape through any unsealed or poorly sealed part of the system. These systems must be designed for ease of maintenance, e.g., proper use of cleanouts. Furthermore, the nature of chemicals or wastes that may enter the systems could require special materials of construction which may in turn require additional maintenance techniques. Any liquid leakage may contaminate the surrounding soil and accumulate in the ground water.

### *8.1.3. Flammable Gas Detection Systems*

Flammable gas detection systems provide early warning of a hazardous condition resulting from a process release. These detection systems may provide time to intervene and prevent escalation of the incident. A properly designed detection system with an adequate number of strategically located sensors can be a valuable fire protection feature.

### 8.1.3.1. Gas Detection Alarm Levels

Flammable gas detection systems are typically used to initiate an alarm at a concentration level below the lower flammable limit (LFL). Two gas alarm levels (low and high) are often utilized to allow early warning prior to taking automatic actions. Detection systems may also be used to stop electrical power and initiate process shutdown. The low alarm setpoint should be ~20% LFL and the high alarm level set point should be between 40%–60% LFL. Where these devices are used to initiate process shutdown or activate fire protection systems, it is common practice to use some form of voting, typically 2 out of 2, such that the frequency of spurious shutdowns or system activation is minimized.

### 8.1.3.2. Gas/Liquid

Once it is determined that flammable gas detection is needed for a processing area, an evaluation of the gas/liquid composition of each credible release can be done. The formation and behavior of gas clouds is primarily governed by whether the material is released as a gas or as a liquid. In order to properly locate detectors, the expected behavior of a gas cloud must be understood. Similar considerations apply to the release of toxic materials and the appropriate location of detectors for these materials.

The release of pressurized gas to the atmosphere and its dispersion can be described in three stages: jet mixing, momentum effects of wind or air currents, and natural diffusion. While the initial properties of the gas at the time of release (e.g. temperature, pressure, density) define the first stage, they have little influence on the second and third stage. The energy associated with the release of a pressurized gas creates a "jet mixing" effect that causes the gas to be diluted in air. Gas releases can rapidly form explosive clouds depending on the rate of release.

A release of a flammable liquid or heated combustible liquid will vaporize at a rate determined by its temperature and vapor pressure. The result is a relatively slowly developing cloud with concentrations decreasing with the height above the liquid surface. To ensure early warning, detector placement may need to be relatively close to the floor or grade level, 3 ft (1 m) or less. The release of liquids under significant pressure creates a spray atomizing effect, somewhat like pressurized gases, resulting in a higher rate of evaporation.

### 8.1.3.3. *Placement of Flammable Gas Detectors*

Placement of detectors should consider the need for protection from physical damage, weather effects (snow, icing, wind-driven rain, or dust), or direct water hose spray during an area wash down. Detector placement should also allow safe access to perform required periodic testing and inspection. Typical approaches to flammable gas detection layout are:

- *Local or point-of-release coverage*—based on fire protection system judgment to predict possible leak sources. Detection should be located in proximity to release source near specific individual equipment items.
- *Spatial coverage* (At-Risk Volume Approach)—based on the correlation between flammable gas accumulation and the resultant overpressure upon ignition. The detection spacing is determined by estimating the size of the cloud that can present a credible escalation hazard.

#### 8.1.3.3.1. LOCAL OR POINT-OF-RELEASE COVERAGE

Local or point-of-release detection is based on estimating credible spills and releases. Location of detectors is based on the likelihood of a spill or release occurring. Usually this results in gas detectors near pumps, compressors, sumps, and trenches. They can also be located at spaced intervals surrounding the perimeter of vessels or groups of equipment items that handle flammable liquids, gases, or heated combustible liquids. Traditional point gas detection is illustrated in Figure 8-4. Example guidance for gas detection for enclosed spaces, rooms, and building is provided in Table 8-2.

#### 8.1.3.3.2. SPATIAL COVERAGE (AT-RISK VOLUME APPROACH)

Spatial coverage is a gas detection layout methodology that utilizes a goal setting or at-risk volume approach (Bond, 1993). The objective is to detect gas accumulations that can cause catastrophic escalation through a pragmatic methodology. For flammable gas hazards, this can be based on detection of

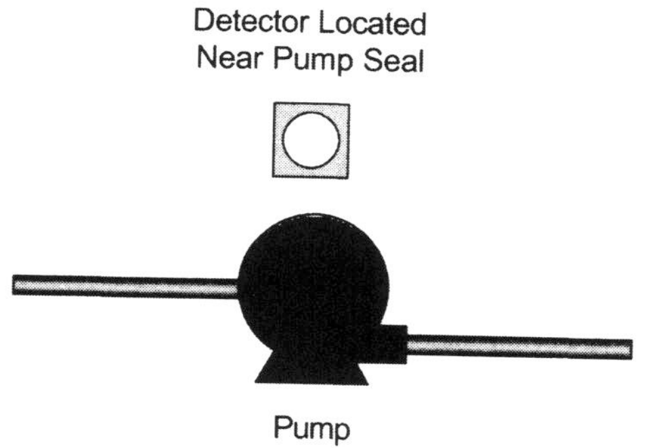


**Figure 8-4.** Traditional Point Gas Detection

**Table 8-2**
*Detection of Flammables in Enclosed Spaces*

| Type Ventilation | Source of Flammables | Ignition Source in Enclosed Space? | Detector Placement | Possible Detection System Actions |
|---|---|---|---|---|
| Natural | Outside the space | Yes | Consider detector(s) at source(s) | ▪ Consider alarm to attended location. |
| | | No | Not normally required | — |
| | Inside the space | Yes | Inside enclosed space | ▪ Multilevel alarm to attended location.<br>▪ Consider automatic equipment shutdowns or other actions. |
| | | No | Inside enclosed space | ▪ Alarm to attended location. |
| Forced | Outside the space | Yes | Ventilation air intake(s) | ▪ Alarm to attended location.<br>▪ Provide automatic shutdown of ventilation system. |
| | | No | Ventilation air intake(s) | ▪ Consider alarm to attended location. |
| | Inside the space | With or without ignition source present. | Exhaust air duct or inside enclosed space | ▪ Multilevel alarm to attended location.<br>▪ Consider automatic increase of ventilation system flow rate. |

accumulated flammable gas and initiating protective actions. The approach recognizes that the hazard associated with flammable gas clouds is the resultant overpressure that can occur upon ignition. An enclosed, ideally mixed flammable gas cloud 16 ft (5 m) in diameter can generate enough overpressure to break small bore piping and instrumentation, which will cause escalation of the event. The detection target for such areas is typically gas clouds of 16 ft (5 m) diameter.

In the at-risk volume approach, a three-dimensional grid of detectors is installed to assure that the hazardous gas cloud can not exist without detection.

Three-dimensional detector coverage, as shown in Figure 8-5, can be obtained by a triangular grid (isometric) pattern of point or open path detectors, with a maximum separation less than the gas cloud diameter that can

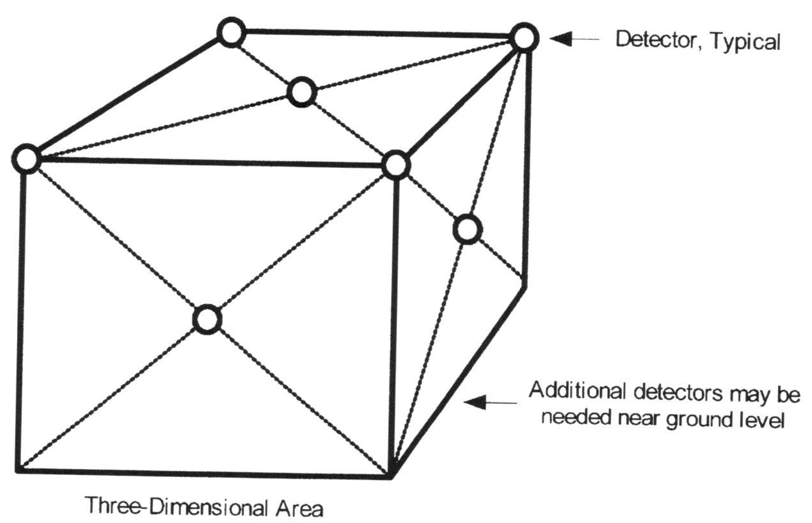


**Figure 8-5.** Three-Dimensional Detector Coverage

cause escalation of the event. Open Path IR gas detection is often utilized to achieve this coverage.

To design general area gas detection, each area should be evaluated based on its volume for credible gas release hazards. The factors to consider when determining credible hazards are:

- Location and elevation of process equipment with gas or flashing liquids.
- Compressed flammable gas storage in cylinders.
- Composition of process stream

8.1.3.3.3. DUCTWORK SUPPLYING NONHAZARDOUS AREAS

Where it is possible for flammable or toxic gas or vapor released within a hazardous area to migrate to the inlets for HVAC systems serving nonhazardous enclosed areas such as control rooms, detection systems should be installed in those HVAC inlets or connecting ductwork. Detection should be provided in HVAC system intakes if the building, room, or enclosure served is not electrically classified and a flammable (or toxic) gas or vapor could feasibly be drawn into the area, either by mechanical ventilation systems or by differential pressures. The detection system should alarm and automatically shutdown the HVAC to prevent gas or vapor concentration in the protected space from reaching the flammable or toxic range.

In addition to ductwork and HVAC air intakes, penetrations in walls, floors, ceilings, or roofs may also allow gas or vapor to be drawn from hazardous areas into areas intended to have nonhazardous atmospheres. Single personnel passage doors and equipment doors can also be a source of unwanted infiltration of gases or vapors. Vestibules with double doors or air-lock doors can provide improved isolation for personnel passageways.

#### 8.1.3.3.4. HYDROGEN GAS DETECTION

Hydrogen is produced or used in gas or liquid form in a number of processes. Hydrogen is a colorless, odorless gas that is easily ignited over a wide flammable range. Hydrogen burns very rapidly and cleanly in air with a nearly invisible clear blue flame. It has flammable limits of 4–74% in air and minimum ignition energy of 0.02 mJ, compared to 0.25 mJ for typical hydrocarbons.

Hydrogen gas detection is normally not provided for equipment located in open air due to hydrogen's normally rapid dispersion and its exceptionally low specific gravity. Hydrogen gas detection should be provided for hydrogen containing systems or portions of systems that are located in fully or partially enclosed buildings, rooms, or enclosures. Detectors should be placed on the ceiling or at other locations where released or leaked gas may accumulate.

Hydrogen gas may accumulate in rooms or areas where multiple batteries are routinely recharged or maintained, such as electric fork-lift truck charging/maintenance stations and storage battery banks for Uninterruptible Power Supplies (UPS). Adequate continuous ventilation (natural or active) should be provided (CGA, 1992; NFPA 50A, 1999; NFPA 50B, 1999).

## 8.1.4. Fixed Fire Detection

### 8.1.4.1. Placement of Fire Detection

There are generally two approaches to fire detection:

- *Point Fire Detection*—based on good judgment and past practices to predict where fire hazards are likely to occur and cause unacceptable damage.
- *Grading Approach Fire Detection*—based on defining fire detection performance requirements for sensitivity, response time, and availability based on risk of fire escalation.

The fire detection grading approach is a coarse hazard assessment, which documents the credible fire hazards and defines and establishes acceptable fire detection performance requirements. The fire detection performance requirements are defined by determining what size of fire presents a credible escalation hazard. The systematic application of the grading process then develops an engineered detection system whose performance can be verified. The general process utilized to develop the detection layout is as follows:

- Review plot plans, PFDs, P&IDs, etc.
- Identify and document credible fire hazards
- Apply the grading process for fire detection
- Determine the detection technology to achieve the performance requirements

Where the need for fire detection is identified, the required performance of the fire detection system is already specified as part of the grading process. Fixed fire detection is typically installed to protect equipment that is high value, long lead time, or likely to be significant fire escalation hazards. The performance specification defines fire size and response time thresholds for alarm and action(s). Fire hazards are defined by radiant heat output (RHO). RHO gives a reasonable indication of the potential damage and the probability that the fire will escalate or cause loss. The RHO should not be used to determine fire thermal loading onto equipment and structures. Table 8-3 compares RHO and flame area for some typical hydrocarbon fires.

### 8.1.5. Fire Protection

The decision to provide or not to provide fixed fire protection systems within process structures, areas, or on specific vessels or equipment is usually based on a qualitative assessment of the following factors such as:

- Quantity of flammables or combustibles present
- Facility/equipment spacing and location
- Business importance of the operation
- Value of vessels or equipment
- Access to manual fire monitors and personnel availability.

Use of fire hazard analysis (FHA) approach will determine the size and expected duration of fires and allow selection of fire protection on a performance basis (see Chapter 5).

**Table 8-3**
*Comparison of RHO and Typical Apparent Flame Area for Hydrocarbons*

| RHO | 10 kW | 50 kW | 100 kW | 250 kW |
|---|---|---|---|---|
| Gas Jet | 3 $ft^2$ (0.3 $m^2$) | 16 $ft^2$ (1.5 $m^2$) | 32 $ft^2$ (3 $m^2$) | 80 $ft^2$ (7.5 $m^2$) |
| Oil Jet | 2.5 $ft^2$ (0.25$m^2$ ) | 14 $ft^2$ (1.3 $m^2$) | 27 $ft^2$ (2.5 $m^2$) | 70 $ft^2$ (6.5 $m^2$) |
| Oil Pool | 5 $ft^2$ (0.5 $m^2$) | 20 $ft^2$ (1.9 $m^2$) | 43 $ft^2$ (4 $m^2$) | 91 $ft^2$ (8.5 $m^2$) |

**Table 8-4**
*Quantity of Flammable Liquids or Liquefied Gases Where Fixed Water Spray Fire Protection Is Recommended*

| How Liquids or Gas Is Stored | Outdoors or Open Process Structure | Enclosed Process Structure or Building |
|---|---|---|
| In the largest vessel | >500 gal (1,893 l) | > 200 gal (757 l) |
| Total in multiple adjacent or connected vessels within 2,500 ft² (232 m²) area | >5,000 to 10,000 gal (18,927–37,854 l) | > 2,000 gallons (7,570 l) |

The quantity of flammable liquids or gases or heated combustibles can be indicative of the need for fixed water spray fire protection. Table 8-4, developed from FM Data Sheet 7-14, suggests quantities of flammables where fixed water spray fire protection should be considered based on one insurance provider's guidance. Some company internal guidelines and practices allow up to 5,000 gal (18,927 l) in the largest vessel before water spray protection is recommended. Additional guidance on fixed fire protection is available in API Publication 2030, *Application of Water Spray Systems for Fire Protection in the Petroleum Industry*, (API, 1998).

In the majority of applications in processing facilities handling flammable gases or liquids, or liquefied flammable gases, fixed water-based fire protection systems will control, but not extinguish, the fire. Foam-water sprinkler systems are an alternative to water sprinkler systems. Both water and foam-water sprinkler systems are discussed in Chapter 7.

In general, fixed water spray fire protection has the two-fold purpose of cooling the affected equipment and flushing any burning liquids from the immediate fire area. This can reduce local damage, limit fire spread, and allow time for other response actions. It should not be expected that these systems will extinguish a fire without the use of other fire protection systems, such as foam, dry chemical, or manual firefighting.

### *8.1.5.1. Area Coverage and Directional Spray*

Fixed water spray fire protection in process structures may be provided as follows:

- Complete area coverage protection (deluge).
- Directional water spray protection on specific equipment (fixed water spray).
- A combination of area coverage with protection of specific equipment.

The arrangement, spacing, and density of vessels and equipment within the process structure usually determine the most practical protection. In making the final choices on active fixed fire protection, consideration should be given to the amount and effectiveness of passive fire protection, such as fire-proofing and other forms of passive fire protection.

When using area coverage protection, additional spray heads should be provided below any obstructions to water distribution that are greater than 3 ft (1 m) in diameter or more than 10 $ft^2$ (0.9 $m^2$) in area. Additional protection should also be provided for equipment located in areas or levels having solid floors and where the height of the nozzles is more than 15 ft (4.6 m) above the solid floor.

In open process structures or grade level process areas, deluge or water spray nozzles should be of the open type to ensure rapid actuation and to combat the effects of wind. For enclosed structures, closed head sprinkler systems can be used.

Fixed fire protection for specific equipment or vessels should be provided to the top of the vessel or for tall vertical vessels, to the greater height of either 10 ft (3 m) above the normal flammable liquid levels in the vessel or to a height of 30–40 ft (9–12 m) above any level, either grade level or a solid floor deck, where a substantial pool fire could develop.

### 8.1.5.2. *Density, Demand, and Duration*

For area protection of process structures, the densities listed in Table 8-5 (NFPA 15) should be provided for the various parts of the building.

Typical fire protection water spray application rates for selected equipment that is typically found in open processing areas or structures are shown in Table 8-6 (API 2030).

Fixed fire protection systems for process structures and areas can be activated automatically or manually. Automatic activation provides quicker and more reliable response than is typically possible with manual activation. Vari-

**Table 8-5**
*Fire Protection of Open and Enclosed Structures*

| | Density, gpm/ft² (lpm/m²) | | |
|---|---|---|---|
| **Protect where...** | **Enclosed** | **Open** | **Demand Estimate** |
| Under roof or ceiling | 0.25 (10.2) | 0.25 (10.2) | ▪ Assume all ground floor system(s) operate.<br>▪ If deluge system is used, assume all systems operate. |
| Below each solid floor level | 0.25 (10.2) | 0.25 (10.2) | |
| Below grated floor levels of 15 ft height or less | 0.15 (6.1) | 0.25 (10.2) | |
| Below grated floor levels of >15 ft height | 0.25 (10.2) | 0.25 (10.2) | |

**Table 8-6**
*Selected Equipment Water Spray Application Rates*

| Item | Application Rate | |
|---|---|---|
| | gpm/ft² | lpm/m² |
| Air-cooled fin-tube heat exchangers | 0.25 | 10.2 |
| Cable trays | 0.30 | 12.2 |
| Compressors | 0.25 | 10.2 |
| Exposure protection | 0.25 | 10.2 |
| Fired heaters | 0.25 | 10.2 |
| LPG loading racks | 0.25 | 10.2 |
| Motors | 0.25 | 10.2 |
| Pipe racks | 0.25 | 10.2 |
| Pressurized storage tanks | 0.25 | 10.2 |
| Pumps | 0.20–0.50 | 8.1–20.4 |
| Electrical switchgear | 0.25 | 10.2 |
| Towers | 0.25 | 10.2 |
| Transformers | 0.25 | 10.2 |
| Turbines and related gear boxes | 0.25 | 10.2 |
| Vessels and heat exchangers | 0.25 | 10.2 |

ous detection systems can be used to activate the protection systems. These detection systems may also be used as alarms only with the activation decision left to an operator. Detection systems are available based on sensing heat, smoke or flame.

### *8.1.5.3. Medium to Small Processing Operations in Buildings*

Medium- to small-scale processing operations handling flammables or combustibles are often housed inside of fully enclosed buildings. In many cases, the buildings may be of relatively conventional construction unlike the special occupancy process structures specifically designed for larger scale processing facility operations.

Typically, medium to small processing facilities are under the jurisdictional coverage of local fire and building codes that require automatic sprinkler protection. These processing operations should be fire protected by automatic sprinklers that extend over all areas that store, process, or transfer flammable or combustible materials. The physical limits of the areas requiring protection can be defined by fire resistive walls of at least a 1-hour fire rating or by curbs or continuous trench drainage systems that are designed to prevent spills from flowing to other areas. Sprinklers protecting curbed areas should extend 20 ft (6 m) beyond the curbing to ensure adequate protection. Care should be taken in the original installation and in subsequent process equipment or facility changes to ensure that sprinkler coverage is not obstructed by ductwork, tanks, or equipment. Obstructions that exceed 3 ft (0.9 m) in width or diameter or 10 $ft^2$ (0.9 $m^2$) in area necessitate the relocation or addition of sprinkler heads to reestablish coverage.

Sprinkler protection may be omitted in areas of a building where the only exposure is from piping running through the area that is of welded metal construction with no flanged joints. Sprinkler protection may also be omitted in buildings of low value that are safely spaced and separated from important buildings and structures. Recommended separation distances can be found in *Guidelines for Facility Siting and Layout* (CCPS, 2003b).

Design guidance for automatic sprinkler systems protecting flammable and combustible liquid processing operations in buildings is provided in Table 8-7 (FM Data Sheet 7-32). Additional guidance can be found in NFPA 30. These systems should be hydraulically calculated. The guidance assumes that there is no potential for three-dimensional fires due to liquid handling equipment on multilevel open grating mezzanines. Foam-water sprinkler systems are an alternative to water sprinkler systems. Both water and foam-water sprinkler systems are discussed in
Chapter 7.

### 8.1.6. Structural Steel Protection

In process structures handling significant quantities of flammables or heated combustible liquids, it should be assumed that the structural steelwork supporting the building, vessels, equipment, and piping can be exposed to substantial jet or pool fire events. The high temperature associated with these fires can cause thermal weakening of steelwork and result in the collapse of structures, failure of equipment, and subsequent escalation of the initial event. Accordingly, it is necessary to suitably protect these structural steel supports to prevent such an event. See Chapter 5, Section 5.13 for details on fire impact to structural steel.

The required protection may be obtained by active, passive, or a combination of both protection systems. For example, steel support located in a fire exposed area within process unit battery limits may be protected by either a fixed water spray system or the application of fire resistant insulating material to the steelwork or possibly both. *Note:* Passive protection is generally the preferable method for protecting structural steel.

**Table 8-7**
*Sprinkler Protection for Medium to Small Processing Facilities*

| Flash Point | Heated To/Above Flash Point | Room or Equipment Explosion Hazard | Sprinkler Density gpm/ft² (lpm/m²) | Sprinkler Temp. Rating °F (C) | Area of Demand ft² (m²) | Manual Hose Streams gpm (lpm) | Duration min |
|---|---|---|---|---|---|---|---|
| Any liquid presenting explosion hazard | | | 0.30 (12) | 286 (141) | 6,000 (560) | 1,000 (3800) | 120 |
| | | | | 165 (74) | 8,000 (740 ) | | |
| <100 °F (38 °C) | Does Not Apply | No | 0.30 (12) | 286 (141) | 4,000 (370) | 5000 (1900) | 60 |
| | | | | 165 (74) | 6,000 (560) | | |
| 100–200°F (38–93 °C) | Yes | No | 0.30 (12) | 286 (141) | 4,000 (370) | | |
| | | | | 165 (74) | 6,000 (560) | | |
| | No | No | 0.25 (10) | 286 (141) | 4,000 (370) | | |
| | | | | 165 (74) | 6,000 (560) | | |
| >200 °F (93 °C) | Yes | No | 0.25 (10) | 286 (141) | 4,000 (370) | | |
| | | | | 165 (74) | 6,000 (560) | | |
| | No | No | 0.20 (8) | 286 (141) | 3,000 (280) | | |
| | | | | 165 (74) | 4,000 (370) | | |

### *8.1.6.1. Fire Resistant Insulating Material*

The purpose of fire resistant insulating material or fireproofing is to limit the temperature of important steelwork or equipment when exposed to high-temperature burning liquids and gases during fire situations. As noted above, this passive protection can prevent the thermal weakening of steelwork, the collapse of structures, and failure of equipment and their supports and thereby reduce or eliminate damage. Fireproofing is more often used on the steelwork of structures and equipment supports, but can also be used on equipment including tanks, vessels, and heat exchangers.

Several different materials and installation systems are available and include various types of proprietary materials, plasters containing perlite or vermiculite, concrete mixtures, or lightweight concretes. The selected material and installation system, encasement, or surface application should provide protection for the expected fire duration. For more information refer to Chapter 7, Section 7.3.2.

The material and its installation should be designed to protect for a specific time period. For hydrocarbon fire exposure materials and their installation are generally tested to either UL 1709 or ASTM E1529 to determine time-to-failure of different thicknesses and installation methods. It should be noted that the fire resistance rating is a measure of the ability of the installed material to withstand a specific "standard" fire. While these conditions closely match those in any given hydrocarbon fire, during actual fires the material may be exposed to conditions that may be more or less thermal intense, thus it can be expected to retain its integrity for a greater or lesser time.

The fireproofing system should also be capable of withstanding exposure to direct jet flame impingement and to dislodgement by direct force of fire water hose streams. The outer surface of the fireproofing material should be suitably protected against weather damage and sealed to prevent ingress of water.

Fire resistant insulating material can provide passive protection for both vertical and horizontal structural steel members. The level or rating of fire resistance should be consistent to the expected fire duration. Where only fire resistant insulating material will be used, the material and its installation system should be specified to have a 2- to 3-hour fire rating (UL 1709). In applications using a combination of fixed water spray or sprinkler protection and fire resistant insulation, a 1- to 2-hour fire rating (UL 1709) is frequently specified for the fireproofing.

Fireproofing should not be confused with thermal insulation applied for process operational or energy conservation purposes. Thermal insulation of the appropriate material can provide some limited fire exposure protection if it is protected with a sheet steel (not aluminum) outer weather jacket and stainless steel banding to hold it in place against fire exposure and hose stream impingement. NFPA 30 and API RP521 both allow a reduction in the size of fire exposure emergency relief venting capacity for tanks and vessels that have the specified thermal insulation (NFPA 30, 2000). See Chapter 7, Section 7.3.2.6.8 for additional information.

#### *8.1.6.2. Application*

Some commonly used fireproofing ratings for structures and equipment are indicated in Table 8-8 (API 2218). Unless otherwise noted, steel structures and

**Table 8-8**
*Typical Fireproofing Ratings for Structures and Equipment*

| Structures and Equipment | Exposure Time to Hydrocarbon Fire |
|---|---|
| Steel structures supporting equipment | 2–3 hours |
| Main pipe rack structures inside process units | 1–2 hours |
| Main pipe rack structures in process units that also support equipment | 2–3 hours |
| Structural supports for process unit transfer lines | 2 hours |
| Piping supported from hangers | 2 hours |
| Flue gas stack supports | 2 hours |
| Vessels containing substantial volumes of flammables or heated combustibles (refinery crude desalters, chemical facility reactors, etc.) | 1–3 hours |
| Equipment, vessel, and column supports (skirts, legs, etc.) | 1–3 hours |
| Main instrument runs | 15 minutes |
| Tubing, conduit, and valve operators<br>*Exception:* For safety or emergency shutdown purposes, some pneumatically operated valves may intentionally use plastic tubing that is intended to fail in fire or high heat exposure in order to ensure that the valve moves to the desired position. | 15 minutes |

equipment supports should be protected with at least a 1-hour, high-rise hydrocarbon fire resistance rating, or equivalent. The results of a specific fire hazard analysis (FHA) and approval by the authority having jurisdiction (AHJ) may allow the use of fireproofing at lower fire resistance ratings than those listed below. For process structures:

- When supporting high fire potential equipment:
  - Support columns should be fireproofed from grade up to the equipment support level.
  - Columns and beams that transmit the equipment load to support columns should also be fireproofed (see Figure 8-6).
  - When low fire potential equipment is located at a higher level, the fireproofing should be extended an additional 40 ft (12 m) above the level at which the high fire potential equipment is supported (see Figure 8-7).

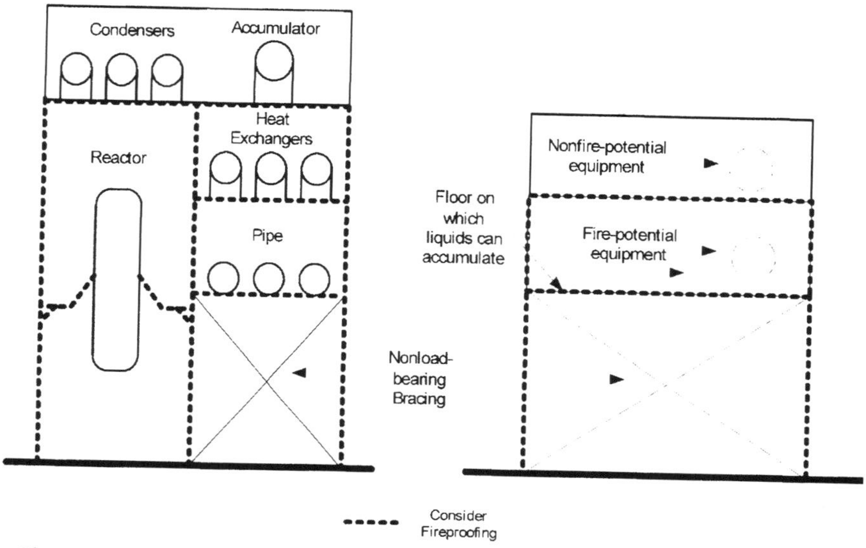


**Figure 8-6.** Structure Supporting Fire-Potential Equipment in a Fire-Scenario Area

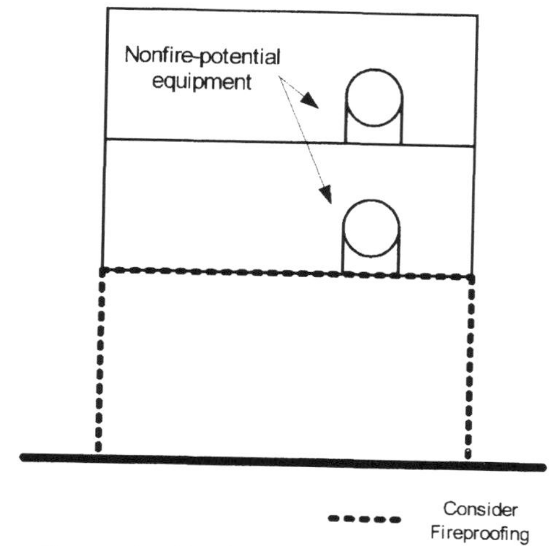


**Figure 8-7.** Structure Supporting Nonfire-Potential Equipment in a Fire-Scenario Area

- When supporting low-fire-potential equipment: Support structure should be fireproofed from its base up to the platform or horizontal bracing plane nearest to an elevation of 40 ft (12 m), but should not terminate below 30 ft (9 m) (see Figure 8-8).
- Knee and diagonal bracing that contributes to the support of vertical loads or stability of support columns should be fireproofed. Knee and diagonal bracing used only for wind loading need not be fireproofed (see Figure 8-6).
- The top surface of a beam that requires fireproofing need not be fireproofed when that beam supports steel flooring or piping.
- Fin fan supports should be fireproofed up to their full load bearing height.

For pipe racks and pipework:

- Main process area pipe racks and interconnecting pipe racks (located in process areas and supporting piping only) should be fireproofed from their bases up to and including the first-level pipe support beams for a 1- to 2-hour rating (see Figure 8-6).
- Wind bracing and structural steel stringer beams running parallel to piping need not be fireproofed.
- Main pipe racks located in process or other fire potential areas that also support equipment such as fin-tube air coolers should be fireproofed with a 2- to 3-hour rating from their bases up to and including the equipment support legs and any horizontal beams transmitting the

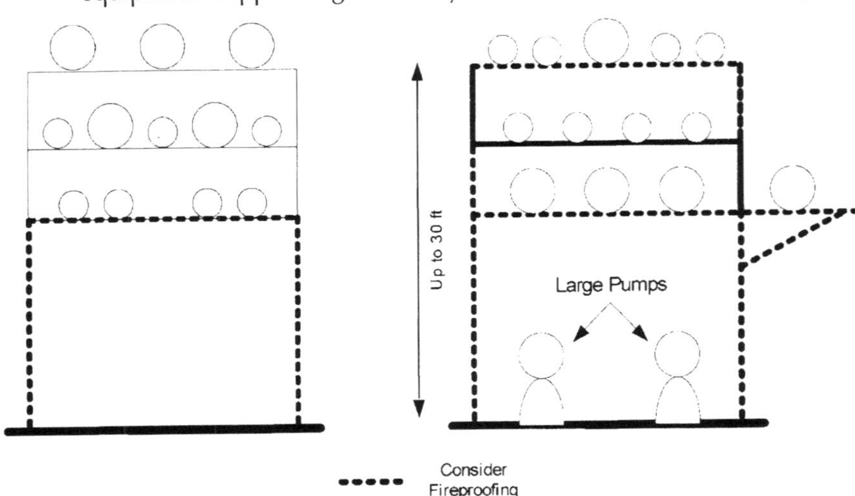


**Figure 8-8.** Pipe Rack with and without Pumps in a Fire-Scenario Area

equipment load to the columns (see Figure 8-6). Short support legs, not over 6 ft (2 m) in height and integral with the fin-tube air cooler should also be fireproofed for a 1- to 2-hour rating.

- Structural steel supports for within-unit process transfer lines should be fireproofed with a 1- to 2-hour fire resistance rating.
- In fire exposed areas, piping supported from spring hangers or rods should be protected against failure of the hangers or rod by a bracket or beam located beneath the pipe (see Figure 8-9). The bracket or beam should be fireproofed with fireproofing material. Sufficient clearance between the bracket and pipe should be provided to permit free movement of the spring hangers.
- For equipment, tanks, and vessels containing flammables or heated combustible liquids, their steel structural supports, legs, and anchors should normally be fireproofed for a 2-hour fire resistance rating. Some typical examples of where fireproofing should be applied are:
  - Skirts supporting columns or towers
  - Support legs for pressure storage spheres
  - Steel saddle supports for horizontal cylindrical vessels when they exceed 2 ft (0.6 m) height, measured at the center
  - Support legs for fin-tube air coolers supported from ground level
  - Support structures for pressure storage cylinders
  - Support legs for fired process heaters

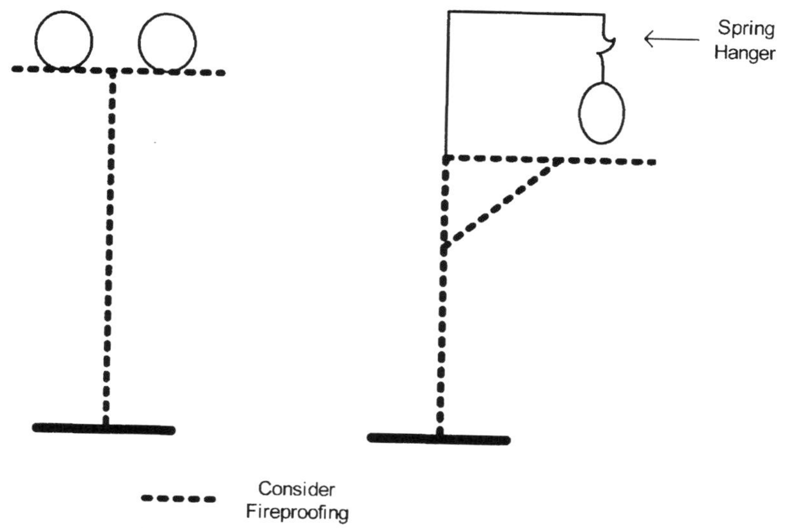


**Figure 8-9.** Transfer Lines in a Fire-Scenario Area

- Special attention is required for the following to ensure that excessive hidden localized corrosion does not occur beneath the fire-resistive insulation due to infiltrated water:
  - Fire-resistive insulation at the bottom of vessels or column skirts should be protected from water absorption by flashing or caulking.
  - Hold-down lugs or anchor bolts for vessels and columns should be coated with corrosion-protective mastic prior to application of the fire-proofing material.

For further guidance on where and how to use fire resistant insulating materials, refer to API Publication 2218, *Fireproofing Practices in Petroleum and Petrochemical Processing Plants* (API, 1999).

#### *8.1.6.3. Water Spray for Structural Steel*

Fire protection for structural steelwork can be provided by water spray that cools and wets fire exposed surfaces. This active form of fire protection can be used on vertical structural steel columns, horizontal supports and other steelwork.

Perimeter vertical columns and other columns that support significant equipment or building loads are usually selected for protection. Typically, water spray is applied to each column with sufficient flow rate to wet and cool the exposed steel surfaces. The water spray is directed onto the column's web at each level of the process structure, or at every 15 ft (4.6 m) of elevation, on alternate sides of the column.

Horizontal, stressed primary structural steel members can also be protected by water spray directed onto the surfaces exposed to fire. Stressed primary horizontal members may include beams connecting perimeter columns as well as those supporting significant equipment loads or connecting to other important vertical members of the process structure.

Fixed water spray systems designed on an area coverage basis may also be used to wet/cool structural steel supports. In this case, the placement of discharge nozzles should be close, usually within 4 ft (1.2 m) of the steelwork being protected. Alternatively additional nozzles or a separate system may be provided.

A generalized comparison of equivalent fire protection for structural steel using water spray rates recommended in NFPA 15 and fire-resistive insulation used alone or in combination is provided in Table 8-9 (NFPA 15).

### *8.1.7. Manual Firefighting Equipment*

Manual firefighting capability is a general requirement for each process structure or area, regardless of any fixed fire protection that may be provided. See Chapter 7, Section 7.4.16.

In process operations where relatively small flammable or combustible liquid leaks are expected or considered possible, hand-held fire extinguishers

**Table 8-9**
*Water Spray and Fireproofing for Structural Steel Protection*

| Type Protection | Vertical Structural Steelwork | Horizontal Structural Steelwork | Comments |
|---|---|---|---|
| Water spray only | 0.25 gpm/ft² (10.2 lpm/m²) of wetted surface | 0.10 gpm/ft² (4.1 lpm/m²) of wetted surface | Active protection; dependent on detection, actuation and fire water supply systems. |
| Fire-resistive insulating material only | 2 to 3-hour rated fireproofing | 2- to 3-hour rated fireproofing | Passive protection; potential for unseen corrosion. |
| Combination water spray with fire-resistive insulating material | 1 to 1½-hour rated fireproofing plus water spray (as above) | 1 to 1½-hour rated fireproofing plus water spray (as above) | Active protection; but the passive fireproofing allows some time of protection in event water system fails. |

of an appropriate type and size should be provided to permit extinguishment of small fires. Hand-held fire extinguishers may also be placed on each level of a multifloor process structure, preferably near the stairways.

Fire monitors should be provided and positioned so that at least two monitor streams can reach each process area, preferably from opposite sides. Often, fire monitors are mounted directly on hydrants. Care should be taken to select and position fire monitors to ensure that their streams can reach elevated equipment. The local fire main's residual pressure under flowing conditions should be determined to ensure it is adequate to deliver effective water streams capable of reaching all elevated equipment. While monitors need to be close enough such that their stream reaches elevated equipment, if they are too close it may not be possible to physically access the monitor to direct its stream due to radiant heat. In these cases, some form of personnel shielding or remote operation is required.

### 8.1.8. Process Vessels

Fire protection can be provided to process vessels with either manual firefighting or fixed water spray systems. Manual firefighting with monitors and hoses can be used to protect process vessels against exposure to fire. See Chapter 7, Section 7.4.3. If water spray is used, it should be applied to all outer surfaces at a rate of 0.25 gpm/ft² (10 lpm/m²) of projected (surface) area. See Chapter 7, Section 7.4.8. Multiple nozzles are typically required. Where water spray from upper nozzles can flow down the sides of the vessel, the nozzles or rows of nozzles can be spaced up to 12 ft (3.7 m) apart vertically to take advantage of this water spray "rundown." Credit for spray rundown cannot be taken for surfaces below the equator of a spherical or a horizontal cylindrical vessel. For spherical or horizontal vessels, additional rows of nozzles should be provided for the lower halves of the vessels (NFPA 15, 2001).

### *8.1.9. Columns, Scrubbers, and Reactors*

#### *8.1.9.1. External Fires*

For those process vessels containing a significant quantity of flammables or heated combustibles, a combination of passive and active fire protection should be considered in addition to manual fire fighting. This additional protection is usually in the form of fireproofing and fixed sprinkler or directional water spray. Additional vessel protection should be applied as follows:

- External sprinkler or directional water spray fire protection at 0.25 gpm/ft$^2$ (10.2 lpm/m$^2$) of exposed surface to:
  - A height of at least 10 ft (3 m) above the vessels internal liquid operating levels (liquid hold-up in distillation column trays is not included)
  - 40 ft (12 m) above any surface surrounding the vessel where a substantial liquid spill may accumulate
- For vertical and horizontal supporting steel members by:
  - Applying fire-resistive material to the steel members from grade level up to and including the level closest to 30 ft (9 m) in elevation
- For the supports of self-supporting skirted columns, towers, and similar vessels by:
  - Appling fire-resistive material to the external skirts and anchor bolts to provide a minimum 1½-hour protection duration
  - If the skirt has more than one access opening or a single opening exceeding approximately 18 in (0.5 m), protect the inside of the skirt by either:
    - Fire-resistive material
    - Water spray at 0.25 gpm/ft$^2$ (10.2 lpm/m$^2$) of exposed surface

Except that skirts less than 4 ft (1.2 m) in diameter with only one access opening need not be fireproofed on the inside surface.

#### *8.1.9.2. Internal Fires*

Columns, scrubbers, reactors, and similar process vessels, particularly those containing packed beds, either structured or random (or loose-fill), can be susceptible to internal fires when they are opened to the atmosphere for internal repair, packing replacement or other work. This susceptibility to fire is clearly a concern for process vessels containing combustible plastic packing. However, fire hazards should also be a concern for vessels equipped with normally non-combustible ceramic and metal packing materials as they may retain combustible, pyrophoric or other process residue that could initiate or fuel a fire.

#### 8.1.9.2.1. PACKING MATERIALS

By design, packing materials have a large surface area in order to perform their process function. These large packing material surface areas may retain a significant quantity of process residues or byproduct coatings that may be difficult to clean and remove. As a fire prevention precaution, packing should always be considered to contain a combustible fuel source. Even new metal packing probably contains a manufacturer's protective oil coating that could ignite.

In addition to process residues, the packing materials themselves may be combustible. Plastic packing materials are clearly combustible, but metal packing should also be considered a potential fuel source. Metal packing is typically manufactured in very thin sections and may be thinned further by corrosion or erosion during service or may even deteriorate into small pieces. The ignition temperature of many metals is significantly lowered as the ratio of surface area to mass increases. The amount of plastic or metal packing materials in vessels with packed beds can be a significant fuel source.

Carbon steel, stainless steel, aluminum, tantalum, titanium, and zirconium are used for both loose-fill and structured metal packing. All have reportedly been involved in packing fire incidents. Titanium, in particular, has been involved in a number of column fires.

#### 8.1.9.2.2. IGNITION SOURCES

Potential sources of ignition for internal fires are often related to ignition of pyrophoric material, hot work during inspection, maintenance, or repair, such as welding, use of cutting torches, grinding, etc. In some incidents, the ignition source could not be clearly determined. Some incident investigation reports and authors have suggested causes related to autoignition and other phenomena, including thermite reactions involving process or byproduct residues or "cool flame ignitions" (Coffee, 1979).

Once ignited by an outside heat source and possibly fueled initially by process residue, metal packing materials can burn with extreme vigor producing temperatures in excess of those from flammable liquid fires. Metal fires (Class D) are difficult to extinguish even with special methods and preplanning. The probability of significant damage is high once a metal packing fire has begun inside a vessel. Metal firefighting methods include preplanned capability for rapid high volume flooding of the interior space with water, steam, or inert gas. Prevention methods include continuous inerting or water flooding over the packing zone and removal of the packing before performing hot work of any type inside of the vessel.

The following are examples of incidents involving metal packing fires in packed columns:

- 01/16/01—a fire in a 73-ft-tall, 42-in-diameter packed column involving titanium packing; column was burned through in two places and destroyed.
- 02/11/01—a fire in a 250-ft-tall, 26–29-ft-diameter packed column during the replacement of carbon steel structured packing; column collapsed.

#### 8.1.9.2.3. PROTECTING AGAINST PACKED BED FIRES

If at all possible, hot work should not be performed above or below any type of packed beds inside of process vessels. Where possible, consideration should be given to removal of the packing prior to performing hot work.

When maintenance involving hot work must be performed inside a vessel with the packed bed in place, the following fire protection measures should be considered:

- Perform a thorough safety and fire protection review including the prevention of and response to metal fires, pyrophorics, and catalysts.
- Adhere strictly to facility hot work procedures with particular attention to any work task that may produce heat or sparks.
- Keep the packing wet with water:
  - If the vessel and its external support structure were designed for water-filled loads, flood the vessel with water before start of hot work to a level just below the work area
  - Provide a well distributed spray or flow of water over the packing below the work area
- If the vessel's internal packing supports can handle the additional weight, create a shallow pool of water above the packing and below the work area by using plastic sheeting appropriately supported by metal sheets cut to fit the interior of the vessel.
- Isolate the packing from the hot work area using physical barriers or noncombustible wetted blankets.
- Position a fire watch at each opened manway into the vessel with a pressurized fire water hose available.
- Work under inert atmosphere.

*Note*: Titanium is reported to burn in pure nitrogen or carbon dioxide atmospheres (NFPA 484).

### 8.1.10. Isolation Valves

Isolation valves should be installed to reduce the amount of material being fed to a fire or released to the atmosphere. Isolation valves should be considered for:

- Flammable liquid volume of 1,300 gal (5,000 l) in columns and reactors (the reduced volume will take into the amount on trays) for C4 and lighter material.
- Flammable liquid volume of 2,000 gal (8,000 l) for accumulators and drums for C4 and lighter material or materials above their autoignition temperatures.
- Flammable liquid volume of 4,000 gal (16,000 l) in process vessels for materials above their flash point.
- Compressors with drivers over 150 kW that handle flammable gases (the compressor driver should be shutdown through the interlock system before the valves are closed).
- Flammable liquid storage vessels that exceed 20,000 gal (80,000 l).

One company has developed a flowchart for determining the need for isolation valves. Their method is shown in Figure 8-10.

Isolation valves should be easily accessible under adverse conditions or valve should be remotely operable. The isolation valve should be fire rated and the actuator and power cables should be fire proofed. Twenty minutes of fireproofing is required when the design is not "fail closed." Isolation valves can serve a dual function, such as equipment isolation. The isolation valves should be located as close to the outlet flange of the vessel as possible.

Normally, isolation valves should be designed to fail closed. However, depending on the process and operating conditions, the valve failing closed may create more problems and the failure position should be reviewed by process personnel. Examples of where the valve should fail open include on:

- Hydrogen recycle line for a reformer unit, to cool the catalyst
- Cooling water systems, to ensure availability of cooling water
- Polymerization reactions, to provide pressure relief

See Section 7.1.2 for additional information.

### 8.1.11. Fired Heaters

Many processes are heat driven, take place at elevated temperatures, or require product drying. As a result, process heaters and dryers are common equipment in processing facilities. Many of these are fired units fueled by a variety of gas or liquid fuels; frequently by natural gas. They may be used to heat a process stream directly, to heat an intermediate heat transfer fluid, or to

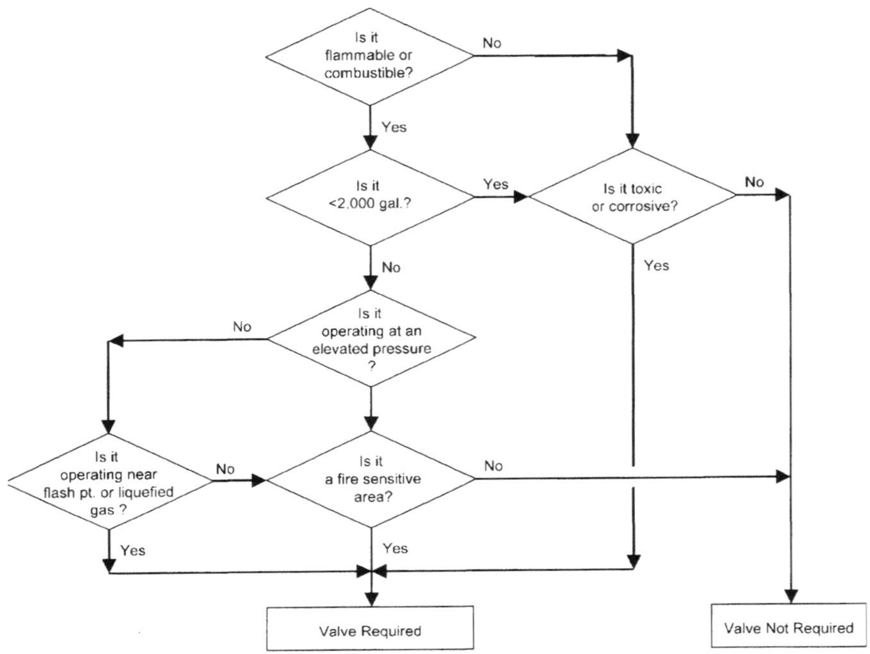


**Figure 8-10.** Isolation Decision Flowchart

generate steam. Secondary heat recovery is common using economizer sections in the heater exhaust gas stream. Statistically fired heaters and boilers have the greatest frequency of fires of all common types of process equipment.

For process liquid and gas stream heating, most designs heat the process stream as it flows through tubes that pass through fireboxes, convection sections, or combustion gas stacks, although a few fire-tube heaters exist.

Fires involving liquid process streams are the most common heater loss. Most involve a ruptured process stream tube leading to a firebox fire or a pool fire under or near the heater. The two most common causes are failure of the tubes due to overheating and rupture of the tubes as the result of a fire box explosion.

When overheated, hydrocarbons tend to break down, leaving carbon residues (coke). This coke builds up on the inside of the heater tubes, slowing the transfer of heat from the tube walls to the product by restricting the flow of product and acting as an insulator. As the control system attempts to maintain the process outlet temperature at the setpoint, the fuel valves will open and the tubes subjected to an increased heat load. With the diminished ability of this heat to be transferred to the process fluid, the temperature of the tubing will increase.

If undetected, the process tube temperature can exceed the design rating of the metal resulting in tube failure and rupture, releasing the process stream into the firebox where it can immediately ignite. The fire damage generally progresses by destroying first the heater tubes, then the stack and shell.

Fire box explosions generally occur in fired heaters during startup as a result of fuel being supplied to the burners that exceeds the safe operating limits. This may be due to failure of the air supply, a sudden increase, or decrease, in fuel pressure, changes in fuel composition or operating the system "in manual." There are number of industry standards available that provide guidance upon comprehensive burner management systems to address such scenarios. Historically, the process industry has been reluctant to adopt the more sophisticated protective measures for reasons of reliability, however there are indications that this attitude is changing with the publication of API RP 556.

Solid materials are often dried or heated using combustion gas exhaust from a fired heater as the material is conveyed through a hot combustion gas zone. Solids handling dryers may take a number of forms, e.g., a rotary kiln. Losses involving dryers usually involve internal fires or explosions.

Increasingly, newer fired process heater installations are adding more fuel-air combustion controls and safety instrumentation systems. However, the decision on the extent of fired heater combustion controls, instrumentation, and safety systems to employ is fundamentally a loss prevention and risk tolerance issue, rather than a fire protection one. The following recommended practices, codes and standards apply to fired heater and dryer controls and instrumentation:

- API RP 556, *Fired Heaters and Steam Generators*, provides current guidance on the recommended practice for controls and instrumentation for fired process heaters for liquid and gas stream heating in the petroleum and petrochemical industries.
- NFPA 86, *Standard for Ovens and Furnaces*, provides guidance for ovens, furnaces, and dryers for solid process materials.
- NFPA 85, *Boiler and Combustion Systems Hazards Code*, provides guidance for steam boilers and similar high reliability automatic combustion systems. In general, NFPA 85 combustion system control and safety instrumentation systems requirements exceed those defined in NFPA 86 and in API RP 556.

#### *8.1.11.1. Fire Protection for Fired Equipment*

Fire-resistive insulation should be provided as passive fire protection for critical load bearing heater supports. Critical supports include those for the firebox, the convection section, any "breeching" (combustion exhaust gas connections to a discharge stack) and adjacent piping such as process inlet and outlet piping located within 25 ft (7.6 m) of the perimeter of the heater. Fire-resistive insulation protection should be applied that is sufficient for 1½ -hour duration protection (per UL 1709) if the structure is fully protected by automatic water spray and equipped with adequate drainage. If either is not fully adequate, fireproofing for 2½-hour exposure protection should be applied.

Steam snuffing can be provided for the interior of process heaters handling flammable or combustible liquid or gas product streams. For steam snuffing to be effective, an adequate supply of high-volume steam must be available. NFPA 86, Appendix E provides guidance on adequate flow rates of steam. At least one steam snuffing connection should be provided for each section or module of the heater. These connections should be installed horizontally and in a manner to preclude direct steam impingement upon the heater tubes and refractory. It is common industry practice to provide the ability to manually activate snuffing steam flow to the fire box from a location at least 50 ft (15.2 m) from the heater.

Steam snuffing in solids dryers can also provide effective fire protection. In some cases, the major concern for a fired product dryer may be explosion involving product dust. Where there is a possibility of an internal dust explosion in a product dryer, a fast response explosion suppression system should be considered in addition to fire protection.

Monitor nozzles should be located so that water streams can be manually applied to any outdoor fired heater from at least two different directions. A flow of at least 500 gpm (1,893 lpm) per monitor nozzle should be provided. In some cases, elevated monitors may be required.

### *8.1.11.2. Isolation and Relief Valves for Fired Process Heaters*

Isolation or emergency shutdown (ESD) valves should be installed to stop fuel flow and the process feed flow into the heater in the event of heater tube rupture. These valves can be automatically actuated by controls or safety interlocks or can be manually operated remotely. Remote actuation can be from a control room console or in the field; field actuation stations should be located at least 50 ft (15 m) from the heater. It is also common to provide a manual block valve, located at least 50 ft (15 m) from the heater, on each of the fuel and process feed lines. These should be accessible to operators in the event of an incident involving the heater.

Where process liquid backflow to a ruptured heater tube is likely, a means to block this flow should be provided. A simple method is to install reliable check valves on the outlet lines. Remotely operated block valves on the outlet side can be substituted where process conditions might render check valves ineffective, e.g., coking, or where higher reliability or more positive shutoff is desired. Where it is possible to block both inlet and outlet lines on a heater, enough pressure could build up to rupture the tubes in an upset condition. Therefore, all heaters capable of being isolated should be provided with relief valves.

In each of these process heater isolation applications, fire-rated valves should be used. Fire-rated valves should meet these specifications:

- API Specifications 6FA and 6FC, *Fire Test of Valves*
- API Specification 6FD, *Fire Test for Check Valves*

### *8.1.11.3. Spacing for Fired Heaters*

Fired process heaters are typically an integral part of a process and are required to be close-coupled to other vessels and equipment. Process heaters are regarded as a prime source of ignition within process units, particularly those employing natural draft combustion. Process fluid stream flow rates, temperatures, and physical size of associated equipment are other important parameters in determining an appropriate spacing of fired heaters to other equipment and structures within a process unit. Thus, it is important to provide as much space as possible between fired heaters and other equipment and structures consistent with process related considerations. Heater spacing and separation with respect to other equipment can often be maximized by locating the fired heater at a corner or on the outside edge of the process unit block and in the prevailing upwind direction

Refer to *Guidelines for Facility Siting and Layout* (CCPS, 2003b) for further spacing guidance.

### *8.1.11.4. Curbing and Drainage around Fired Equipment*

Provisions should be made to direct flammable or combustible liquid spills away from fired heaters. Spills from other equipment that flow into or under fired heaters can and have been ignited by the hot surfaces or flames of fired heaters. Conversely, in the past, spills from ruptured tubes in process heaters have allowed burning liquid to flow around and damage other equipment. Both process heaters and nearby equipment should be protected from each other due to possible spills of flammable or combustible liquids using one of the following methods:

- Slope or grade the area to direct liquid spills away from fired equipment and an open area if adequate space is available or to a trapped drainage system.
- Provide curbing with trapped drains around each heater to prevent a liquid spill fire from involving other equipment. Drains should be sized according to heater liquid holdup to minimize pooling of the spill within the curbed area.

- No open process or chemical sewer drains should be under or in the drainage zone from a fired process heater since the heater could ignite flammable vapors present in the process/chemical sewer. Where process/chemical sewer drains are required at a fired heater during cleanup and maintenance, the drain openings should be sealed by a gasketed and bolted cover prior to startup of the heater.

### *8.1.12. Heat Exchangers*

Many different heat exchange requirements exist in processing facilities, involving process-to-process streams, gas-to-liquid streams, and utility streams exchanging with process streams. The design and locations of heat exchangers are equally diverse. Heat exchangers may be mounted horizontally or vertically, at ground level, elevated in process structures, or attached to the process equipment served.

Heat exchangers mounted at or near ground level can be exposed to a spill fire. Their supports are often made of concrete to maximize structural integrity in the event of a spill fire beneath the exchanger.

A similar spill fire exposure may exist for heat exchangers mounted above solid floors in a process structure where the heat exchanger is supported by the process structure steelwork or separate steel legs.

Other heat exchangers are mounted on the item of process equipment being served by the exchanger. Fire protection here is typically directed by the requirements for the major process equipment being served.

For shell-and-tube heat exchangers, limited fire protection may be provided during fire exposure by the liquid in the shell that provides some internal cooling for the shell wall. However, if a spill fire exposure exists and the shell side liquid has low heat capacity, then either water spray or fire resistive material should be applied to the surface of the exchanger itself.

Use of manual fire monitors or area coverage water spray or deluge systems are typically sufficient for the fire protection of heat exchangers. However, where exchanger support saddles exceed 12 in (30 cm) in height, the supports should be protected by either water spray at a density of 0.3 gpm/ft$^2$ (12 lpm/m$^2$) or application of fire resistive material.

Maintenance work on shell-and-tube heat exchangers often involves cleaning the tubes by mechanical drilling or high-pressure water. Some of these drilling practices can cause residue materials in the tubes to ignite. There is also the possibility of the presence of pyrophoric material. Keeping an appropriate liquid flowing on the shell side and having manual firefighting capability available is usually sufficient to control and quickly extinguish any ignition that might occur during maintenance.

Finned-tube, air-cooled heat exchangers are usually placed in elevated locations, often above pipe racks. Finned-tube heat exchangers that are fan-forced are particularly vulnerable to spill fires from below, since the fans can draw the heat up and into the exchanger. Exchanger fans should be provided with automatic shutdown by high air temperature sensors if doing so will not create other process hazards. In order to protect the exchanger from fire related damage, it is often necessary to provide water spray protection under the exchangers. Care should be taken to ensure that the piping on the pipe racks supporting air-cooled exchangers is all welded, and where possible, no equipment handling flammable material is located under fin fans.

### 8.1.13. Pumps

Pumps handling flammable materials represent a significant potential for spill and subsequent fire. This is due to damage to seals and failures of other potential leak points. The first consideration in fire protection for pumps is their location relative to other equipment, vessels, process structures and buildings housing personnel, and key control or utility systems. When locating a pump, consideration should be given to the size, properties of material handled, temperature, and pressure.

In general, pumps handling flammable materials should be separated from:

- Other types of process equipment.
- Fired heaters, field located emergency shutdown or activation switch stations, and unit electrical substations.
- Cooling towers and utility areas (boilers, air compressors, power generation).

Pumps handling flammable materials should not be located under pipe racks as an ignited spill could rapidly allow fire to rise and envelop the overhead pipework. Finned-tube, air-cooled heat exchangers, often fan-forced, are frequently mounted above pipe rack structures.

Refer to *Guidelines for Facility Siting and Layout* (CCPS, 2003b) for further spacing guidance.

The fire potential for pumps in flammable or heated combustible liquid service is influenced by certain design features. Providing pumps with double or tandem seals with leak detection in between the seals can provide early warning that the primary seal has failed. The potential for seal leaks associated with centrifugal pumps can be avoided by utilizing special pump designs that employ a seal-less design, such as canned pumps. Canned pumps are totally enclosed with the drive element internal to the pump housing. Similarly, magnetically coupled pumps avoid the use of seals. This design employs a rotor/impeller enclosed in the pump housing and closely spaced across a thin stainless steel membrane, thus it can be magnetically coupled to a drive rotor on its external electric motor. Canned pumps or magnetic driven pumps do not require water spray protection.

Water spray for fire control should be provided on pumps based on their location or on the materials handled as follows:

- LPG, LNG
- Hydrocarbons at temperatures of 500°F (260°C) or higher
- Hydrocarbons at pressures higher than 500 psig (3,447 kPa)
- Hydrocarbons heated above their autoignition temperatures
- Class I liquids [flash point at or below 100°F (37.8°C)]
- Class II liquids heated above their flash point

The design of fixed water spray fire protection over pumps should provide one or more spray nozzles positioned so that all parts of all objects within the protection pattern are thoroughly wetted and enveloped by the spray. The recommended water spray densities based on the horizontal coverage area is a minimum of 0.5 gpm/ft$^2$ (20 lpm/m$^2$) (API 2030).

The coverage area for water sprays protecting pumps should include the pump casing and the horizontal area 2 ft (0.6 m) from the periphery of the casing. The suction and discharge parting flanges, check valves, gage connections, balance lines, and pump seals should also be included in the coverage area.

### 8.1.14. Compressors

Compressors are of two general classes; *positive-displacement* or *centrifugal*. The positive-displacement compressor simply compresses an initial gas volume to a final smaller volume in a closed containment; a centrifugal compressor compresses gas by changing its velocity. Each of the common types of compressors has their advantages and applications. Drivers for compressors may be either electric motors, steam turbines or gas turbines.

From a fire protection perspective, the primary issues are the gas or vapor handled, their size, operating temperature and pressure, auxiliary systems (lubricating oil), and value or importance to the process they serve.

In general, compressors, like pumps, handling flammable materials should be provided with passive fire protection by spacing and layout by separating the compressors from:

- Other types of process equipment
- Fired heaters, field located emergency shutdown or activation switch stations and unit electrical substations
- Cooling towers and utility areas (boilers, air compressors, power generation)

Refer to *Guidelines for Facility Siting and Layout* (CCPS, 2003b) for further spacing guidance.

Compressors should preferably be located outdoors or, alternatively, in buildings of totally noncombustible construction with adequate ventilation. Concrete or concrete-on-protected-steel should be used for a compressor system supporting structure. Exposed steel construction is acceptable if protected by water spray or deluge system.

Compressors should be provided with a system for emergency isolation/shutdown (Section 8.1.10) and possibly blowdown, both as a form of machinery protection and as part of the overall fire protection of the unit. Release and ignition of lubricating oil is a recognized hazard with compressors. Accordingly, it is important to provide a means of confining lubricating oil in event of a pipe break or other accident. Relief valves on the lube oil pump should discharge to a safe location.

### *8.1.14.1. Fire Protection for Compressors*

In addition to the provision of manual fire protection, water spray or deluge sprinkler protection may be considered for compressors where one or more of the following conditions exist:

- Adjacent equipment handles flammable or combustible liquids or gases
- Compressor has a rating of 150 HP or greater handling flammable gases
- Compressor has an external lubrication system with:
  - Oil flow rate exceeding 25 gpm (95 lpm/m$^2$)
  - Oil capacity greater than 100 gal (378 l). For multiple compressors, the capacity should be considered the aggregate total for all compressors within 25 ft (8 m)

*Note:* Use of water spray is strongly recommended for compressors located in enclosures.

*Note:* Where normal process and compressor operating temperatures are low and there are no nearby hot surface ignition sources, the probability of fire may be low. In these cases, specific compressor water spray protection may not be needed.

There has been a reluctance to provide water spray on compressors or turbines that typically tend to run "hot," with respect to the risk that a sudden deluge of cold water will result in severe thermal shock and cause cracking of the compressor case. However, studies indicate that this hypothesis is not valid (EPRI, 1985). These studies support the use of water spray protection where the fire hazard so warrants. Water mist systems are now being approved for gas turbines and compressors.

For compressors in buildings, the ceiling sprinklers should extend 20 ft (6 m) beyond the compressor and over any part of an associated lubricating oil system.

Manual firefighting capability should be provided by monitor nozzles located within 50 ft (15.2 m) and hose stations within 100 ft (30 m) of the compressor(s). Hose stations are particularly important for compressors located in a building or enclosure.

### 8.1.14.2. *Flammable Service Compressors*

The hazards specific to the handling of a flammable gas in a compressor system are internal and external explosions and jet fires caused by ignition of gas escaping from a leak or break:

- An internal explosion is possible if air is drawn into the system through leaking packing glands, fittings or valves under conditions of negative pressure produced by a suction line obstruction. These explosions frequently lead to external fires.
- The release of pressurized material from a compressor or its associated piping may be subject to delayed ignition, resulting in an explosion (if located in a building) or a flash fire (if located outdoors).
- A high-pressure leak, subject to immediate ignition, could result in a jet fire impinging on other adjacent equipment or piping leading to subsequent failure of those systems.

Where multiple compressors are involved in the same service or duty, separation should be provided between compressors to reduce mutual exposure. Compressor buildings housing flammable service compressors should be provided with a combustible-gas detection system. The system should alarm at a concentration of 20–25% of the LFL and shutdown the compressor at 40–50% LFL (see Section 8.1.3). The shutdown should include closing all inlet and discharge process lines. For reciprocating compressors in flammable service, explosion vents on the crankcases should be provided.

### 8.1.14.3. *Oxygen Service Compressors*

When oxygen compressors are located in a building or enclosure, piping leaks could result in an oxygen-enriched atmosphere with the potential for easier ignition and an intense fire. Other equipment, compressors, or pumps handling flammables should not be located in the same area or building with oxygen compressors. It is advisable to consult with manufacturers of oxygen compressors and other oxygen handling equipment regarding appropriate location and spacing to other equipment and facilities.

Compressor components and downstream piping may burn in an oxygen atmosphere. An internal fire, regardless of origin, could work its way through the system causing extensive damage to the compressor and downstream processing equipment.

Petroleum-based lubricating oils form explosive mixtures with oxygen; they are also highly susceptible to spontaneous ignition in an oxygen atmosphere. Although petroleum-based lubricating oils are not to be used directly in oxygen service, accidents have occurred when the wrong oil was mistakenly used. If a reciprocating compressor is used for oxygen, fires and explosions can occur if petroleum crankcase oil leaks into the compressor cylinder.

### *8.1.15. Cable Trays*

Fire damage to cable trays can be caused by exposure to flames and heat from spill or pool fires below, falling burning liquids from above, thermal radiation from an adjacent fire, or fire originating among the cables themselves. Cable trays and other grouped cable, wire, and nonmetallic tubing runs should be evaluated to determine the potential for fire exposure where warranted by their size and cost or safety-related importance of their service.

Cable trays outside of process areas or unit battery limits are normally not provided with either passive or active fire protection regardless of size. Small cable trays are usually not fire protected regardless of location when they provide service for ancillary equipment or for equipment designed to fail to a safe state on loss of power, control signal or communication.

Primary cable trays inside of process areas, process structures, or unit battery limits should be reviewed to determine if their routing or location presents a potential fire exposure. Potential external fire exposures to cable trays are typically at those locations where flammable or combustible liquids or gases are handled, such as:

- Connections to process structures and for a short distance from the structure, 20–50 ft (6–15.2 m)
- Distribution runs inside or through process structures and areas
- In proximity to fired heaters
- Above pumps/compressors

Cable trays may be protected by water spray or fire resistant material. Large or fully filled cable trays, particularly those carrying power cables, should also be reviewed for cable overloading that could result in wire overheating and internal fire.

#### *8.1.15.1. Water Spray Protection for Cable Trays*

NFPA 15 provides guidance on water spray protection for cable trays and other grouped cable, wire, and nonmetallic tubing runs. These options are summarized in Table 8-10.

In all of the cases where protection from a spill or pool fire is justified, the structural supports for the cable trays also require fire protection. Protection

**Table 8-10**
*Protection Options for Cable Trays Exposed to Spill Fires*

| Source of Exposure | Type of Protection | Sprinkler Density (Net Rate) | Where Applied |
|---|---|---|---|
| Spill fire (no protection) | Water spray cable tray only | 0.30 gpm/ft$^2$ (12.2 (lpm/m$^2$) | Top and bottom (or both sides) |
| | Flame shield below tray plus water spray above | 0.15 gpm/ft$^2$ (6.1 (lpm/m$^2$) | Top only |
| Spill fire (with its own separate water spray protection) | Water spray cable tray | 0.15 gpm/ft$^2$ (6.1 (lpm/m$^2$) | Top only |
| Fire originating in cable tray itself | | | |

can be passive fire-resistive insulation which provides protection for 1½ to 2½ hours, applied to the support structure up to the level of the cable trays or, active fixed water spray.

For very large cable tray arrays and multiple trays and levels, additional guidance can be found in NFPA 850, Appendix C-4, Grouped Cable Fire Tests.

### *8.1.15.2. Passive Fire Protection for Cable Trays*

Passive fire protection for electrical power cables, instrument cables and lines, wiring, and nonmetallic tubing can be provided in a number of ways, including:

- Fire resistive cable insulation
- Cable wrap insulation systems
- Spray on insulating coatings after installation
- Routing underground, buried or in duct banks
- Routing away from potential fire exposures
- Use of flame shields

Flame shields on the underside of cable trays have been used effectively to deflect flames or heat emanating from fires below. Flame shields should be fabricated of $\frac{1}{16}$in (1.6-mm) thick steel plate or equivalent mounted below the cable tray and extending a minimum of 6 in (152 cm) beyond the tray side rails. These shields can improve tray survivability usually in concert with water spray systems. Flame shields alone (with no water spray) can provide only brief fire protection and are not normally used in this manner. Flame shields coated with an approved fire-resistive material designed to achieve a specific fire exposure duration, have been employed in some applications.

Cable tray fires have resulted from the accumulation of process leakage, residues or combustible dusts or debris on top of cable trays with densely packed wires and cables. Covering shields above cable trays have been used to minimize such accumulations; however, such shields should be used with caution as they can limit air circulation and increase cable and wire temperatures in densely loaded trays particularly those carrying power cables. Covering shields may also block fixed fire protection water spray or manually applied firefighting water spray from reaching the cables.

Main banks of aboveground instrument runs in cable trays (such as those coming from the control room and supported on process unit pipe racks) that are located inside process unit battery limits should be considered for fireproofing by one of the following passive methods:

- *Insulation*—Perforated and solid metal instrument trays should be completely enclosed with 1 in (25 mm) minimum thickness insulating block or board. The insulation should be secured with stainless steel bands and should be weatherproofed with vinyl-coated galvanized steel sheet. The bands and sheet should be in conformance. The metal jacket should be secured with stainless steel, sheet metal screws or bands.
- *Intumescent Mastic*—The following techniques may be used:
    - For solid galvanized steel sheet metal trays, Method A or Method B:
      *Method A*—Solid galvanized sheet metal trays that completely enclose instrument cable and tubing should be coated on all sides with an approved intumescent mastic ¼ in (6.4 mm.) dry thickness. This exterior coating should be reinforced with wire mesh in accordance with manufacturer's recommendation. Prior to application of the mastic, new galvanized trays should be mechanically abraded by grit blasting or other means to ensure adhesion of the prime coat.
      *Method B*—This is the same as Method A except that 1 in (25 mm) thick mineral wool insulating board, with a density of 6 to 8 lb/ft$^3$ (96 to 128 kg/m$^3$), should be glued to the bottom of each tray to separate the instrument lines from the sheet metal tray. Also, the outside of the tray should be coated with approved intumescent mastic $\frac{3}{16}$in (4.8 mm) dry thickness.
    - For perforated, solid, and ladder-type trays, a box with a removable cover should be fabricated with 0.04 in (1.1 mm) minimum diameter, galvanized hardware cloth with ½ in by ½ in (12.5 mm by 12.5 mm) openings. The box should be lined on all sides with 1 in (25 mm) thick mineral wool insulating board with a density of 6 to 8 lb/ft$^3$ (96 to 128 kg/m$^3$). The outside of the mesh should be sprayed with mastic $\frac{3}{16}$in (4.8 mm) dry thickness. The entire assembly should be prefabricated,

sprayed, and field installed. Prior to installation, the instrument cable or tubing should be wrapped with 1 mil thick aluminum foil.

- *Subliming Mastic*—Solid galvanized steel sheet metal trays that completely enclose instrument tubing can be coated with mastic for a 15-minute test rating. Two weather-protective top coatings should be applied. Prior to the application of the subliming mastic, the galvanized steel sheet metal should be primed with one coat of primer. The mastic coating should be reinforced on all four corners of the cable tray by application of a 5½ in (140 mm) wide strip of open-mesh fiberglass cloth between coats of mastic.
- *Mineral wool/cement board*—Perforated and solid sheet metal and ladder-type cable trays can be completely enclosed with 1 in (25 mm) minimum thickness mineral wool/cement board panels, or equivalent. The panels should be secured with stainless steel screw and stainless steel bands ¾ in (19 mm) wide by 0.02 in (0.5 mm) thick, located no more than 18 in (450 mm) apart. In freezing climates the fireproofing panels should be given a weatherproof coat of paint. The structural strength of the metal cable tray should be adequate to support the weight of the fireproofing.
- *Prefabricated fireproof wrap-type systems*—Prefabricated fireproof wrap-type systems that meet the 15 minutes protection may be considered for use on instrument runs. These systems should be composed of high temperature insulating material that is wrapped around the instrument runs and is readily removable for maintenance activities.
- *MI electric cable*—For critical service applications, mineral insulated (MI) instrument cable may be used in lieu of fireproofing. Magnesium oxide insulation should be stable at 3,000°F (1,650°C). An outside metal sheath of copper or stainless steel should be provided. High-pressure, liquid-tight fittings should be used.

### 8.1.16. Pipe Racks and Piping

Piping in processing facilities may be routed and supported in several ways; the most common are:

- Overhead pipe support structures or pipe racks
- On grade supported by short supports or sleepers
- Below grade in open pipe trenches
- Buried subsurface

Grade level piping that is supported by sleepers has the potential to be immersed in a pool fire during an incident. Choice of routing and provision of good drainage away from the piping are the most effective means of protection. Fire protection for a pipe rack is usually based on an assessment of hazards and risk. Protection is often provided on an individual basis for important pipe racks in the vicinity of the process structures, buildings or major process vessels they serve. Where fire protection for a pipe rack is determined to be needed, the protection can be based on water spray, fire resistive coatings or a combination. Pipe racks outside of process areas or unit battery limits are normally not provided with either passive or active fire protection regardless of their size or number of lines carried.

Pipe racks and load-bearing piping supports located in areas with a high fire exposure can be protected by one or more of the following passive or active features:

- Drainage to prevent flammable or combustible liquids from accumulating under the pipe rack
- Fire-resistive construction
- Water spray systems designed and installed in accordance with NFPA 15

### *8.1.17. Pipe Trenches*

A "pipe trench" is an installation of process and utility piping on supports within a below grade, open excavated trench. Pipe trenches may provide some protection from damage due to explosion overpressures or explosion fragments or missiles.

Pipe trenches have been used in large multiunit processing facilities for multiple runs of larger size process piping connecting between units, leading to or from tank farms, and for utility services. Pipe trenches should be provided with drainage; they should not themselves be used as a part of the drainage system for large process spills, fire water, or rain water since such use could expose the piping contained in the trench.

Processing facilities have experienced several serious pipe trench fires. Contributing to the size of the fires were inadequate or plugged drains, lack of isolation valves, pipelines on the ground, or inadequate fire stops along the length of the trenches. Pipe trench fires can result in significant business interruption.

The surface below pipes in a pipe trench should have a 1% slope toward one side of the trench and toward a drain point so that spills will quickly drain away from the pipes. A drainage channel 10 ft (3 m) from the edge of a pipe trench is desirable for fire protection.

Where it is necessary for a drainage channel to cross under pipes in a pipe trench, the drainage channel should be covered to prevent burning liquids in the channel from endangering the piping. Additionally, a common practice is to fireproof the pipework supports for 20–40 ft (6–12 m) on each side of the drainage channel crossing if failure of the supports would result in piping damage.

Solid transverse barriers should be installed as fire stops at 300–500 ft (91–152 m) intervals in main below grade pipe trenches to prevent spills from leaking or broken lines from spreading to the entire pipe trench. If a spill ignites, the barrier can prevent the spread of fire to other sections of the pipe trench. A drain inlet should be provided in each section of the pipe trenches to carry away the flow of leaks and fire water, if a fire should occur.

## 8.2. Storage

### *8.2.1. Storage Tanks*

Tank storage of flammable and combustible materials normally represents the largest inventory of such materials in a process facility. History indicates that once a fire occurs in storage facilities, the event can be catastrophic and the loss substantial. For this reason substantial effort needs to be devoted to both minimizing the possibility of such events occurring and ensuring that the appropriate resources are available should a fire occur. The main components of this prevention and mitigation strategy are:

- Spacing and layout.
- Dikes, drainage, and remote impounding
- Selection of storage tank type to minimize potential for fire
- Fixed active fire protection systems incorporating both water spray and foam application
- Proper maintenance of tank roofs and drains
- Manual firefighting, using both water and foam application

Table 8-11 (API 2021) provides guidance by tank type on the potential types of fires and some general comments.

#### *8.2.1.1. Dikes, Drainage, and Remote Impoundments*

Dikes, drainage, and remote impounding should be provided in accordance with NFPA 30. Typically, companies will use either dike or remote impounding for controlling spills of flammable materials. Spills that become fires need to be controlled to limit the fire spread to other storage areas and process units.

Remote impounding is preferred over diking, however, in most cases sufficient real estate is not available for remote impounding. If diking is chosen, then drainage away from the tanks should be provided.

**Table 8-11**
*Tank Types and Fire Potential*

| Tank Type | Potential Type(s) of Fire | Comments |
|---|---|---|
| Fixed (cone) roof tanks | ▪ Vent fire<br>▪ Overfill ground fire<br>▪ Unobstructed full liquid surface area<br>▪ Obstructed full liquid surface fire if frangible roof remains partially in tank | ▪ For volatile liquids, the rich vapor space typically prevents ignition within the tank.<br>▪ Environmental regulations typically prevent storage of Class I flammable liquids in larger fixed roof tanks. |
| Vertical, low-pressure fixed roof tanks without frangible roof seams | ▪ Vent fire<br>▪ Overfill ground fire<br>▪ Tank explosion and failure with subsequent ground fire | ▪ Rich vapor space inside of tank typically prevents ignition within tank.<br>▪ Lack of frangible roof seam can result in failure of tank at bottom or side, resulting in significant or total loss of tank integrity, and launching of tank. |
| Internal (or covered) floating-roof tanks | ▪ Vent fire<br>▪ Overfill ground fire<br>▪ Obstructed rim seal fire<br>▪ Obstructed full liquid surface fire | ▪ Many fires in this type of tank occur as a result of overfilling.<br>▪ Tank will be extremely difficult to extinguish if entire liquid surface becomes involved.<br>▪ Fires in tanks with pan type covers can be expected to develop into obstructed full liquid surface fires. |
| Domed (or covered) external floating-roof tanks | ▪ Vent fire<br>▪ Overfill ground fire<br>▪ Obstructed rim seal fire<br>▪ Obstructed full liquid surface fire | ▪ Fires in this type of tank most often occur as a result of overfilling.<br>▪ Tank will be extremely difficult to extinguish if entire liquid surface becomes involved. |
| Open floating-roof tanks | ▪ Rim seal fire<br>▪ Overfill ground fire<br>▪ Obstructed full liquid surface fire<br>▪ Unobstructed full surface fire | ▪ Application of fire water to the roof area should be carefully controlled to prevent overloading and sinking the roof when fighting a rim seal fire. |

Drainage to a remote impounding area may be used to control spillage by directing flammable or combustible liquids away from the tank and to a safer location, if spacing permits. Such systems, per NFPA 30, should provide a slope of not less than 1% away from the tank for at least 50 ft (15 m) toward the impound area and should have capacity of not less than the volume of the largest tank plus anticipated fire water that can drain into it. The drainage system should be routed such that other tanks or adjacent property are not exposed to significant thermal radiation in the event of burning liquid being run to the remote impoundment.

Dikes are provided around most flammable liquid storage tanks. The dike capacity should be sized for at least 100% of the largest tank for multiple diked tanks. Individual dikes should be provided for each tank that is 150 ft (46 m) or greater in diameter that stores crude oil, hot oil over 200°F (93°C), or slop oils. Individual dikes should be provided for each tank that is 200 ft (61 m) or greater in diameter that stores Class I liquids. For tanks with individual dikes the capacity of the dikes area should be sized at 100% of the tank volume. For tank groups, intermediate dikes that are a minimum of 18 in (50 cm) high should be provided. When calculating the dike capacity relative to the spill from a particular tank, it should be noted that the spillage volume that the dike area needs to contain is the maximum volume of the tank less that volume of the tank up to the same level as the top of the dike.

Where tank diameters exceed 150 ft (46 m) or more, the base of a dike should be located a minimum of 50 ft (15 m) from the tank's shell. Minimum height for dikes should be 3 ft (1 m) with a maximum height of 6 ft (2 m). Where this height is exceeded, firefighting may be restricted and additional fire protection may be required. It may be necessary to provide roads on the top of earthen dikes.

Tanks may be equipped with a "water draw" that empties into a catch basin within the dike and adjacent to the tank. To limit exposure of tank valves to this potential fire hazard, the "water draw" catch basin should not be placed directly below or adjacent to tank valves or nozzles.

Drainage of diked areas should be provided to the plant drainage systems and be provided with fire traps. The drainage lines connecting the outlet from the tanks to the plant drainage systems shall be designed in accordance with that system and incorporate the appropriate sealed and vented manholes to act as fire seals. Sump pumps may be necessary where the diked area is below the water drainage level. Normally, all drainage valves are located outside of diked areas, maintained in the closed position, and are the indicating type.

Dike walls should be constructed of compacted earth, concrete, or solid masonry, and designed to be liquid tight. The dikes should be designed to withstand full hydrostatic head, fire, earthquake, wind, and rainfall exposure. Piping penetrations should be provided with liquid-tight expansion joints and sleeves or packing.

Pumps and other equipment are generally not provided inside a storage tank area, but are located in their own curbed area adjacent to the dike.

#### 8.2.1.2. Layout and Spacing

An important consideration for reducing the hazards of tanks fires is escalation of a tank fire to adjacent tanks. One method of reducing escalation is layout and spacing. Generally, storage tanks are not placed higher than process units, so that a spill will not run downhill and impact process units, creating a significant fire hazard.

Spacing between tanks and other tanks, units, buildings, and property lines is another important consideration. NFPA 30, *Flammable Liquid Code*, provides guidance on minimum tank spacing. Additional guidance and discussion is provided in CCPS *Guidelines for Facility Layout and Spacing* (CCPS. 2003b).

Tanks should be arranged so that every tank with a diameter greater than 50 ft (15 m) is directly accessible for firefighting from an access road on at least one side. Tanks 150 ft (45 m) in diameter or larger should be accessible for firefighting from access roads on at least two sides.

#### 8.2.1.3. Types of Storage Tanks

##### 8.2.1.3.1. ATMOSPHERIC STORAGE TANKS

Atmospheric storage tanks are operated at or slightly above atmospheric pressure and are commonly used for the storage of flammable and combustible liquids. A number of designs are in use for above ground storage of liquids at atmospheric conditions; the most common are of a vertical cylindrical design—cone roof, open top, floating roof, covered floating roof, and cone roof with an internal floating cover.

In general, atmospheric storage tanks are limited to a maximum operating pressure of 0.5 psig (3.5 kPa) with the exception that those designed and constructed according to API Standard 650, Appendix F are permitted to operate up to 1.0 psig (6.9 kPa) (NFPA 30, 2000).

Other designs and configurations are occasionally used in small size tanks. The fire protection issues presented in this section have specific or general application to all atmospheric storage tanks. Underground storage tanks will not be addressed in this book, since the processing industry's experience and insurer loss records indicate that these tanks do not present a fire loss or exposure concern and are seldom used in processing facilities.

The principal design standards for above ground atmospheric tanks in common use are:

- API Specification 12B, *Bolted Tanks for Storage of Production Liquids*
- API Specification 12D, *Field Welded Tanks for Storage of Production Liquids*

- API Specification 12F, *Shop Welded Tanks for Storage of Production Liquids*
- API Standard'650, *Welded Steel Tanks for Oil Storage*
- UL 58, *Standard for Steel Underground Tanks for Flammable and Combustible Liquids*
- UL 80, *Standard for Steel Inside Tanks for Oil Burner Fuel*
- UL 142, *Standard for Steel Aboveground Tanks for Flammable and Combustible Liquids*
- UL 2080, *Standard for Fire Resistant Tanks for Flammable and Combustible Liquids*
- UL 2085, *Standard for Protected Aboveground Tanks for Flammable and Combustible Liquids*

#### 8.2.1.3.1.1. Floating Roof Tanks

Floating roof tanks have become increasingly popular as a means of satisfying environmental and safety concerns over cone roof tanks in Class 1 liquid service. The evolution of vapor from the liquid surface of atmospheric storage tanks can present significant environmental air emission concerns and, in some cases, fire safety concerns. Fixed roof tanks, whether vented straight to atmosphere or through a pressure–vacuum conservation vent device, have vapor losses to the atmosphere due to day–night thermal breathing as well as internal liquid level increases. Vapor emission losses from open top tanks are even greater. The US Environmental Protection Agency regulations require that many hydrocarbon liquids be stored in tanks equipped with one of the following:

- Floating roof
- Vapor recovery system
- Equivalent system

Generally, the floating roof tank is considered the least costly and most practical option.

A floating roof can be designed into vertical cylindrical tanks of either open-top or external fixed roof design. The internal roof floats on the liquid surface and is prevented from rotating horizontally by fixed guides attached to the tank wall. The annular space between the floating roof and the tank shell is sealed with flexible, movable seals to control vapor losses and emissions. The upper or secondary seal is generally a wiper type. For the primary or lower seal, there are two designs in common use:

- Pantograph
- Tube-type

There are several designs for the floating element itself. API 650 and NFPA 30 differentiate between these in several ways based on whether the design allows a vapor space beneath the float and on their survivability under severe fire exposure.

- *Contact*—the full bottom surface of the float is in contact with the liquid, examples are:
    - Double-deck and pan types
    - Closed cell honeycomb floating cover of metal construction
- *Noncontact*—vapor spaces exist under the float, examples are:
    - Liquid-tight pontoon and buoy types

Metal double-deck or pontoon type floats designed according to API 650 have sufficient buoyancy to prevent sinking when half of their support chambers or pontoons are punctured or the access hatches are opened.

For reasons of corrosion in the vapor spaces and for ease of construction and cost, some floating cover designs have used nonferrous metals, primarily aluminum. Some designs have used encapsulated plastic flotation materials. These designs are generally considered more vulnerable to damage in a fire than those fabricated of metal components. API 650 and NFPA 30 classify these different designs as:

- *Floating roofs*—designs using all welded steel construction meeting the design requirements in API 650 (Figure 8-11)
- *Floating covers*—designs that use materials other than metal

Tanks that use the floating cover design are not considered true floating roof tanks. Fire protection and firefighting for tanks with floating covers should be planned for full tank involvement.

Typical fires involving floating roof tanks occur at the annular seal where the surface of the flammable liquid may be exposed or vapors may escape. Application of foam to smother a fire at the seal is a proven, effective fire extinguishment method.

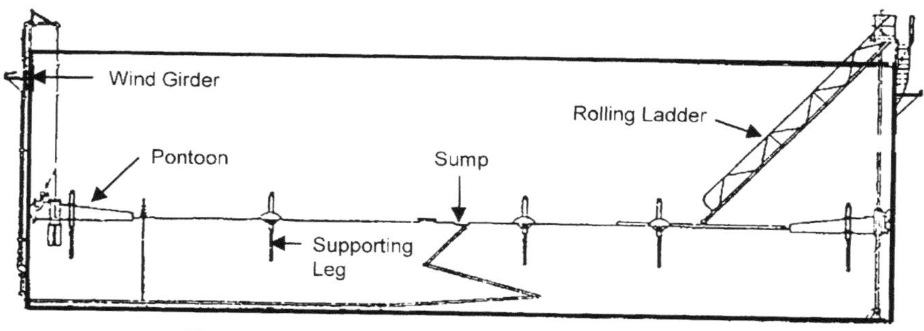


**Figure 8-11.** Open-Top Tanks with a Floating Roof

However, a floating roof can sink or become upset resulting in a very large exposed liquid surface area with evolution of much more vapor product and potential for a much larger fire than could occur in the seal area alone. Thus, the importance of a roof that can maintain buoyancy, such as the pontoon and double-deck design is beneficial. A floater sinking event may require application of much more foam (entire area of the tank) to either suppress vapor and reduce probability of ignition or to control a fire once started. The causes for sinking of floating roofs include:

- Leaks caused by internal corrosion of roof compartments.
- Filling of roof compartments with inadequately drained rainwater (open top tanks).
- Becoming jammed in one position during filling or empting of the tank (open or fixed roof tanks).

#### 8.2.1.3.1.2. Fixed Cone Roof Tanks

The main safety concern with fixed roof tanks is related to the potentially large volume contained between the roof and the surface of the liquid. The potential exists for there to be a flammable mixture present in this space, which, if ignited, will generate significant overpressure due to its confinement. Experience indicates that unless appropriate protective features are provided, the overpressure will cause the tank's flat bottom to assume a convex shape, lifting the walls off the foundation and overstressing the bottom-to-wall joint, leading to its catastrophic failure with sudden release of the tank's entire contents. These, in turn, are usually ignited, resulting in a significant fire. While all such tanks should be provided with a pressure/vacuum vent or flame arrestor (NFPA 30), there are two primary ways of reducing the risk associated with such tanks: incorporation of a weak seam roof design and provision of an internal floating roof.

#### 8.2.1.3.1.3. Weak-Seam Roof Tanks

Fixed cone roof vertical cylindrical tanks in diameters larger than 30–50 feet (9–15 m), should have a weak seam or weak seam roof. These are intended to respond to an internal overpressure by being blown off and allowing the tank to retain its integrity of content containment. Design of an effective weak seam roof becomes increasingly difficult as the diameter decreases.

A weak seam roof design allows tank failure to occur at a predictable overpressure and in a controlled mode at the roof-to-shell or wall joint. In a weak seam roof design, the tank walls are anchored by bolting to the foundation (or by their weight) and an intentionally designed weak seam is made at the roof-to-wall joint. The weak roof-to-wall seam becomes the point of failure allowing the roof to separate, leaving the walls intact to retain the contents. The resulting fire involves the tank's exposed liquid surface rather than a spill fire of its contents. The references cited do not suggest that the weak seam roof design provides overpressure relief for internal tank deflagrations. However, industry experience suggests that this design may provide sufficiently rapid overpressure protection at the roof seam to prevent failure of other tank seams. Further information on weak-seam tank roofs can be found in:

- API Standard 650, *Welded Steel Tanks for Oil Storage*
- UL 142, *Standard for Aboveground Atmospheric Storage Tanks for Flammable and Combustible Liquids*
- NFPA 30, *Flammable and Combustible Liquids Code*

#### 8.2.1.3.1.4. Cone Roof Tanks with Internal Floating Roofs

An internal (or covered) floating roof tank (Figure 8-12) is essentially a conventional fixed conical metal roof vertical cylindrical tank with an internal floating roof. The fixed outer roof provides weather protection for the floating internal roof and its seals extending their life and the outer roof also can significantly reduce the solar heating of the tank's contents. Additionally, a fixed roof tank with an internal floater does not require complicated provisions for draining rainwater and snow melt from the floating roof. The outer roof also eliminates the occurrence of lightning strikes igniting seal fires that occur with some frequency on open top floating roof tanks.

The internal floating roof tank's head space between the floating roof and the fixed roof is naturally ventilated by openings around the perimeter of the tank high on the side walls at the roof seam line and at the top center high point of the conical roof. Under most operating conditions, studies have shown that the head space is sufficiently ventilated to keep the flammable vapor concentration acceptably below the LEL of the liquids stored. However, during the

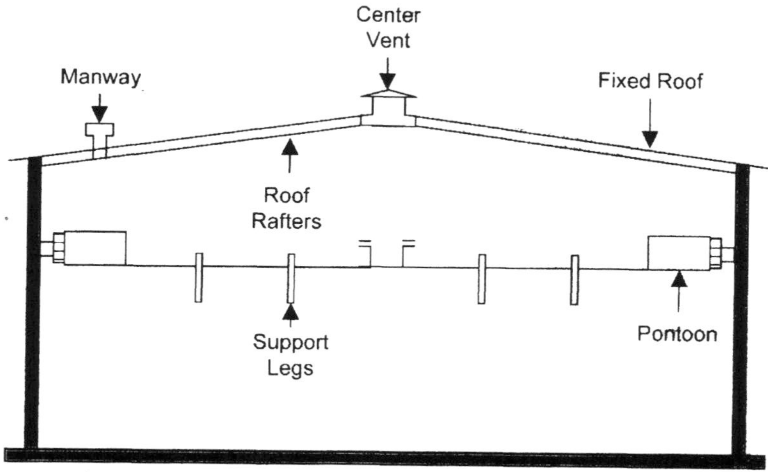


**Figure 8-12.** Conical fixed roof tank with internal floating roof

time that an empty tank is refilling with its floating roof still sitting on its internal support legs, the vapor rich atmosphere under the float is vented into the head space above the floating roof. During this time, head space vapor concentrations can be in the explosive range. As a result, the external fixed roof for conical roof internal floating roof tanks should be of the weak seam design.

#### 8.2.1.3.2. PRESSURIZED STORAGE VESSELS

Pressurized storage vessels are either refrigerated tanks (low pressure) or horizontal vessels and spheres for storing lighter material under pressure so they remain liquids.

For low-pressure tanks <15 psig (103 kPa) the following resources should be considered:

- API 620, *Recommended Rules for the Design and Construction of Large, Welded, Low-Pressure Storage Tanks*
- ASME *Code for Unfired Pressure Vessels*, Section VIII, Division I

For pressure vessels >15 psig (103 kPa) the following resources should be considered:

- Fired pressure vessels should be designed and constructed in accordance with Section I (Power Boilers), or Section VIII, Division 1 or Division 2 (Pressure Vessels), as applicable, of the ASME *Boiler and Pressure Vessel Code*.
- Unfired pressure vessels should be designed and constructed in accordance with Section VIII, Division 1 or Division 2, of the ASME *Boiler and Pressure Vessel Code*.

#### 8.2.1.3.3. REFRIGERATED STORAGE TANKS

A number of flammable liquids and gases used in processing facilities are stored in refrigerated vessels. Common among these are liquefied gases, such as liquefied natural gas (LNG) and anhydrous ammonia, and a number of reactive or self-polymerizing liquids, such as acrylic acid and organic peroxides.

The outside surfaces of refrigerated storage vessels are thermally insulated to assist in maintaining the desired storage temperature. Thermal excursions increase product vaporization, raise vessel pressure, and increase the amount of gas relieved to atmosphere or to a relief vent collection system. Fire exposure of a refrigerated vessel can increase product temperature and vessel pressure, possibly exceeding the capacity of relief valves or relief vent collection system, and could result in vessel rupture with major fire or explosion consequences.

Spacing is the primary means of fire protection for most refrigerated flammable liquid or gas storage vessels. Refer to CCPS *Guidelines for Facility Siting and Layout* (CCPS, 2003b).

Refrigerated vessels handling flammables should also be provided with fixed fire protection in a manner similar to that of other storage tanks and vessels. Water spray protection or fireproofing of vessel surfaces that could be fire exposed should be considered. Where fireproofing is used, it should be specified for a fire endurance rating of 1½ hr.

#### *8.2.1.4. Isolation Valves*

Consideration should be given to the use isolation valves as an aid in the control of spills external to the tanks (pipe leaks, etc.). Types include:

- Remotely operated isolation valves may be installed at the piping connection to the vessel in place of manually operated valves. The advantage of remotely operated isolation valves is that they can be quickly activated when so required.
- Excess flow valves provide automatic isolation when pipe failures occur. For these valves to be effective, the downstream piping must have a flow capacity greater than the set point of the excess flow valve. The main advantage of these valves is that they operate automatically to stop leaks and do not require fire conditions to be present for them to close. For example, in LNG service, excess flow valves can limit the maximum possible leak rate to protect nearby areas.
- Heat-activated valves or other types of valves that close automatically when exposed to fire ensure that the tank will be isolated from the piping during a major fire. They operate regardless of leak rate and whether or not the pipe in which they are installed is the source of a spill. Additional advantages are that they require no instrumentation, utilities, or operator intervention and can be very reliable. The main disadvantages are that they do not operate until the fire is already in progress, and they may shut off against incoming pumped flow with resulting pressure surges unless designed for a timed close rate.

#### *8.2.1.5. Storage Tank Fire Protection*

Firefighting foam is the extinguishing agent typically utilized for flammable and combustible liquid storage tank protection. Protection is through fixed and manual systems for tanks. Such systems require:

- An adequate water supply
- An adequate supply of foam concentrate with proportioning device(s)
- A means of foam application to the tank in the proper amount and proportion

The specific design of each system will vary based upon the size and type of tank being protected, the type of system (fixed or semi-fixed) on the tank, and the product being stored in the tank. The recommendations of NFPA 11 should be followed in the design and installation of foam systems for tank protection. Refer to Chapter 7 for design information on foam systems for storage tanks.

The latest research on storage tank fire protection is the LASTFIRE Project conducted in the UK. This joint industry project reviewed the fire-risk of large, open top floating roof storage tanks. For additional information on this project, refer to http://www.resprotint.co.uk/.

#### 8.2.1.5.1 ATMOSPHERIC STORAGE TANKS

Table 8-12 summarizes the protection measures that should be provided for vertical atmospheric storage tanks to protect them against fire risks. Fixed or semi-fixed fire protection for cone-roof tanks should be provided as follows:

- For tanks containing liquids with flash points greater than 140°F (60°C), no protection is required unless, under normal operation, the material entering the tank or in the tank is heated to within 30°F (15°C) of its flash point.
- For tanks containing liquids with flash points less than 140°F (60°C), a subsurface foam system should be considered.

Use of subsurface foam delivery systems for fixed roof vertical tanks is an option but there are limitations, refer to Section 7.4.10, *Foam Systems*.

A fixed roof tank with an internal floating roof should have fixed fire protection only for the floating roof seal area when the tank design satisfies the following criteria:

- Tank has fixed metal roof with side wall and roof top ventilation that meet API Standard 650.
- A closed-top pontoon or double-deck metal internal floating roof that meets API Standard 650.
- A metal floating cover supported by liquid-tight metal pontoons or floats that provide sufficient buoyancy to prevent the tank's liquid surface from being exposed when half of the flotation is lost.

A cone roof tank with an internal floating roof that does not meet these criteria or uses plastic foam for flotation, even if encapsulated in metal or fiberglass, should be fire protected by side wall foam chambers suitable for the full (surface) area of the tank. This type of internal floating-roof installation is not recommended, particularly for flammable liquids.

Fixed monitor nozzles can be considered as the primary means of protection for fixed-roof tanks up to 60 ft (18 m) in diameter. Foam hand lines should not be considered as the primary means of protection for fixed-roof tanks over 30 ft (9 m) in diameter or those over 20 ft (6 m) in height.

**Table 8-12.** *Protection Measures for Atmospheric Storage Tanks*

| Tank Type and Diameter ft (m) | Number of Hydrants Required | Foam Protection Recommended | Tank Shutoff Valves Required | Lightning Protection | Dike for Liquid Volume of Largest Tank | More than Two Tanks within a Common Diked Area | Overpressure Protection |
|---|---|---|---|---|---|---|---|
| *Cone Roof (Storing combustible liquids >30°F below flashpoint)* | | | | | | | |
| <120 (<37) | 4 | No foam system recommended. Fixed foam system may be provided if desired. | Yes | Provide bonded metal roof or use rods, masts, or conducting wires. | Yes | Provide a drainage channels or curbing if largest tank is 150 ft (46 m). If tank diameter exceeds 150 ft (46 m), provide intermediate dike with capacity of 10 percent of largest tank volume for Class I, II, and IIIA liquids. | |
| 120 - 200 (37 - 61) | 8 | Foam protection for Class I and II liquids. | Yes | | Yes | | |
| >200 (>61) | 16 | | Yes | | Yes | | |
| *Cone Roof with Internal floating roof** | | | | | | | |
| <120 (<37) | 4 | No foam system recommended. Fixed foam system may be provided if desired. | Yes | Provide bonded metal roof or use rods, masts, or conducting wires. | Yes | Same requirement as for cone roof tanks. | Provide weak roof seam or vent according to API Std 650. |
| 120 - 200 (37 - 61) | 8 | Foam chambers for Class I and II liquids. | Yes | | Yes | | |
| >200 (>61) | 16 | | Yes | | Yes | | |
| *Floating Roof* | | | | | | | |
| <120 (<37) | 4 | Portable hose lines for seal fire protection. | Yes | Bond and ground all metal tank parts. Provide shunts where required. | Yes | Same requirement as for cone roof tanks. | None required |
| 120-150 (37-46) | 4 | Portable hose lines from tank standpipe for seal fire protection. | Yes | | Yes | | |
| 150-200 46-61 | 4 | Foam chambers for seal fire protection. | Yes | | Yes | | |
| >200 (>61) | 8 | | Yes | | Yes | | |
| *Covered Floating Roof* | | | | | | | |
| <120 (<37) | 4 | No foam system recommended. Fixed foam system for seal fire may be provided if desired. | Yes | Tank design provides lightning protection | Yes | Same requirement as for cone roof tanks. | Vent according to API Std 650. |
| 120 - 200 (37 - 61) | 4 | | Yes | | Yes | | |
| >200 (>61) | 8 | | Yes | | Yes | | |

*When internal consist of either a thin metallic skin on floats or an aluminum honeycomb sandwich panel, foam protection guidelines for covered floating roof tanks apply. The exception is that aluminum thin-skin covers on floats require full liquid surface area foam application when foam protection is provided.

Where water supplies are inadequate for conventional firefighting and foam making, automatically activated fixed clean agent or $CO_2$ systems may be considered for seal fire protection. The agent should be discharged into the seal area below the secondary seal. Fire detection options for these automatic systems include reusable thermal wires and pneumatic tube devices. Thermal wire is typically the more economical choice.

#### 8.2.1.5.2. PRESSURIZED VESSELS

The effect of fire exposure is predictable for pressure vessels, such as, spheres, spheroids or horizontal vessels. If no fire protection is provided or is not adequate or inoperative, the vessel will probably fail catastrophically in a prolonged fire. Vessel failure typically results from excessive metal temperature weakening the tank wall above the liquid level of its contents. This weakening can occur within a few minutes if the initial liquid level is significantly below the maximum flame height and the flames impinge on the shell.

Three basic fire emergency situations that deserve special consideration regarding pressurized vessels are shown in Table 8-13. They are listed in increasing order of severity.

Water applied to the outside of a vessel exposed to a fire serves the following purposes:

- Slows the loss of product vapor thus retaining a higher liquid level in the vessel and delays the risk of direct flame impingement on metal vessel walls unwetted by liquid contents.
- Directly cools the vessel metal wall by spray impingement and rundown and, in sufficient quantity, can prevent vessel failure by reducing excessively high metal wall temperatures even in an empty, or nearly empty, vessel.

For exposure protection, the decision whether or not cooling water is needed for a pressure storage vessel should be based on the amount of radiant heat it will receive from the adjacent fire of a given diameter and the maximum temperature that the unwetted shell will reach if it is not cooled.

**Table 8-13**
*Situations Requiring Special Consideration*

| Type Fire Exposure | Water Application Rate |
|---|---|
| Radiant heat only with no flame contact | 0.1 gpm/ft$^2$ (4 lpm/m$^2$) |
| Direct flame contact | 0.25 gpm/ft$^2$ (10 lpm/m$^2$) |
| High-velocity jet flame contact | 250–500 gpm (950–1,900 lpm) at point of jet flame contact |

For direct flame impingement, as in the case of a fire around a vessel, there is much greater heat input to the vessel. Therefore, higher water application rates are required to control the rate of temperature rise of the liquid contents and limit the relief valve from lifting or the shell above the liquid level from being weakened by high metal temperatures.

Jet flame contact on the shell of a vessel makes water spray cooling ineffective. The momentum and velocity of medium to large jet flames is such that they will deflect any water spray pattern and thus prevent the local application of cooling to the vessel's shell. The concentrated application of fire water by monitors can provide adequate cooling. Unwetted steel shell subjected to a jet flame can be expected to fail within 10 minutes, thus there are practical difficulties in being able to bring the necessary resources to bear in sufficient time to be effective. Therefore, fireproofing and separation distance are the fire protection options for jet fires.

Spheres, spheroids, and horizontal pressure vessels should be actively protected with cooling water at an application rate shown in Table 8-13 over the entire tank surface including the structural supports and the underside of the tank, including leg area. Adequate coverage should also be provided for the ends of horizontal vessels. This may be accomplished by a water spray, water distribution weirs, monitors or combination of all three. Monitor nozzles and manual hose streams should be provided to supplement the fixed water spray fire protection on the vessel.

Fireproofing may also be used as passive protection for pressure vessels. Fireproofing reduces the fire exposure heat input to the protected vessel and the rate of increase of the vessel wall temperature. Outside surfaces of vessels that may be exposed to fire should be covered with a fireproofing material having a fire endurance rating of 2 hours. Refer to Chapter 7 for additional information on fireproofing.

Where fireproofing is applied using the above guidance, some facilities choose to reduce or eliminate water spray systems where assured effective manual firefighting is available. However, if the fire cannot be extinguished within a reasonable period of time (duration rating of the fireproofing) or the fireproofing fails, then escalation of the incident should be expected.

Information and guidance on minimizing the risk of fires in pressurized vessels is available in:

- API Publication 2510A, *Fire Protection Considerations for the Design and Operation of Liquefied Petroleum Gas (LPG) Storage Facilities*
- NFPA 58, *Liquefied Petroleum Gas Code*
- NFPA 59, *Utility LP-Gas Code*

#### 8.2.1.5.2.1. Burying and Mounding

Another option that has been successfully used for pressurized storage fire protection is burying or mounding.

LPG vessels may be directly buried below grade or mounded above grade to reduce exposure to an external fire. Both of these methods require special precautions, careful preparation, and special design features, since they introduce other substantial hazards. These hazards include undetected corrosion, undetected leaks, and the potential for mechanical damage when the vessel is unearthed for inspection.

#### 8.2.1.5.3. STORAGE TANK FIREFIGHTING

Pre-fire planning and drilling is an important part of any emergency response plan but is particularly critical for ensuring appropriate response to large storage tank fires where there is a great likelihood that substantial support will be required of external resources and responders. Pre-fire plans and drills allow timely identification and resolution of firefighting details such as hose connection compatibility, prearranged staging areas, how and where to run hose etc. See Chapter 11 for additional information.

A combination of low product levels and partial or full surface fires can result in an intense fire which may create up-drafts creating a chimney like effect that may make a portable foam monitor or foam hose streams ineffective. These updrafts will prevent sufficient foam from reaching the product surface and forming a foam blanket. Foam should be applied continuously and evenly to establish a foam blanket. Foam streams should be directed against the inner tank shell so that the foam flows gently onto the burning liquid surface. This may be difficult to accomplish in windy conditions because, depending on velocity and direction, wind reduces the effectiveness of foam streams. Care should be exercised when foam is applied by monitor or hose stream to tanks containing crudes, heavy fuel oils, and similar materials that can develop a boil-over (API 2021, 2001).

Foam management can be a critical part of the firefighting equation. Pre-planning is essential as effective extinguishment can be limited by the amount of foam onsite at the time of an incident. Knowing the amount of foam concentrate necessary to extinguish a given fire, tank specific firefighting tactics, and where to obtain more concentrate on a 24/7 basis are elements of foam management. It is important to have the entire quantity of foam necessary prior to applying any foam to a fire. Attempting an attack with less than the required amount only wastes foam and prolongs the fire.

Portable foam monitors or foam hose streams may be used to extinguish tanks fires. These devices can be used for protection of cone roof tanks not over 50 ft (15 m) in diameter or 30 ft (9 m) high. Foam hose streams are effective in extinguishing floating roof tank rim fires. In tanks up to 200 ft (60 m) in diameter and 42 ft (13 m) high, high-capacity nonaspirating foam monitors have been successful in extinguishing fires that involved the entire tank. However, use of large foam monitors should not be the primary means for extinguishment of large cone roof tank fires.

Despite the limitations, there are sufficient advantages in the use of portable foam monitors to justify consideration for tank fire extinguishment. Some foam trucks are equipped with a 1,000 gpm (3,800 lpm) foam monitor or snorkel articulated boom with a 500 gpm (1,900 lpm) monitor. In addition, a 1,000 gpm (3,800 lpm) portable monitor is recommended and can be towed behind the foam truck for use on tank or spill fires. Monitors have the advantage of fire extinguishment without close approach, resulting in less personnel hazard than the use of foam hose lines.

Also, less time is required to prepare for foam application, resulting in faster extinguishment, especially for numbers of closely spaced tanks up to about 50 ft (15 m) in diameter. Foam monitors are effective in fighting large spill fires, such as those in tank impounding basins. However, industry experience indicates that foam monitors and foam hose streams are ineffective in fighting tank fires in water-soluble materials. It has been determined that foam is only effective on these materials if applied very gently by fixed application devices or hand hose.

Considerable experience has been gained from past incidents of storage tank fires. API 2021 *Management Atmospheric Storage Tanks Fires* contains information useful for tank firefighting.

### 8.2.2. Unstable/Reactive Material Storage

Firefighting for unstable materials is based on generally accepted fire service tactics, provided the unique characteristics of the material are considered. If the material is extremely flammable, then additional spacing distances should be implemented. If it reacts violently with water, other means of extinguishment should be considered. Table 8-14 provides a listing of several unstable, reactive, and water-reactive materials.

*Guidelines for Safe Storage and Handling of Reactive Materials* (CCPS, 1995) provides essential information to assist in materials assessment and analysis as well as general and specific design considerations for storage and handling of reactive materials and the appropriate fire protection.

### 8.2.3. Outdoor Storage

Outdoor or yard storage is frequently used for those materials or containers that are not significantly damaged by exposure to the weather or sunlight and that fit one of the following groups:

- Low fire hazard requiring no protection even if stored indoors.
- Low value does not justify use of building space.
- High fire hazard but low value does not justify costly indoor fire protection.
- Very large size or bulk quantity not practical to store in a building.

**Table 8-14**
*Unstable, Reactive, and Water-Reactive Materials*

| Unstable or Reactive | Water Reactive |
|---|---|
| Acetyl Peroxide | Acetic Anhydride |
| Acrolein | Acetal Chloride |
| Acrylic Acid | di and tri-Chlorosilane |
| Tert-Butyl Hydroperoxide | Diborane |
| Butyl Nitrate | di-Chloro-acetyl-chloride |
| Tert-Butyl Peracetate | Diethyl-aluminum Chloride |
| Tert-Butyl Perbenzoate | Diethyl-aluminum Hydride (*) |
| Tert-Butyl Peroxypivalate | Diethyl Carbamyl Chloride |
| Chloro Nitropropane | Diethyl Zinc (*) |
| Cyanamide | Di-isobutyl Aluminum Hydride (*) |
| Cumene | Di-propyl Aluminum Hydride (*) |
| *p*-Nitroaniline | Ethyl Aluminum Di-chloride (*) |
| *p*-Nitro Chloro Benzene | Ethyl Aluminum Sesqui-chloride (*) |
| Nitroethane | Iron Carbonyl |
| Nitroglycerine | Isobutric Anhydride |
| Nitromethane | Methyl Aluminum Sesqui-bromide (*) |
| *m* and *o*-Nitrotoluene | Methyl Aluminum Sesqui-chloride (*) |
| 2-Nitro-p-toluidine | Methylene Di-isocyanate |
| Peracetic Acid (diluted with 60% acetic acid) | Methyl Isocyanate |
| Piperidine | Sulfur Chloride |
| Propyl Nitrate | Tri-ethyl Aluminum (*) |
| Tetraethyl and Tetramethyl Lead | Tri-methyl Aluminum (*) |
| Tetrafluoroethylene | Tri-isobutyl Aluminum |
| Vinyl Acetylene | |

* *These materials also react with water-based foams and halogenated agents*

Outdoor storage related to processing facilities can encompass a wide-range of chemicals including raw materials, intermediates, finished, and off-specification product in a variety of container types such as drums, portable and tote tanks, and intermodal tank containers. Materials, including spare, surplus, or salvaged equipment, are also stored outdoors.

Failure to properly locate outdoor storage areas can result in fire loss of the stored goods or result in a low value outdoor storage placing a high value processing operation or other important building in danger of fire exposure. Common examples of this are:

- Storing large numbers of wood pallets or empty plastic drums too close to a process structure, packaging, or drumming building or to a warehouse.

- Staging waste solvent or spent pyrophoric catalyst in drums immediately adjacent to the involved operating process structure or area.
- Staging empty metal drums containing residual flammables too close to a waste disposal incinerator, fired process heater or other ignition source.

### *8.2.3.1. Storage Arrangements*

From a fire protection view, all proposed or existing outdoor storage areas should be reviewed, assessed and controlled as to type and quantity of material stored and that the storage area boundaries are well defined. The review and assessment should consider the fire or explosion hazards that the outdoor storage presents to nearby operations or that may be presented by those operations to the storage area itself.

#### 8.2.3.1.1. IDLE PALLETS

Large numbers of wood pallets can present a serious fire threat if stacked outdoors, too close to processing structures and areas, cable trays, pipe racks, tanks, power lines, buildings, and warehouses. Some commonly used separation distances are shown in Table 8-15.

#### 8.2.3.1.2. DRUMS AND OTHER PORTABLE CONTAINERS

When the outdoor storage of significant numbers of drums, portable and tote tanks or intermodal tank containers becomes necessary, the following should be considered in addition to a review and assessment of mutual fire and explosion hazards:

- Hard surface roadway access to permit safe truck, fork-lift truck, and emergency vehicle access.

**Table 8-15**
*Separation Distances for Outdoor Idle Pallet Storage*

| | Minimum Separation Distance ft (m) | | |
|---|---|---|---|
| **Exposed Item** | **<50 pallets** | **50–200 pallets** | **>200 pallets** |
| Masonry wall with no openings | 2 (0.6) | 2 (0.6) | 2 (0.6) |
| Noncombustible wall with no openings | 10 (3.0) | 20 (6.0) | 50 (15.0) |
| Open structure | 15 (4.5) | 30 (9.0) | 90 (27.0) |
| Other similar stacks of pallets | 7.5 (2.3) | 15 (4.5) | 45 (13.5) |

- Hard surface storage pad of adequate size with drainage and curbing to control and direct spills and runoff.
- Containment of runoff as required environmentally or for fire water control.

*8.2.3.2. Manual Firefighting Equipment*

If the outdoor stored material consists of flammables or combustibles in containers or combustible bulk piles, an adequate number of hydrants should surround the area with approximately 200 ft (61 m) between each other and the material to be protected to allow manual firefighting.

*8.2.3.3. Protection from Exposures*

Outdoor storage areas are seldom provided active fire protection by fixed systems to protect them from exposure by events in adjacent areas. However, on a case-by-case basis, specific fixed fire protection systems, often fire monitors, may be added where the outdoor stored material:

- Is combustible
- Is of sufficient value
- Could create exposure to other facilities
- Could harm to the environment or community
- Protection to Exposures

Protection of facilities that are exposed to a fire hazard from an outdoor storage area is preferably achieved passively by adequate separation distance to the potentially exposed facilities. When the quantity of stored material is large, it should be separated into smaller groups or piles with adequate aisle space maintained to prevent uncontrolled fire spread within the defined storage area. An often used spacing guide for stacked or piled materials is aisle width should equal stack height.

## 8.3. Buildings

*Guidelines for Evaluating Process Plant Buildings for External Explosions and Fires* (CCPS, 1996c) has been published to provide a practical approach to identify, evaluate, and manage the explosion and fire impacts to plant buildings and occupants resulting from events external to the building. Using these or other analysis methods, it is recommended that a facility siting study be conducted to assess the external fire and explosion risks for all buildings routinely occupied by plant or other personnel.

### 8.3.1. Control Buildings

Control rooms have evolved into control buildings containing support services and people that may include a break area, restrooms, locker rooms, small laboratories, maintenance shops, conference/training room and offices. These control room/buildings continue to grow in size due to the consolidation of the control of several process units into one control room and the inclusion of their support facilities and staff.

These support areas should be considered as a possible fire exposure to the control room especially since many of these areas are not continuously occupied. This section considers the control room and its support facilities together.

#### *8.3.1.1. Control Building Location*

The location of a control building should be selected with care to ensure that it is exposed to minimal hazards from the processing facility it serves as well as all adjacent units and external facilities.

Additional analyses are required and special building design features may be needed if the control building in its planned or existing location is exposed to more severe hazards, such as:

- Thermal radiation or fire spread resulting from a spill fire hazard involving flammables or combustibles.
- Explosions or ruptures of vessels or process equipment from internal deflagrations, runaway reactions or pressure explosions with possible damaging shock wave and missile ejection.
- Vapor cloud explosion resulting from release of flammable vapor.
- Toxic gas release.

Control rooms serving more than one process unit do not present unique hazards when considering the proper location of the building. However, they can significantly increase the size of a loss because of their multiple uses. For example, an incident in one unit could damage the central control room and impact the ability to operate other units. The business interruption financial impact is likely to far outweigh the actual damage repair costs.

Refer to API RP 752 and *Guidelines for Facility Siting and Layout* (CCPS, 2003c) for additional guidance on control building location.

#### *8.3.1.2. Heating, Ventilating, and Air Conditioning*

The heating, ventilating, and air conditioning (HVAC) system design for any control rooms near process units handling flammable or toxic materials should include these features:

- A manual HVAC emergency shutdown switch in the control room to stop all air handlers and, preferably, close tight-fitting dampers on air intakes and exhausts. (Exhaust fans or blowers in break areas and restrooms should be included in this shutdown and activated by the same emergency shutdown switch).
- A system to maintain a positive pressure of approximately 0.1 inch of water (25 Pa) and provide an alarm for loss of this positive pressure.
- Flammable/toxic gas detection in the air intake.
- Fresh air intake from a safe location to minimize the infiltration of low-lying plumes of released gas or unfiltered air carrying dust. Under most conditions, this elevation should be at least 25 ft (7.5 m) above ground.

#### *8.3.1.3. Smoke, Fire, and Toxic Detection*

If the potential exists for a large process release or spill and the separation distance to the control room building is such that the building is likely to be rapidly engulfed by the resulting flammable or toxic gas plume, then the appropriate type of detectors or continuous analyzers should be installed in the fresh air intakes. The detection system should alarm in the control room. It is advisable that the detection system should automatically close the inlet dampers and put the HVAC into total recycle or stop all blowers. The detection system's logic can be set to provide a warning alarm at a low concentration and automatically perform an HVAC system shutdown at a higher concentration.

Smoke/fire detection should be provided in areas of the building surrounding the control room. Preferably, detection should be provided throughout the building housing the control room. The detection system should alarm locally and in the control room to alert operators of possible fire during off-hours when other building occupants are not present. If the I/O and DCS hardware area is to be served by an HVAC system also serving other control building areas, the system should have a smoke detection system monitoring the return air to the air handler.

The detection system should sound an audible alarm in the control room and be interlocked to shutdown the air handler in the event smoke is detected. When these areas are not served by an independent HVAC system, an approved 1-hour fire damper needs to be installed where the duct penetrates the room closure.

Early detection is essential. The placement of detectors inside cabinets can significantly reduce the time to detection. An early warning high sensitivity smoke detector (HSSD) smoke detection system can be used to detect incipient stage fires and allow planned repairs prior to equipment failure. Hence, fire suppression may not be required.

In plants having a central plant-wide fire and emergency alarm center (fire house or guard control station), consideration should be given to sending the control room/buildings smoke/fire alarms to the central location, as well as to the control room. An alternate alarm monitoring location may be warranted if a control room is expected to have some periods when it will not be occupied, such as during a process shutdown or if the control room has limited staffing and the control operator(s) have some outside duties.

The location of the smoke detectors should be based on an engineering survey of the area to be protected. Factors such as air flow, proximity to air-handling system diffusers and other physical features of the installation need to be taken into account. Smoke tests can be run to verify that the air flow within the protected area favors the smoke detectors.

Additional general guidance on the placement of detectors can be found in FM Data Sheet 5-28, *Smoke Detectors* and in NFPA 72, *National Fire Alarm Code*.

### 8.3.1.4. Control Building Construction and Layout

Only noncombustible materials should be used in the construction of control room buildings. While walls of masonry construction offer greater protection from external fire exposure, control buildings can be of pre-engineered construction if fire or explosion exposure is minimal. When the possibility of damaging explosion overpressure exists, the entire building design must be carefully evaluated.

Exposed ordinary combustibles (Class A) should not be permitted to accumulate in the control room area. Process data printouts, batch records, shipping documents, and other paper in the open should be minimized, preferably limited to one-day's output. Where longer-term storage of such paper files or storage of supplies is required, closed-door metal cabinets should be provided. Metal file cabinets should be provided to store drawings, electrical diagrams, manuals, equipment catalogs, etc.

#### 8.3.1.4.1. INTERIOR LAYOUT AND SEPARATIONS

Interior partition walls should be noncombustible construction, such as fabricated of drywall (gypsum board) on metal studs or equivalent. Wood or plywood wall paneling or other combustible construction materials should not be used. The use of wallpaper or other thin wall coverings is acceptable.

To protect the control room area from incidents occurring in the support areas, the control room area should be separated from the support areas by minimum 1-hour fire-rated partition walls. All penetrations of the walls, floors, ceilings, and roof for cables, etc. should be closed with an approved sealant having a fire resistance rating of at least one-hour. Approved ¾-hour rated (Class B) fire doors equipped with self-closing devices should be installed on all openings between the control room and other portions of the building.

Restrooms, laboratories, and uninterruptible power supply (UPS) systems and battery rooms should have once through ventilation. Laboratories and battery rooms should only be accessed from the outside.

Separate rooms should be provided for I/O wiring panels, the process distributed control computer hardware and related systems, Motor Control Centers (MCC), UPS, and battery rooms. A minimum 1-hour fire-rated partition wall should be provided to ensure isolation of I/O and MCC areas from control rooms.

##### 8.3.1.4.2. RAISED PANEL FLOORS

If feasible, raised panel computer-room type floors should not be used. Where raised floors must be used to accommodate wiring and cables, they should be constructed entirely of noncombustible materials and provided with floor panels easily removable by hand. Power distribution in the subfloor space should be in conduit. Redundant data highway cables should be run as remotely from each other as possible and not mixed with power cables.

The void created by raised panel floors should be provided with smoke detectors and considered as a separate detection zone. The actual design and detection method used depends on several variables including ventilation and routing of electrical/data cables. Passive or active protection may be considered based on the results of a fire hazard analysis.

#### *8.3.1.5. Control Building Fire Protection*

Active external fire exposure protection for control buildings is seldom needed since the primary protection is the passive spacing distance between the building and the process areas and the building's noncombustible construction.

Since control rooms are normally constantly manned, it is considered unlikely that a fire could progress undetected to a hazardous size. Fixed fire protection systems, whether manually or automatically activated, are seldom installed in control rooms of processing facilities, mainly due to the fact that they are normally constantly manned. As such, it is considered unlikely that any fire that does occur would progress undetected to a size that cannot be extinguished by manual intervention (fire extinguishers, hose reel, etc.).

An appropriate number, type, and size of hand-held fire extinguishers should be provided throughout the building to handle ordinary combustibles and electrical fires in the control room and related areas. Typically, clean agent or carbon dioxide fire extinguishers should be provided for electrical and electronic equipment. Dry chemical extinguishers should be avoided because of equipment contamination with powder. For use on ordinary combustible fires in the associated areas water or multipurpose dry chemical fire extinguishers should be provided.

### 8.3.1.6. *Unattended Process Control Equipment Areas*

The distributed control system (DCS) hardware areas are often referred to as "process computer rooms." I/O Rooms contain the incoming and outgoing wiring, cables and data highway links, and often small transformers and other related electrical equipment. Often, additional space is needed for a master process engineering computer terminal/work station for process control system changes and for critical safety instrumented systems (SIS) for interlocks and emergency shutdowns.

The potential for a prolonged process unit outage exists in the event of fire, smoke, or heat damage to the process control system hardware, connecting wiring and data highways, and related systems. While these areas represent a significant capital investment, the business interruption losses associated with their damage could be much greater.

Rooms housing cable and wiring marshalling panels, input/output devices, and process control computers should be isolated from all other building areas by partitions with a minimum of 1-hour fire resistance rating, extending from the concrete floor to the bottom of the roof. Any penetrations of the walls for cables, etc. should be sealed with an approved sealant system having a fire resistance rating of at least 1-hour. All entrances from other building areas should have approved ¾-hour fire rated doors (Class B) with self-closing devices.

Where practical, the distributed control system hardware areas should be provided with an independent HVAC system.

Instrument cabinets should not have pockets, shelves or other devices that encourage the storage of drawings or other combustible materials inside the cabinets. Metal file cabinets or shelves preferably in the control room area should be provided for the storage of drawings, manuals, etc. Closed-door metal cabinets should be provided for the storage of any spare parts that must be kept in the area.

Where large cable trays feed into the control building or I/O room, these openings or wall penetrations should be sealed against fire and smoke. Penetration seals should be provided that are or should meet the appropriate test requirements of ASTM E 814, *Standard Method of Fire Tests of Through-Penetration Fire Stops* or other test methods, such as IEEE 634, *Testing of Fire Rated Penetration Seals*.

## 8.3.2. *Computer Rooms*

Computer rooms considered in this section are those installations of substantial size and importance providing primarily data processing and network support service for multiple large process units or for an entire plant site. These installations are typically much larger than ones for an individual processing facility's dedicated process control computer.

### 8.3.2.1. Detection

An approved smoke detection system should be provided throughout the computer equipment room(s) and record storage spaces, with alarms locally and to another constantly attended location. The smoke detection system can be used to actuate the fire protection system.

An early warning HSSD smoke detection system can be used to detect incipient stage fires and allow manual intervention prior to significant equipment damage. Hence, fire suppression may not be required.

Smoke detection systems should be extended to include the areas under raised floors and above suspended ceilings that contain grouped cables or a significant quantity of ungrouped cables.

The location of the smoke detectors should be based on an engineering survey of the area to be protected. Factors such as air flow, proximity to air-handling system diffusers and other physical features of the installation need to be taken into account. Smoke tests can be run to verify that the air flow within the protected area favors the smoke detectors.

Additional general guidance on the placement of detectors can be found in FM Data Sheet 5-28, *Smoke Detectors* and in NFPA 72, *National Fire Alarm Code*.

### 8.3.2.2. Fire Protection

Fire protection options for computer rooms include:

- None, manual firefighting only
- Pre-action sprinkler system
- Water/mist system
- Gaseous total flooding system

A gaseous total flooding extinguishing system using an approved clean agent can be provided to protect the equipment in a computer room where the thermal and nonthermal damage from a fire starting in the room can result in very high loss expectancy. For further details, refer to NFPA 2001, *Standard on Clean Agent Fire Extinguishing Systems*.

Automatic sprinkler systems are generally considered more effective in suppressing a fire and more reliable in operation than gaseous total flooding systems. However, sprinkler use in computer rooms is increasing. Control of the following conditions is often used to justify a gaseous total flooding system over a sprinkler system:

- Construction is noncombustible
- Equipment enclosures are metal
- There is minimal and controlled use of paper and other combustibles in the room

However, for high-value computer centers where high reliability fire protection is determined to be necessary, both a pre-action sprinkler and a gaseous flooding system should be installed. Each system should be provided with an independent detection system for the automatic actuation of their respective isolation valves. A smoke detection system is recommended for the gaseous flooding system and a heat detection system for the pre-action sprinkler system. This arrangement is intended to ensure that the gaseous flooding system will activate first and avoid the need for the pre-action system to be utilized. It would be possible to substitute an additional smoke detection system for the heat detection system, but this must be entirely separate from that associated with the gaseous flooding system.

If a sprinkler system is installed in a computer room or similar area, provisions must be made to automatically de-energize all electrical power to the room and equipment, except power to lighting, in the event of sprinkler operation. Ensure that de-energizing activity leads to a fail-safe condition. Preferably, this should take place prior to water application to minimize damage to exposed electronic circuits. This can be accomplished automatically by smoke detection systems. Manual activation is tolerable for constantly attended locations. Where automatic sprinklers are installed in areas containing minimal combustibles as described above, a sprinkler density of 0.10 gpm/ft$^2$ (0.38 lpm/m$^2$) should be provided. Refer to NPFA 75.

#### *8.3.2.3. Manual Firefighting Equipment*

An appropriate number, type, and size of hand-held fire extinguishers should be provided to deal with ordinary combustibles and electrical fires in the I/O room and related areas. Typically, 10–15-lb (4–6-kg) clean agent or carbon dioxide fire extinguishers should be provided for electrical and electronic equipment. Dry chemical extinguishers should be avoided.

### 8.3.3. *Laboratories*

Laboratories are classified as Class A (High Fire Hazard), Class B (Moderate Fire Hazard), Class C (Low Fire Hazard), or Class D (Minimal Fire Hazard), according to the quantities of flammable and combustible liquids each is allowed to have. Table 8-16 and Table 8-17 (NFPA 45) describe the flammable and combustible liquids limitations and the requirements involving automatic sprinkler protection and fire-resistive partition wall separations. The tables apply to laboratories less than 10,000 ft$^2$ (929 m$^2$) area and should be suitable for most labs in processing facilities.

For the purposes of determining laboratory fire hazard classification, liquefied flammable gases should be treated as Class I flammable liquids.

**Table 8-16**
*Maximum Quantities of Flammables and Combustibles in Unsprinklered and Sprinklered Laboratories—Areas <10,000 ft² (929 m²)*

| | | Without Sprinklers | | With Sprinklers | |
|---|---|---|---|---|---|
| Laboratory Unit Fire Hazard Class | Flammable and Combustible Liquid Class | Maximum per 100 ft² (9.3 m²) of Laboratory Unit gal (l) | Maximum per Laboratory Unit gal (l) | Maximum per 100 ft² (9.3 m²) of Laboratory Unit gal (l) | Maximum per Laboratory Unit gal (l) |
| A | I | Not permitted | Not permitted | 20 (76) | 1,200 (4,540) |
| | II, IIIA | Not permitted | Not permitted | 40 (150) | 1,600 (6,060) |
| B | I | Not permitted | Not permitted | 10 (38) | 600 (2,270) |
| | II, IIIA | Not permitted | Not permitted | 20 (76) | 800 (3,028) |
| C | I | 4 (15) | 150 (570) | 4 (15) | 300 (1,136) |
| | II, IIIA | 8 (30) | 200 (760) | 8 (30) | 400 (1,515) |
| D | I | 2 (7.5) | 75 (284) | 2 (7.5) | 150 (570) |
| | II, IIIA | 2 (7.5) | 75 (284) | 2 (7.5) | 150 (570) |

**Table 8-17**
*Fire Protection Requirements for Unsprinklered and Sprinklered Laboratories—Areas Less Than 10,000 ft² (929 m²)*

| | Fire Separation–Fire-Resistive Partition Rating | |
|---|---|---|
| Laboratory Unit Fire Hazard Class | Without Sprinklers | With Sprinklers |
| A | Not permitted | 2 hours |
| B | Not permitted | 1 hour |
| C | 1 hour | Not required |
| D | Not required | Not required |

### *8.3.3.1. Fire Protection for Laboratories*

All laboratory units should be provided with fire protection appropriate to the fire hazard, as follows:

- Automatic sprinkler systems
- Fire alarm systems
- Standpipe and hose systems
- Portable fire extinguishers

8.3.3.1.1. AUTOMATIC SPRINKLER SYSTEMS

Automatic sprinkler system protection for Class A and Class B laboratories should be in accordance with NFPA 13, *Standard for the Installation of Sprinkler Systems,* for Ordinary Hazard (Group 2) occupancies. Automatic sprinkler system protection for Class C and Class D laboratories should be in accordance with NFPA 13, *Standard for the Installation of Sprinkler Systems,* for Ordinary Hazard (Group 1) occupancies.

Where water will create a serious fire or personnel hazard, a suitable non-water automatic extinguishing system should be considered. Penetrations through fire-rated floor, ceiling, and wall assemblies by pipes, conduits, bus ducts, cables, wires, air ducts, pneumatic tubes and ducts, and similar building service equipment should be protected in accordance with NFPA 101®, *Life Safety Code*. All floor openings should be sealed or curbed to prevent liquid leakage to lower floors. Door assemblies in 1-hour rated fire barriers should be ¾-hour rated. Door assemblies in 2-hour rated fire barriers should be 1½-hour rated.

8.3.3.1.2. FIRE ALARM SYSTEMS

Class A and Class B laboratory units should have a manual fire alarm system installed and maintained in accordance with NFPA 72, *National Fire Alarm Code*. The fire alarm system, where provided, should be designed so that all personnel endangered by the fire event should be alerted. The fire alarm system should also alarm to an attended location to alert emergency responders or the public fire department.

8.3.3.1.3. STANDPIPE AND HOSE SYSTEMS

For large laboratory buildings of two or more stories, standpipes should be installed in accordance with NFPA 14, *Standard for the Installation of Standpipe, Private Hydrant, and Hose Systems*. Hose lines of an approved type should be provided.

8.3.3.1.4. PORTABLE FIRE EXTINGUISHERS

Portable fire extinguishers should be installed, located, and maintained in accordance with NFPA 10, *Standard for Portable Fire Extinguishers*. For sizing and placement purposes, Class A laboratory units should be rated as extra (high) hazard, and Class B, Class C, and Class D laboratory units should be rated as ordinary (moderate) hazard.

### 8.3.4. MCCs, Substation Rooms, and Buildings

A motor control center (MCC) contains a number of enclosed motor control switches and breakers on a common power bus. MCCs should be of noncombustible construction and be located so that they are protected from physical damage. Two exits, remote from each other, should be provided.

Wall penetrations for cables and conduits should be sealed with an approved fire retardant sealant. If conduit is used that subsequently passes through flammable areas, appropriate conduit seals must be used at the MCC terminus.

If the MCC will be located within an electrically classified area, the ventilation system must be designed, operated and maintained to create a positive pressure inside the MCC building and the air intake located outside of the electrically classified area. The MCC should be equipped with smoke detectors and alarmed to a constantly attended location. $CO_2$ fire extinguishers are the suggested protection for MCC rooms. The storage of combustible materials inside an MCC must be prohibited.

Substations should be located so they are well separated from process areas and protected from damage. They should be located in a similar manner as control rooms. An enclosed substation should be equipped with smoke detectors and alarmed to a constantly attended location.

In larger MCCs, if the combustible loading gets extensive, then passive fire protection, e.g., fire-resistant cables or passive fire protection should be considered. The practices of using multiple tiers of cable trays or stripping the outer jacket insulation (to make the cable easier to handle after it enters the room or cabinet) are not recommended.

### 8.3.5. Clean Rooms

Clean rooms should be located in areas of fire resistant construction and equipped with full automatic fast-response sprinkler systems. These rooms should be cut off from other areas by fire-rated construction having a minimum rating of 1-hour. Because of the high values generally found in clean rooms, additional compartmentation is highly desirable. Ventilation systems should be independent of the general building system as well as adjoining compartments.

Recirculating ventilation systems should be designed for shutdown in the event of a fire or chemical spill. Exhaust systems should continue to run during the incident to facilitate the removal/ treatment of potentially toxic materials. Caution must be exercised when considering the placement of automatic sprinklers inside exhaust ducting. The flowing sprinklers have been shown to greatly reduce the capacity of the exhaust system.

Fire extinguishers and hose stations should be provided in accordance with NFPA 10, *Standard for Portable Fire Extinguishers*. Carbon dioxide ($CO_2$) extinguishers are the most common in clean rooms. The use of dry-chemical extinguishers is not advisable due to the corrosive nature of the chemicals and the expense of cleaning the area and replacing high-efficiency particle air (HEPA) filters after discharge.

## 8.3.6. Warehouse Protection

The approach to fire protection for onsite warehouses should be influenced by a number of factors, some of which may be overlooked in reaching a decision on whether or not to provide fixed fire protection and, if so, to what extent.

Without dependable fixed fire protection, fire mitigation will rely only upon compartmenting and spatial separation to limit the magnitude of the fire. Table 8-18 summarizes these points.

Further guidance on fire protection of warehouses and their contents can be found in the book *Guidelines for Safe Warehousing of Chemicals* (CCPS, 1998).

### 8.3.6.1. Rack Storage

Open frame steel rack storage arrays allow for greater exposed surface area of the product and greater pile stability during a fire than solid pile and palletized storage arrays. For these reasons, rack storage arrays result in more severe fires than other types of storage arrays under similar conditions.

In-rack sprinklers can result in smaller more easily controlled, suppressed, or extinguished fires, with fewer operation sprinklers. This can have the added benefits of reduced quantity of contaminated water runoff and reduced product damage.

### 8.3.6.2. Automated Warehouses

Automated warehouses can present unique conditions not found in warehouses where material handling is done manually. Typically, these facilities use rack storage heights that are much higher and present a greater fire challenge

**Table 8-18**
*Warehouse Fixed Protection Considerations and Options*

| Considerations | Decision: Protect or Not? | Basic Design Options |
|---|---|---|
| Hazards of Materials Involved<br>Quantities of Materials Involved<br>Exposure of Other Operations<br>Emergency Response Capabilities<br>Community Exposures | No | Accept risk (do nothing)<br>Provide separation distance<br>Compartmentation |
| Environmental Exposures<br>Investment at Risk<br>Insurance Considerations<br>Business Continuity<br>Local Codes and Authorities | Yes | Control<br>Suppression<br>Extinguishment |

requiring special fire protection considerations. Their automation reduces the presence of personnel and may limit opportunities for early detection of spills and other fire precursor events. An automatic rack retrieval system has the potential to inadvertently move burning or leaking product to other parts of a warehouse. Consideration should be given to automatically return the retriever apparatus to the staging area and shutdown upon fire detection.

### 8.3.6.3. *Refrigerated Warehouses*

In some facilities, there is a need for refrigerated storage of one or more flammable or combustible raw materials, catalysts, intermediate products, or finished goods. These materials may be in solid or liquid form in containers, boxes, drums, or small portable tanks. Depending on quantity of materials to be stored and required temperature, these facilities can range from cold storage warehouses for storage of food products to walk-in freezer rooms for pharmaceutical materials.

The thermal insulation materials used in the walls, ceilings, and floors of refrigerated storage are usually combustible, such as polyurethane and polystyrene foam, and could present an additional fire hazard. The insulation should be covered with a noncombustible barrier material; a ½-in (1.3-cm) thickness of Type-X gypsum board or cement plaster is often used for this purpose. Where the floor is insulated, the floor insulation should be covered with a layer of concrete.

Sprinkler protection should be provided as for any storage facility handling similar material products. In storage areas that are always above freezing, pre-action type systems electrically activated by a separate fire/heat detection system are often used.

Where storage temperatures are below freezing, the sprinkler system has to be designed to prevent accidental freeze-up. One method is to use a pre-action sprinkler system. The pre-action valve is electrically activated by a separate fire/heat detection system and has a ⅛-in (0.3-cm) orifice in a bypass line that maintains pressure on the preaction valve, keeping it closed. The pressurizing air should have a low dew point to avoid frosting and possibly plugging the system over time; one option is to draw compressor intake air from the cold space itself. Additional information and a description of the above method can be found in NFPA 13.

### 8.3.6.4. *Warehouse Construction*

Generally, chemical warehouses whose contents classify them as hazardous occupancies should be constructed of fire resistive or noncombustible materials. Fire resistive construction should consist of materials that will withstand the anticipated fire exposure of a given duration without structural failure.

When designing for fire safety in a chemical warehouse, building area fire separations should limit the risk or size of loss by:

- Separating personnel facilities from warehouse areas
- Segregating different stored materials due to their hazardous properties
- Increasing the likelihood of control in a fire situation by exposing a limited area that is considered manageable by automatic and manual suppression efforts, or by a building's containment/drainage design
- Meeting building or fire code requirements for building area limitations, separation of occupancies/hazards, travel distance considerations, or other provisions

Fire-related barriers or partitions are those walls that are supported in some manner by the building's structural frame, floor, ceiling, or roof. These walls rely upon the integrity of the supporting construction to remain in place and to withstand a given fire duration. Refer to Section 7.3.1.1.

#### 8.3.6.5. *Fire Protection*

Fire propagation through a warehouse is a function of a number of conditions, including the intrinsic fire hazard of the product, storage height, and storage array configuration. Chapter 4 of NFPA 30 contains many schemes for protecting flammable and combustible liquids in containers in warehouses.

Fire control can be achieved for a number of hazardous chemicals by an automatic sprinkler system using deluge sprinkler heads operating directly over the origin of the fire and the surrounding area. Water that is discharged from sprinklers extinguishes burning materials and pre-wets the surrounding product to prevent or slow the spread of fire. This is most effective with material that is water absorbent such as products in cartons. For other products, such as flammable and combustible liquids in containers or drums, sprinklers protect by cooling to prevent overpressure and container rupture.

The convective updraft from severe fires, frequently involving flammable and combustible liquids may prevent water droplets from standard ½-in (1.3-cm) orifice sprinklers from reaching the seat of the fire and product may continue burning until it has been consumed or manually extinguished. Therefore, the variety of new high performance, large orifice, large droplet, or early suppression fast response (ESFR) sprinkler systems should be considered.

Palletized or solid pile storage arrays can be protected with ceiling mounted sprinkler systems. Depending on storage conditions and system design, rack storage arrays can also be protected with ceiling mounted or combination ceiling mounted and in-rack sprinkler systems.

Fire extinguishment can be attained with certain products by using a sprinkler system that discharges a low expansion foam or AFFF agent, typically at concentrations of 1%, 3%, or 6% with water. Foam–water sprinkler systems have been successfully tested with ordinary products and containerized storage of flammable and combustible liquids. The cooling capabilities of foam–water are the same as ordinary water. Foam-water can extinguish burning flammable and combustible liquid pool fires while cooling exposed containers. Foam-water also acts as a wetting agent with ordinary products.

#### 8.3.6.6. Manual Firefighting Equipment

Manual fire suppression is not normally dependable as a primary fire protection strategy for chemical storage warehouses. Fires may grow to uncontrollable size before effective manual response can be employed, and may pose severe risks to firefighters. In warehouses, manual suppression usually involves far more water application than automatic systems, aggravating problems of disposal of fire water runoff.

Fire extinguishers should be provided throughout the warehouse according to requirements specified in NFPA 10. Even though flammable liquids may be stored in the warehouse, warehouses typically are a mixture of types of fire hazards. Therefore, Class ABC extinguishers are generally recommended. Extinguishers for special hazards may be required in some areas.

### 8.3.7. Temporary Buildings and Office Trailers

Temporary, nonpermanent structures may be used at processing facilities. The most common are mobile office trailers used during construction or periodic major unit overhaul or turnaround. A common practice is that these temporary offices are located near processing areas for convenience and are not removed on completion of the job; thus, they transition from temporary to semipermanent.

Office trailers frequently have combustible framing, flooring, and interiors. Their exteriors may have a technically noncombustible skin, but with little or no flame or thermal radiation resistance. They are susceptible to windstorm damage even when properly anchored. Accounting for these features, NFPA 241 and 80A suggest minimum separation distances to minimize exposure to other structures. Greater spacing may be required for larger groupings. Some vendors now have temporary buildings that have fire and explosion ratings.

If trailers remain in service after startup and are located next to a flammable and combustible processing area, the personnel using these trailers do not have the same level of protection from events in the processing area as do those in permanent buildings. Office trailer installations should be included in the facility siting study (API RP 752; CCPS, 1996c).

## 8.4. Loading Racks and Marine Terminals

Loading racks and marine terminals associated with chemical, petrochemical and hydrocarbon processing facilities vary widely in the type of materials and quantities handled. Specific and detailed requirements exist in the regulations, particularly for land-based loading racks handling hazardous, flammable, toxic, or environmentally harmful liquids and liquefied compressed gases. Marine liquid bulk terminals and their operation are closely regulated and regularly inspected by authorities.

### 8.4.1. General

#### 8.4.1.1. Fire Prevention

Fire prevention efforts for handling flammables and combustibles rely heavily on the mechanical integrity of the transfer system, the containment of small spills that do occur, and the control of ignition sources. While operator presence in the immediate area of the rack or wharf is recommended for any loading or unloading operation, full-time surveillance at the unloading spot is recommended.

#### 8.4.1.2. Flow Induced Static Generation

When loading into a compartment where a flammable atmosphere could exist, any liquid splashing or free-fall could create a static discharge with potential ignition. Dip pipes are required and, in addition, the loading pump and transfer system must be designed to allow a low flow rate start. The initial velocity of the incoming liquid stream should be 3 ft (0.9 m) per second or less until the inlet pipe is sufficiently submerged to prevent splashing and potential static generation (NFPA 30).

Filters, strainers, pumps, and other devices can produce static charges through normal flow turbulence with a number of nonconductive flammable liquids. Loading system pipe run lengths from such potential static-generating elements should be so located to allow a minimum of 30 seconds relaxation time prior to discharging into the vessel's tank compartment (NFPA 30 and NFPA 77).

#### 8.4.1.3. Bonding and Grounding

Bonding and grounding should be provided to minimize the accumulation of static electrical charges in the liquid. The integrity of the grounding conductors and connectors should be inspected annually. The integrity of the bonding system should be inspected before each use.

#### 8.4.1.4. Emergency Shutdown

An automated, emergency shutdown system is strongly recommended to isolate and stop the loading or unloading operation in the event of overfilling, failure of a transfer hose/arm, or manifold valve. However, manual shutdown activation by the operator in attendance is permitted and requires that the manually activated device(s) are well marked and easily accessible during an emergency. A distance of 50 ft (15 m) is typically used as the distance to locate an emergency shutdown from the hazard.

Where a loading system's emergency shutdown system closes a valve on gravity or pipeline fed transfer systems, care should be taken to ensure the line is protected against pressure surges or hydraulic hammers which may cause gasket blowout or line failure (NFPA 30).

### *8.4.2. Loading Racks*

Loading and unloading of tank cars and tank trucks represent an increased potential for spill or release. Design and operation must be conducted to minimize the chance for spill, release, or fire and to ensure protection of personnel and facilities.

One alternative to the typical practice of loading rail tank cars individually is the *unit train*. A unit train can be practical when there are large volume regular shipments of the same product to the same destination requiring a number of cars. A unit train consists of a number of tank cars semipermanently coupled together and interconnected by flexible transfer lines and isolation valves. All tank cars in the unit train can be loaded or unloaded with only one set of line connections to one car. A unit train offers advantages in reducing the likelihood of spills during loading/unloading, the likelihood of personnel fall injury, and the ability to provide improved fixed fire protection and spill control at the single loading/unloading rack connection point.

For flammable liquids, top loading or unloading of tank cars or tank trucks by pump typically has the lowest risk of spill or release. Metallic piping is recommended for loading/unloading arms and piping. Use of hose should be minimized and, where possible, metallic hose specified. Standard loading racks should be provided that permit safe access by personnel to the rack itself and onto the tops of the cars or trucks. Fall protection is a key issue.

Loading/unloading arms from the loading rack should be extended to within 4 in (10 cm) of the bottom of the vehicle to prevent splashing on loading or drainage back after loading or unloading.

#### *8.4.2.1. Truck Bonding and Grounding*

Tank trucks, where flammable vapors are likely to be present, should be electrically bonded via the downspout, piping or steel loading rack as shown in Figure 8-13 (API 2003, 1998). If the bonding occurs through the rack, then the downspout, piping, and rack must be electrically interconnected. The connection should be made before the dome cover is opened and remain in place until the dome cover is closed.

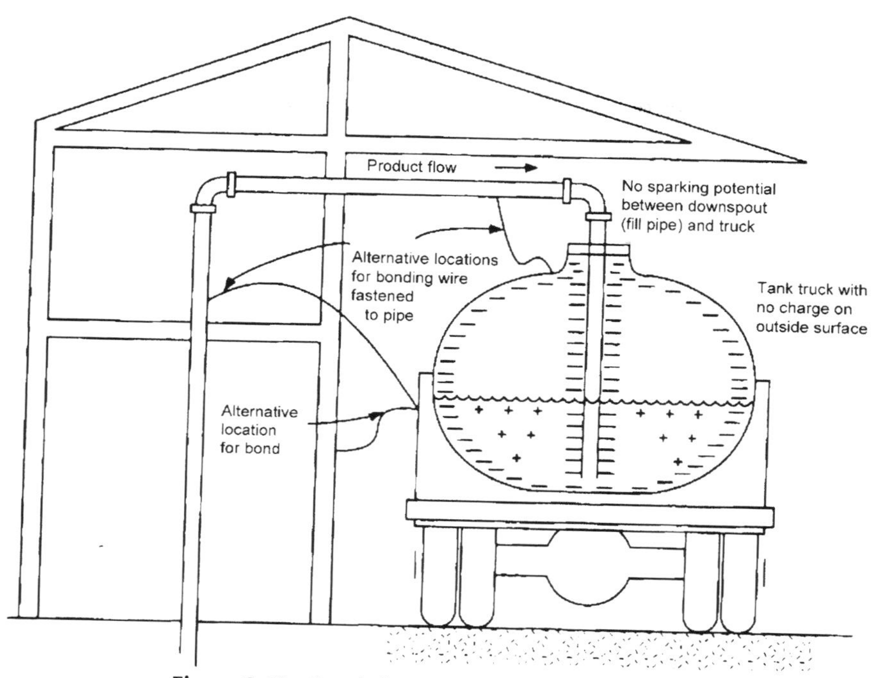


**Figure 8-13.** Bonds for Top Loading of a Tank Truck

Bottom-loading tank trucks must also be electrically bonded. Bonding generally occurs through the use of a metal jacketed hose and metallic couplings.

### 8.4.2.2. Drainage

Loading/unloading racks should be protected from physical damage, particularly from collision during spotting of trucks for unloading to minimize the potential for accidental releases. Spill collection and containment is required around pump manifolds and other areas to prevent the spread of liquids. Means should be provided to manage small spills that may occur during hose system connection, disconnection or draining.

Since spill potential is increased with bottom unloading, containment of a potential spill should be considered in the design of the unloading location. Separation from other loading/unloading spots is recommended.

### 8.4.2.3. Fire Protection

Where unloading of flammable liquids must be done from the bottom of the tank car or tank truck, a remote actuated unloading valve equipped with a fusible link for detecting high temperature should be provided.

Minimum recommended fire protection is fixed foam monitor nozzles, either hydrant mounted or elevated, located such that their streams can be directed to both sides of tank cars or trucks at the rack. This foam–water protection may also be augmented by portable fire extinguishers and larger wheeled units. Depending on the nature of the material being unloaded, its behavior in a spill or release, and its location relative to other operations or site boundaries, automatic foam–water spray protection may be provided over the tank car or tank truck spots at a density of 0.25 gpm/ft$^2$ (10.2 lpm/m$^2$). If unvented roofs are placed over loading/unloading stations, automatic deluge protection is recommended.

Some products are unloaded by nitrogen or other inert gas pressurization of the tank car or truck. The use of air for unloading flammable or combustible liquids is *not* recommended.

### 8.4.3. Marine Terminals

Marine transfer hoses tend to be large and heavy, particularly when filled during transfer. As a result, hose handling systems and winches are frequently required. The design of the hose handling system becomes more critical for those terminals that experience large water level variations such as encountered on larger inland rivers subject to major floods. Facilities on the Mississippi and Ohio rivers in the United States commonly experience seasonal water level changes of 25 ft (7.6 m) or more.

The transfer connections from the wharf piping to the vessel manifold system are typically made using one of the following:

- Hoses
- Articulated swivel-joint transfer arms
- Hoses in combination with one or more swivel joints

All loading/unloading transfer hoses and articulated arms in hazardous liquid service should be inspected and pressure tested annually along with their couplings. The test date should be documented and marked on the hose or arm. Equipment showing deterioration, signs of leakage or fatigue or damage to couplings must be removed from service or repaired before continuing transfer operations.

The swivel joints used in articulated arms or with hoses should be of a design that the mechanical strength of the joint will not be impaired if its packing materials should fail by exposure to fire. In areas of strong tidal effect or river current, movement alarms should be provided along with a method for quick uncoupling of loading arms.

Transfer handling conditions can affect mechanical integrity and lead to premature catastrophic hose failure. These potentially damaging conditions include the following:

- Bend radius too small for the type and size of hose
- Localized misalignment at hose-to-fixed pipe connection due to arrangement of hose handling guide or winch
- Hydraulic hammer from sudden closing of valve downstream of hose (primarily when loading with substantial gravity head or at high flow rates)
- Inadequate support of hose or loading arm and its weight during transfer

#### *8.4.3.1. Marine Bonding and Grounding*

Pipelines on wharves that handle Class I or Class II liquids must be adequately bonded and grounded. Bonding between the wharf and the vessel is not allowed. Insulating flanges or joints should be installed. If stray current isolation is used, the grounding connection(s) must be accessible for inspection (NFPA 30, Section 5.7.10).

#### *8.4.3.2. Drainage*

Spill containment is required around the marine vessel during loading and unloading to prevent the spread of an accidental release. Means should be provided to manage small spills that may occur during transfer arm or hose system connection, disconnection or draining.

#### *8.4.3.3. Fire Protection*

In general, the extent of fire protection for marine terminal wharves is related to the materials handled, plant site and local emergency response capability, terminal size, location, size of ships or barges serviced, frequency of use, and adjacent exposures..

A fire water main should be provided to the shore terminus area of the wharf. Hydrants and monitor nozzles should be located so that effective fire water streams can be remotely applied to any berth or loading/unloading manifold from two directions. Where the wharf is of such length that onshore monitors cannot adequately cover the berths, the fire water main should be run onto the wharf to permit the required monitor and fire hose coverage. Monitors located on marine wharves may be remotely operated from onshore and use of elevated monitors is common. The offshore segment of the main can be a dry system. In all cases, isolation valves and fire department connections should be provided at the wharf-to-shore connection (ISGOTT, 1996).

Where no fire water main is provided, at least two 150 lb (68 kg) wheeled dry chemical extinguishers should be provided and located within 50 ft (15 m) of pump or manifold areas and easily reached along emergency access paths.

## 8.5. Utilities

### 8.5.1. Cooling Towers

Cooling towers can be either the victim of an outside ignition source or they can be the culprit causing propagation of a fire to other nearby processing, storage, or utility operations or facilities. In fire protection terms, they may present an "exposure to" or they may suffer "exposure from" other facilities or hazards. Some of the situations and events that have led to fires in cooling towers of combustible construction are:

- Proximity to process equipment handling flammables or combustibles, including vents, flares, incinerators or thermal oxidizers, where either close horizontal distance or where process elevation and trajectory can result in burning material contacting the cooling tower
- Nonuse periods when tower surfaces and internals are dry and more susceptible to ignition such as during a cold season, idle shutdown, or periodic unit maintenance turnaround
- During maintenance involving hot work on or near the cooling tower, particularly the fan deck
- Contamination of cooling tower water with flammable or combustible process leaks potentially resulting in a flammable atmosphere in and around the cooling tower
- Lightning strike

Fire prevention may include providing degassing boots/vent stacks to the top of risers where flammable combustible vapors can be entrained in the water system, i.e. water pressure lower than process pressure. Combustible gas detection at the top of the vent stacks can be used to detect flammable material presence in cooling tower. However, the hostile environment and difficult to access location makes the detection challenging to maintain.

In general, if any of the cooling tower's major components are combustible materials, then the following factors should be considered in determining the fire protection required for that particular cooling tower:

- Location of cooling tower with respect to exposure to or from nearby processing, storage, or utility operations or facilities
- Size and materials of construction
- Importance to continuity of operation
- Type of tower
- Value of tower

If the above indicates the need for additional fire protection, then appropriate passive and active fire protection should be applied.

Passive fire protection for cooling towers involves increasing spacing distances and using noncombustible materials of construction. For cooling towers of totally noncombustible materials of construction, there are no fire protection requirements. Noncombustible means that the cooling tower's structure, fan and distribution decks, louvers, and fill materials must all be noncombustible materials.

Some manufacturers of internal cooling tower components, specifically fill material and drift eliminators, have products produced from less easily ignited plastic that have been tested by a nationally recognized testing laboratory and determined to have sufficient fire resistance or reduced flame spread ratings that when, and only when, used in an otherwise noncombustible cooling tower, do not require fixed automatic fire protection.

A summary of cooling tower fire protection guidance from NFPA 214 is provided in Table 8-19.

Cooling towers spacing and layout should follow the criteria found in CCPS book *Guidelines for Facility Siting and Layout* (CCPS, 2003b). If the cooling tower's combustible exterior is protected for exposure with a sprinkler system, it may (on a case-by-case basis) be located less than 100 ft (30 m) from the hazard.

Active fire protection of cooling towers is primarily provided by sprinkler systems.

This guidance applies equally to induced, fan-forced, or natural draft types; however, there are specific fire protection system design detail differences for cross-flow versus counter-flow cooling towers. Fire protection systems used on counter-flow cooling towers include closed- or open-head systems, wet-pipe, dry-pipe, pre-action, or deluge systems. For cross-flow cooling towers, deluge open-head systems are generally recommended to maximize the water distribution and heat detection activation.

**Table 8-19**
*Cooling Tower Construction, Protection, and Spacing*

| Type Construction | | |
|---|---|---|
| **Exterior** | **Interior** | **Fire Protection System** |
| Totally noncombustible (all major components) | Totally noncombustible (all major components) | Not required |
| Noncombustible surfaces | Approved low flame spread components | None |
| Noncombustible surfaces | Combustible components | None<br>Interior protection |
| Combustible surfaces | Combustible components | Interior protection and all exterior surfaces |

NFPA 214 provides specific guidance regarding fire protection of cooling towers including recommended water spray application rates. The detailed design of fixed fire protection systems for cooling towers should follow the requirements of NFPA 13 or NFPA 15.

Refer to Appendix A, Case History 4, for an example cooling tower fire that resulted in over 2 million dollars damage (1997 dollars).

### 8.5.2. Air Compressors

Air compressors supply compressed utility air and, if damaged, their function is quickly replaceable with rental compressor units. On the other hand, an air compressor may supply compressed air for a process stream, be very large in size, and not quickly replaceable. Air compressors that supply instrument air are critical to plant operations and must be installed, maintained, and protected to maximize reliability.

There have been fires and explosions in compressed air systems. The fuel is a combustible lubricant that has entered the air system from the compressor. Although maintenance of the check valves will minimize the hazard, the valves can stick open from scale or other causes that can affect even recently serviced valves. The explosion potential can be minimized by replacing the combustible lubricants with noncombustible lubricants.

Like flammable gas compressors, the external lubrication system for large air compressors can present a fire hazard. Oil flow rates above 25 gpm (95 lpm) and or oil capacity greater than 100 gal (379 l) may require fire protection.

If an air compressor is located such that a flammable release from nearby equipment could envelope the air compressor, then automatic deluge or sprinkler protection may be required to protect the air compressor from fire damage.

### 8.5.3. Electric Generators

Electric generators should be separated from other equipment and facilities as determined by the fire risk assessment to limit the spread of fire, protect personnel, and limit resulting damage to the plant. For guidance on spacing distances, refer to *Guidelines for Facility Siting and Layout* (CCPS, 2003b).

Fire protection for generators is based on size, type, location, and criticality. The water supply for fire protection systems should be based on providing a 2-hour supply (NFPA 850, 2000) for:

(1) Either a) or b) below, whichever is larger:
   a) The largest fixed fire suppression system demand
   b) Any fixed fire suppression system demands that could reasonably be expected to operate simultaneously during a single event
(2) A hose stream demand of not less than 500 gpm (1,890 lpm)

When working in areas such as combustion turbine compartments where actuation of the fire protection system could affect personnel safety, the fire extinguishing system must be designed to allow the system to be locked out. Evacuation of a protected area is recommended before any special extinguishing system discharges. Alarm systems that are audible above machinery background noise, or that are visual or olfactory or a combination, should be used where appropriate. A trouble indication should be provided when the system is locked out.

### *8.5.4. Boilers and Thermal Oxidizers*

Boilers and thermal oxidizers should be located outside to minimize containment of releases of hazardous materials, such as fuel gas. Locating these units away from other process structures is also recommended to protect the boiler/thermal oxidizer from damage.

Automatic sprinkler protection should be:

- Provided if the boiler/thermal oxidizer is installed in a room or building of combustible construction or the building or room contains significant quantities of combustible materials.
- Installed if the boiler or thermal oxidizer is burning flammable liquids and a significant quantity could be spilled in the building or room in the event of a release.
- Provided if the boiler/thermal oxidizer is located near potential release sources of flammables that could envelope the building and equipment in fire.

### *8.5.5. Transformers*

Outdoor transformers should be physically separated from adjacent structures and from each other by firewalls, spatial separation, or other means for the purpose of limiting the damage and potential spread of fire from a transformer failure.

The appropriate physical separation to be used should be based on consideration of the following:

- Type and quantity of oil in the transformer
- Size of a potential oil spill (surface area and depth)
- Type of construction of adjacent structures
- Power rating of the transformer
- Fire suppression systems provided, if any
- Type of electrical protective relaying provided

Unless consideration of these factors indicates otherwise, NFPA 850 recommends that any oil-insulated transformer containing 500 gal (1,893 l) or more of oil be separated from adjacent structures by the distances in Table 8-20. If separation can not be obtained, then either a fire barrier or water spray system designed for 0.25 gpm/ft$^2$ (10.2 lpm/m$^2$) should be provided.

A 2-hour fire rated wall may also be used to achieve the needed separation. Where a firewall is provided separating a structure or an area from a transformer, the wall should extend vertically and horizontally a sufficient distance to provide line-of-sight shielding of the transformer to the exposed structure. Where the space separation can not be achieved an automatic water spray system can be used in lieu of the fire barrier.

Dry-type transformers are preferred for installations inside buildings. Where oil-insulated transformers are installed indoors, they should be separated from adjacent areas by fire barriers of 3-hour fire resistance rating if the transformers' capacity or rating exceeds either of the following:

- Oil capacity greater than 100 gal (379 l)
- Rating greater than 35 kV and the oil is a low- or nonflammable fluid

### 8.5.6. *Waste Handling*

Waste handling facilities have become an essential function for operation of many process systems. Historically, waste handling facilities are often overlooked and should be included in a fire hazard analysis. If the waste handling/treatment system cannot operate, then a plant or unit shutdown may be required.

If a flammable process spills significant quantities of flammable material, this material may flow to the waste handling area. Waste digestion systems can be shocked into shutdown if large quantities of material are suddenly introduced. Also, a large release may occur and could result in a fire that may damage equipment that cannot be quickly replaced.

The potential for significant quantities of flammables reaching the waste handling system should be examined in a process hazards analysis. If the poten-

**Table 8-20**
*Outdoor Oil Insulated Transformer Separation Distance*

| Transformer Oil Capacity | Minimum (Line-of-Sight) Separation Distance |
|---|---|
| < 500 gal (1,893 l) | No requirement |
| 500 to 5,000 gal (1,893 to 18,925 l) | 25 ft (7.6 m) |
| > 5,000 gal (18,925 l) | 50 ft (15 m) |

tial is significant and the vulnerable equipment not quickly replaceable, then extension of the fire water system to the waste handling area and the installation of water or preferably foam–water monitor nozzles should be considered. These monitor nozzles should be located so that their streams can, from two directions, reach and protect essential equipment associated with above-ground open top digester tanks or with ground level treatment lagoons where a spill could be concentrated.

# 9

# INSTALLATION OF FIRE PROTECTION SYSTEMS

This chapter discusses the various phases of installation of fire protection systems. Fire protection projects vary from new construction projects to retrofitting an existing system in a facility. Managing a fire protection system project is very much like any other construction or capital improvement project. The management process starts with the development of an installation plan that identifies the various phases of the project.

The project plan should encompass all aspects of a fire protection system, such as the underground fire water distribution system, fire pumps, aboveground water header, valving and standpipes, structural support, and detection and alarm systems. All work on the fire protection system must be coordinated with other work activities at the site or in the operating unit. The recommended installation practices for the different types of fire protection systems are covered in consensus standards, such as NFPA. The installation process is illustrated in Figure 9-1.

## 9.1. Approval Process

### *9.1.1. External*

Before beginning the installation of a fire protection system, it may be necessary to secure approval from local, state or federal agencies that govern fire protection installations. The facility insurance carrier may request that they be notified of any new fire protection construction or retrofitting of existing systems so that they may offer assistance in managing the change.

***NOTE: It is important to verify if drawings require approval by company fire protection personnel, regulatory agencies or other authority having jurisdiction (AHJ). Drawing approval may have to be completed by a licensed professional engineer certified in fire protection engineering.***

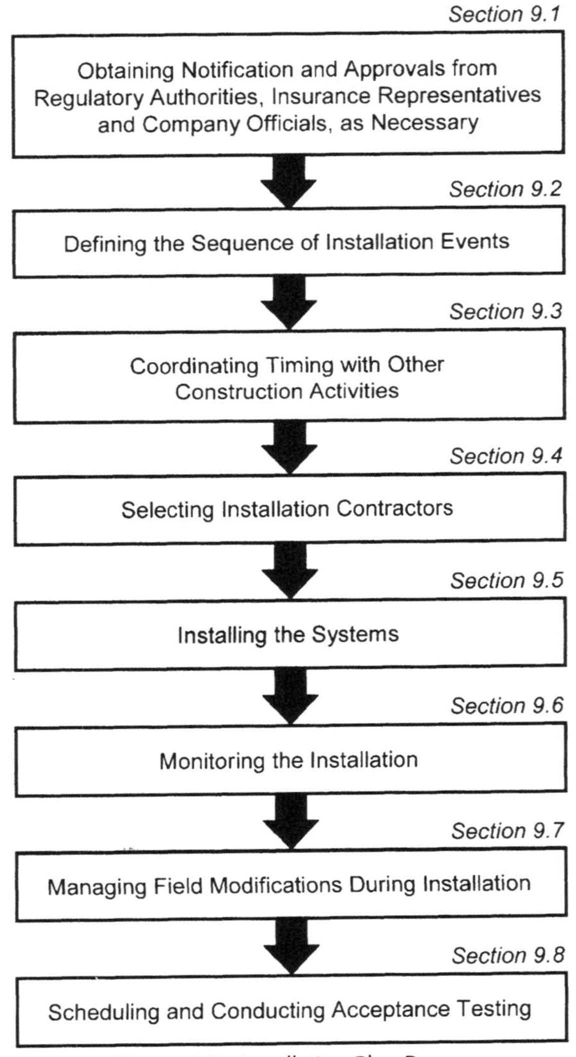


**Figure 9-1.** Installation Plan Process

### 9.1.2. Internal

Company management may require risk managers and fire protection engineers to approve and inspect all fire protection projects. Affected local operations managers, maintenance managers, and emergency responders should be notified of changes to the fire protection systems in their units/areas of responsibility. It is not unusual for a company to identify one person as a single point of contact for obtaining and coordinating required approvals. Sufficient time must be allotted up front to ensure that the schedule is not impacted.

## 9.2. Sequence

Sequencing is important to ensure the system meets the construction timetable and to minimize impact on existing systems and project cost. Sequencing ensures an efficient installation of the fire protection system and assists in meeting the project schedule. For example, fire water distribution systems should be installed early, so that it can be used during construction for firefighting. Sequencing is also important to minimize downtime when protection systems are being impaired. It is important to sequence those operations that require protection systems to be impaired such that their down time is minimized. For example, tie-ins to an existing fire water system should be scheduled as convenient, prior to the commencement of the main construction activity. This will minimize the impact of parallel activities and the potential for errors leading to unnecessary downtime.

The sequence of the installation details the order each part of the fire protection system will be installed. The following is a typical sequence for installing a fixed water protection system (sprinkler system):

- Tie-in piping and alarms to existing systems, where applicable
- Construct underground piping system
- Construct water supply header system
- Install sprinkler piping
- Install electronic monitoring system
- Perform acceptance testing

## 9.3. Timing

The timing of installation of fire protection systems is important because it must be integrated with other construction activities. Fire protection is generally a small but important element of the overall project. To ensure a timely and efficient installation, the person responsible for scheduling the installation must work closely with other project schedulers. Many companies subcontract the installation of fire protection systems to qualified fire protection service companies. Using subcontractors adds another degree of difficulty in managing the project. For example, if the sprinkler heads are installed before painting, then the sprinkler heads could be damaged and have to be replaced, delaying the completion of the project.

Typically, large fire protection piping is installed early in the construction process and smaller piping later in the project. A later installation may result in rerouting of piping that changes the hydraulic characteristics of the system. Another example is that fireproofing on structural steel is completed very early in the project; however, the steel is used for attachment of conduit, control system wiring, and unplanned equipment supports that will damage the fireproofing. Thus, the fireproofing will need to be repaired at the end of the project.

## 9.4. Selection of Installation Contractors

Fire protection engineering, architectural, or engineering companies should provide documentation of experience and qualifications, along with evidence of knowledge of the most current hardware technologies. Prospective fire protection service companies should be required to demonstrate qualifications and work experience. Examples of these qualifications include licensed Professional Engineers or the National Institute for the Certification of Engineering Technicians (NICET) guidelines. Professional engineers chosen should be licensed in fire protection engineering. Fire protection companies must have sufficient staff of NICET technicians or licensed Professional Engineers (PE) to successfully staff fire protection projects of the size they are bidding. The use of qualified and competent contractors is essential in developing fire protection systems that are appropriate for the hazards. The use of contractors with well qualified and competent staff should ensure that designs incorporate the latest and most appropriate fire protection technology and hardware, thus achieving the most cost-effective performance.

## 9.5. Installing the System

**Only after acceptance/approval has been obtained in writing from all necessary external and internal agencies should the installation begin.** The failure to do so could result in delay of the project, major changes after the system has already been installed, or inadequate protection. The contractor would likely wish to pass any additional costs along to the owner.

## 9.6. Monitoring of the Installation

The project manager should assign either inspection staff for new facilities, the facility fire protection staff, or appropriate personnel at existing facilities the responsibility of monitoring the activities of the contractor installing the system. These duties should include:

- Monitoring contractor work activities
- Performing safety inspections and audits
- Performing performance evaluations to ensure installation follows specifications and work plan
- Ensuring the terms of the contract are being met

Fire safety practices such as housekeeping, hot work permits, and elevated work should be in place during installation. The owner's representative should make periodic progress inspections to check for nonconformance issues and good work practices.

## 9.7. Managing Field Modifications During Installation

The assigned staff should monitor the installation to ensure that no unauthorized changes are being made and that installation meets the design specification. All field modifications must be approved before making the changes. The requested changes must be evaluated and approved following the company's Management of Change process. Changes may need to be approved by external agencies and insurance representatives.

## 9.8. Acceptance Testing

Acceptance tests should be performed on all new or modified fire protection systems. The goal of the final acceptance test is to determine if the fire protection system performance meets the design specification. The installation contractor is responsible for the acceptance test and remains the owner of the fire protection system until the acceptance test has been successfully completed.

The project manager should coordinate the acceptance test with installation contractors and appropriate facility personnel. The project manager should develop the testing protocol and the expected results. A review of the testing protocol and requirements should be shared with the groups involved in the testing.

It is responsible stewardhip for facility personnel to witness all performance tests and make detailed examination of the piping installation to verify the quality and adequacy of the system.

Acceptance tests should be documented. Local regulations, insurance carriers, and company requirements may require that officials from these groups witness the acceptance testing. Acceptance testing typically includes:

- Verifying installation complies with working drawings
- Conducting pressure test
- Flushing and hydrostatic testing of underground piping
- Flushing and hydrostatic testing of system piping
- Trip-testing of systems
- Verifying operation of detection, alarm bells, and notification signals
- Ensuring the contractor provides:
  - Contractor's Material and Test Certificate
  - Complete set of as-built drawings, including piping, electrical, and hydraulic calculations
  - Operation and maintenance manual for all system components

### 9.8.1. Water Supply Systems

Flushing activities and flow testing can be conducted on sections of the system as they are completed, but a full system test is still required. Details of the flushing and hydrostatic test requirements can be found in NFPA 24.

***WARNING: Make sure you know where water runoff will be discharged before starting any flow test or flushing of piping.***

#### *9.8.1.1. Flushing*

Flushing is performed on new piping to remove material that was trapped during construction. Typical materials include dirt, stones, and small pieces of concrete; however, hardhats, lumber, and tools have been found during actual tests.

Flushing of new underground piping system, including branches from existing or new underground mains to devices, should be wit-

**Table 9-1**
*Acceptance Test Requirements*

| System | Criteria | NFPA Reference |
|---|---|---|
| Sprinkler | Flushing | 13 |
| | Hydrostatic test | |
| | Installation verification | |
| | Component tests | |
| Water Spray | Flushing | 15, 16 |
| | Hydrostatic test | |
| | Flow test | |
| Foam | Flushing of underground supply and aboveground system piping | 11, 11A |
| | Hydrostatic test of piping | |
| | Flow test | |
| Carbon Dioxide | Verifying system is properly installed and functions per design | 12 |
| | Discharge test to ensure design and function | |
| Dry Chemical | Verifying system is properly installed and functions per design | 17 |
| | Discharge test to ensure design and function | |
| Clean Agent | Visual inspection | 2001 |
| | Pneumatic test of piping | |
| | Flow test of the piping | |
| | Enclosure integrity test | |
| | Electrical systems operation properly | |
| | Discharge test | |
| Fire Water Distribution Piping | Flushing | 24 |
| | Hydrostatic test | |
| Fire Pumps | Flushing of new suction piping | 20 |
| | Hydrostatic test of suction and discharge piping | |
| | Flow test | |
| | Controller test | |

**Table 9-2**
*Flushing Requirements*

| System | Criteria | NFPA Reference |
|---|---|---|
| Sprinkler | New lines supplying the system should be flushed based on rates in Table 9-3. | 13 |
| Water Spray | New lines supplying the system should be flushed based on rates in Table 9-3. | 15, 16 |
| Foam | New lines supplying the system should be flushed based on rates in Table 9-3. Additionally, the above ground system piping should be flushed. | 11, 11A |
| Carbon Dioxide | None. | 12 |
| Dry Chemical | None. | 17 |
| Clean Agent | None. | 2001 |
| Fire Water Distribution Piping | Flush new lines based on rates in Table 9-3. | 24 |
| Fire Pumps | Flush new suction piping<br>Hydrostatic test of suction and discharge piping<br>Flow test<br>Controller test | 20 |

nessed by facility personnel. The system should be thoroughly flushed-out under pressure through hydrants or blow-offs before connections are made to sprinkler risers. The branches should also be thoroughly flushed before connecting the water system risers. The velocity of the flow should be at least 10 ft/s (3 m/s). The rates specified in Table 9-3 will produce the minimum flow rate.

The flushing should continue until the system is clean. This is accomplished by flowing a hydrant that will flush the new main. The flow should be through a length of 2½-in (6.4-cm) or larger hose with a burlap bag tied over

**Table 9-3**
*Minimum Flushing Rates*

| Line Size, in (cm) | Flow Rate, gpm (lpm) |
|---|---|
| 6 (15.2) | 880 (3,331) |
| 8 (20.3) | 1,560 (5,905) |
| 10 (25.4) | 2,440 (9,236) |
| 12 (30.5) | 3,520 (13,325) |

the end of the hose. After 2 minutes of flowing, the hydrant should be shut-off to examine the interior of the bag for rocks or other possible obstructing material. If foreign material is found, the underground mains should be flushed until they are clear.

### 9.8.1.2. *Hydrostatic Testing*

Hydrostatic testing is performed to ensure the integrity of the system. Following the hydrotest, and before the system is put in service, a leak test is performed to ensure that the assembled piping will not leak. Testing is generally performed before the underground piping is buried, nozzles are added in sprinkler systems, and acceptance testing with agents is performed. Table 9-4 provides general guidance on hydrostatic testing requirements.

**Table 9-4**
*Hydrostatic Test Requirements*

| System | Criteria | NFPA Reference |
|---|---|---|
| Sprinkler | The test should be made at no less than 200 psi (1,379 kPa) for 2 hours or 50 psi (345 kPa) above static pressure, where static pressure is in excess of 150 psi (1,034 kPa) for 2 hours. Additionally, dry pipe systems require an air test after the hydrostatic test. The air test is at an initial air pressure of 40 psig (276 kPa) with loss not exceeding 1.5 psig (10.4 kPa) in 24 hours. | 13 |
| Water Spray | The test should be made at no less than 200 psi (1,379 kPa) for 2 hours or 50 psi (345 kPa) above static pressure, where static pressure is in excess of 150 psi (1,034 kPa) for 2 hours. | 15, 16 |
| Foam | The test should be made at no less than 200 psi (1,379 kPa) for 2 hours or 50 psi (345 kPa) above static pressure, where static pressure is in excess of 150 psi (1,034 kPa) for 2 hours. | 11, 11A |
| Carbon Dioxide | None. | 12 |
| Dry Chemical | Not required, if a pressure test is conducted, then a dry gas should be used. | 17 |
| Clean Agent | Pneumatic test of piping at 40 psi (276 kPa) for 10 minutes. | 2001 |
| Fire Water Distribution Piping | The test should be made at no less than 200 psi (1,379 kPa) for 2 hours or 50 psi (345 kPa) above static pressure, where static pressure is in excess of 150 psi (1,034 kPa) for 2 hours. | 24 |
| Fire Pumps | Suction and discharge piping should be hydrostatically tested to 200 psi (1,379 kPa) or 50 psi (345 kPa) above the design pressure for 2 hours. | 20 |

Underground mains should be tested before joints are covered. Water leakage through the joints should be measured by pumping at the specified test pressure from a calibrated container into the section of pipe being tested. Care must be taken to expel all entrapped air and ensure the main is completely full of water. Permissible leakage per 100 joints per hour is 2 quarts (1.9 l) (NFPA 24).

Hydrostatic testing carries with it the possibility of releasing high-pressure fluids. Precautions should be taken during hydrostatic testing, such as:

- Ensuring all air has been removed from the system
- Securing all lines to prevent movement in the event of a rupture
- Ensuring personnel stand clear of the operation

### 9.8.2. Fire Water Pumps

New fire water pump installations require acceptance testing as defined in NFPA 20. This section highlights some of the important tests required.

All fire water pump piping requires a hydrostatic test to ensure no leakage or loss of pressure. The test should be made at no less than 200 psig (1,379 kPa) for 2 hours or 50 psig (345 kPa) above static pressure, where static pressure is in excess of 150 psig (1,034 kPa). In systems with differential dry pipe valves, the clappers should be left open during these tests to prevent damage.

The fire water pump should be flow tested to determine its performance. The flow test should validate the certified fire water pump curve provided by the manufacturer. The fire water pump should run for at least 1 hour.

All actions required of the controller must be function tested. The test requires starting the fire water pump six times automatically and six times manually. Following each start, the pump must be run at rated speed for at least 5 minutes.

### 9.8.3. Water Tanks

The acceptance test for the water storage tank should follow the requirements of the design specification.

### 9.8.4. Sprinkler Systems

Sprinkler system acceptance test requirements are detailed in NFPA 13. This section highlights some of the important factors in acceptance test of sprinkler systems.

#### 9.8.4.1. Installation

The completed installation should be checked against the approved working drawings to verify that the installation is in accordance with the design specification. This is accomplished by checking each sprinkler head, pipe diameter, and hanger in the system against the approved working drawings.

#### 9.8.4.1.1. DESIGN COVERAGE

The design coverage of the sprinkler system should be verified to ensure all areas requiring protection are protected. In many cases, the design of the sprinkler system has been completed before construction began. There have been instances where additional sprinklers were required, usually along walls, because of incorrect building dimensions on drawings.

Design coverage is the square foot coverage per sprinkler and is the product of the distance between branch lines times the distance between sprinklers on branch lines. Also, the distances from end branch lines and end sprinklers on branch lines to wall must be checked to verify these distances do not exceed one-half the distance between branch lines or sprinklers on branch lines. This has a direct bearing on the coverage at the perimeter of the building.

#### 9.8.4.1.2. PIPE PITCH

To ensure proper drainage, inspection personnel must examine overhead piping for proper pitch. Sprinkler systems are designed to drain back to the sprinkler riser, where a 2 in (5.1 cm) valve drain outlet is provided. When this is not possible and some piping is trapped, supplementary low point drain valves should be provided. Low-point drains should be shown on the system drawings.

The presence of adequate pipe slope may be obvious by virtue of the roof slope or may be determined by sighting down the piping. Where there is any doubt concerning pipe slope, a spirit level should be used. Design slope is ¼ in (0.64 cm) per 10 ft (3 m) of pipe for feed mains for dry pipe sprinkler systems, and ½–¾ in (1.3–1.98 cm) per 10 ft (3 cm) of branch line piping for dry pipe sprinkler systems. Proper drainage is extremely important for dry pipe systems since these systems are utilized in areas subject to freezing temperatures. Wet pipe sprinkler systems piping can be level (NFPA, 13).

#### 9.8.4.1.3. OBSTRUCTIONS

It is important to check the sprinklers for possible obstructions. Typical obstructions are building framing, wind bracing, piping, process equipment, partitions, platforms, and ducts. Sprinkler spacing is based on an effective discharge radius of 7.5 ft (2.3 m) and maximum coverage of 130 $ft^2$ (12.1 m) per sprinkler. If a sprinkler is located adjacent to a framing member, up to 50% of the discharge pattern and sprinkler coverage for that sprinkler could be affected. Specific information relative to minimum clearance from joists, beams, girders, and trusses is in NFPA 13. Obstructions are especially critical in early suppression fast response (ESFR) sprinkler installations.

9.8.4.1.4. SPRINKLER TEMPERATURE RATING

Sprinklers are available in various temperature ratings. It is important to check for proper temperature rating of sprinklers in close proximity to hot surfaces, such as steam pipes, ducts, unit heaters, or hot equipment. For example, high temperature sprinklers should be used when located within 30 in (76 cm) vertically and 12 in (30.4 cm) to the side of a steam pipe or under skylights.

Sprinkler frames are color coded to identify the temperature rating as described in Table 9-5.

Further detail concerning proper temperature rating can be found in NFPA 13.

9.8.4.1.5. PIPE HANGERS

Facility personnel should check sprinkler pipe hangers for proper type, spacing, and installation. The intent is to use heavy duty hangers with relatively close spacing so that the sprinkler piping will withstand considerable shock without failure. Sprinkler piping should be supported from the structure and not from process piping.

### *9.8.4.2. Alarm Tests*

A number of agencies (security, control room, public fire department , etc.) may be able to monitor the alarm system via direct connections. These agencies should be notified prior to a test commencing and upon completion of the test. Receipt of each alarm should be confirmed by those agencies.

The alarm is tested on wet pipe systems by fully opening the inspector's test connection valve. This test connection is piped to outdoors and fitted with

**Table 9-5**
*Sprinkler Color Coding and Temperature Rating*

| Rating | Maximum Temperature at Sprinkler Level °F (°C) | Rated Temperature of Sprinkler °F (°C) | Frame Color | Glass Bulb Color |
|---|---|---|---|---|
| Ordinary | 100 (38) | 135–170 (57–77) | Uncolored | Orange/Red |
| Intermediate | 150 (66) | 175–225 (79–107) | White | Yellow/Green |
| High | 225 (107) | 250–300 (121–149) | Blue | Blue |
| Extra High | 300 (149) | 325–375 (163–191) | Red | Purple |
| Very Extra High | 375 (191) | 400–475 (204–246) | Green | Black |
| Ultra High | 475 (246) | 500–575 (260–302) | Orange | Black |

an orifice equal to the smallest sprinkler head in the system. Water discharge through this orifice simulates operation of one sprinkler and should be sufficient to activate the alarms within 90 seconds.

Alarms on dry pipe and deluge systems are activated when the valves are tripped. These systems are equipped with permanently piped direct water bypasses to permit testing alarms without tripping the valves.

#### *9.8.4.3. Restoring System*

Final acceptance should include a turn test of all valves in the system to ensure they are operable and wide open. After verifying all valves are wide open, they should be sealed or locked open.

### *9.8.5. Water Spray Systems*

Supplementing the general details of test and examination of new installations outlined in Section 9.8.1, it is required to test deluge systems. These tests verify operation of automatic release mechanism, deluge valve, and water spray coverage. These tests should be witnessed and test data obtained during the contractor's test of the system. The deluge system should be tested in the following order:

1. Check number, spacing, and piping of pilot heads, heat actuating devices, or other types of initiating devices against the working drawings.
2. Replace water gauge at deluge valve with test gauge. Remove highest water spray nozzle in system and install test gauge.
3. Check the area to determine that the water discharged will not adversely effect the immediate area and the surface drainage will not adversely effect the surrounding area, including open excavations. Cover equipment and systems considered to be sensitive to water.
4. Trip the deluge valve manually or by activating an initiating device, such as a heat detector. Record the elapsed time between application of heat to the detector and opening of the valve (acceptance criterion is less than 40 seconds). Observe whether alarms operate satisfactorily. Record flowing pressure at deluge valve and highest nozzle.
5. Check all nozzles to ensure they are discharging in a uniform pattern and are not obstructed.
6. Check positioning of nozzles and spray patterns on protected equipment. The entire surface of the protected area should be covered by the direct impingement of the spray pattern. If portions of the protected equipment are dry or minimally wetted, the system must be readjusted as necessary and then retested.

7. Determine that each circuit of the automatic release mechanism is provided with an accessible heat sensing device for testing purposes.
8. Test low-pressure alarm on automatic release mechanism by loosening the plug at one of the heat actuating devices or a fitting on the tubing system. Retighten following alarm.
9. Test all manual trip stations, including remote stations.
10. If the system is tested with sea water, consider flushing system with fresh water.
11. After completing these tests, return the system to service in accordance with the manufacturer's instructions.

### 9.8.6. Carbon Dioxide Systems

The acceptance test of a carbon dioxide total flooding or local application consists of:

- Thorough visual inspection of the system including piping and piping supports, nozzle locations, placement of signs, and openings that would let carbon dioxide escape.
- A discharge test to ensure the design concentration is achieved and maintained for duration specified (generally 20 minutes).

NFPA 12 should be followed for the acceptance testing of carbon dioxide systems.

> ***WARNING: Personnel should not enter the test area due to the possibility of asphyxiation. The test area should be treated as an IDLH atmosphere until testing proves otherwise.***

### 9.8.7. Foam–Water Sprinklers and Water Spray Systems

The acceptance test for foam–water sprinkler and water spray systems is similar to systems without foam. Tests include:

- Flushing—see Section 9.8.1.1.
- Hydrostatic test—see Section 9.8.1.2.
- Discharge test to ensure the hazard is fully protected. The flow should ensure that the foam concentrate injected into the system is within the design specification.

### 9.8.8. Clean Agent Systems

The acceptance test of a clean agent system consists of:

- Thorough visual inspection of the system including piping and piping supports, nozzle locations, containers properly mounted, and placement of signs.
- A pneumatic test of piping should be conducted at 40 psi (276 kPa) for 10 minutes. After 10 minutes the pressure drop should not exceed 8 psi (55.1 kPa).
- A flow test of the piping to ensure the piping is continuous should be conducted using air or nitrogen. Note that oil may accumulate on the interior of the piping and may come out and make a mess during the flow test.
- The enclosure for the clean agent system should be examined for openings and tested to ensure that the agent concentration can be maintained. Currently, the preferred method is using a blower door fan unit and smoke pencil.
- All electrical systems, including detection, alarms, supervisory systems, and manual activation systems, should be function tested to ensure proper operation.
- A discharge test should be conducted to ensure the design concentration can be achieved and maintained for the minimum required duration.

NFPA 2001 should be followed for the acceptance testing of clean agent systems.

### 9.8.9. Foam Systems

The acceptance test for foam systems is defined in NFPA 11. Highlights of the acceptance requirements include:

- Flushing—see Section 9.8.1.1.
- Hydrostatic test—see Section 9.8.1.2.
- Flow test for deluge and water spray—see Section 9.8.3. Additionally the flow test should ensure that foam coverage meets the design requirements.
- Proportioning system test—The flow should ensure that the foam concentrate injected into the system is within the design specification.

Not all foam systems can be tested. There are some conditions where discharge of foam in the protected area during testing is undesirable, such as for subsurface injection or placing foam on an internal floating cover.

# 10

# INSPECTION, TESTING, AND MAINTENANCE

The inspection, testing, and maintenance of fire protection systems by qualified personnel is essential to ensure the operability of the systems during an emergency.

This chapter covers key elements of an effective inspection, testing, and maintenance program:

- Establishing ownership for the inspection, testing, and maintenance of fire protection systems.
- Qualifying personnel performing the inspections, testing, and maintenance of fire protection systems.
- Developing an inspection, testing, and maintenance program.
- Developing inspection and testing requirements.

## 10.1. Ownership of Fire Protection Systems

A qualified member of the management team should be responsible for managing the fire protection program and assuring that an adequate budget is provided for fire protection maintenance. The assignment of this responsibility may be dictated by the size of the facility and the staffing levels, for example:

- *Large facilities* may have a Fire Protection Department led by a member of the management team. Facilities with Fire Protection Departments could manage the inspection, testing, and maintenance of fire protection systems in different ways:

- Personnel assigned to the Fire Protection Department may be responsible for all elements of the fire protection program, including inspection, testing, and maintenance activities.
- The Fire Protection Focal Point may assign responsibilities to different groups at the facility. The maintenance group may be assigned the responsibility for performing a weekly test of fire pumps, conducting repairs, or scheduling repairs by qualified fire protection service companies. The operations group may be assigned the responsible for daily or weekly visual inspections of the fixed fire protection systems.
- The Fire Protection Focal Point may utilize a service company to conduct periodic inspections, flow testing, repairs, and maintenance of the systems.

■ *Medium to small facilities* may assign responsibility for the fire protection system to a manager who has additional areas of responsibilities. The manager may assign the various program elements to different groups or to one selected employee. Minor maintenance activities may be performed by the maintenance department or by a fire protection service company. Periodic inspection, flow testing, and major maintenance activities may be assigned to a fire protection service company.

No matter how the responsibilities for completing the different elements of the fire protection program are assigned, one person must be ultimately responsible for the management of fire protection equipment and systems. This person has the overall responsibility for managing the fire protection equipment and systems. This responsibility includes the following activities:

- Develop and manage the program to insure the integrity of fire protection equipment and systems.
- Identify the fire protection equipment and systems that must be inspected tested and maintained.
- Develop inspection criteria.
- Determine frequency of inspection and testing.
- Monitor inspections, testing, and maintenance.
- Review inspection and test data.
- Maintain documentation of inspections, testing, and repairs.
- Qualify fire protection service companies to perform work at the facility.
- Develop projects to improve the system based on results of the inspection and testing programs.
- Develop and manage the fire protection equipment and systems impairment program.
- Notify management and affected units when the fire protection equipment and systems are impaired.

The person assigned the responsibility to manage the fire protection equipment and systems must approve all decisions that affect the integrity of the system.

## 10.2. Qualifications of Personnel

This section deals with the qualifications of personnel who manage the fire protection equipment and systems and conduct the inspections, testing, and maintenance. This includes qualifying service companies and qualification of service company employees.

### *10.2.1. Fire Protection Focal Point*

The person assigned the responsibility of managing the integrity of fire protection equipment and systems must have a general knowledge of the facility and its fire protection needs, be familiar with the facilities fire protection systems, and have access to resources to ensure the equipment and systems are operable.

### *10.2.2. Inspection Personnel*

Facility personnel who perform inspections on fixed fire protection systems must be knowledgeable of the systems and have received training on the inspection protocols.

### *10.2.3. Testing and Maintenance Personnel*

Facility personnel who perform testing and maintenance on fire protection equipment and systems must be experienced and knowledgeable in the systems and on the protocols for testing and maintenance. Knowledge can include work history, educational experience, craft certification, manufacturer certification, field verification, and job assessment and testing. Testing and maintenance personnel may be pump mechanics, pipe fitters, instrument technicians, electrical technicians, millwrights, fire protection personnel or other qualified personnel.

### *10.2.4. Fire Protection Service Companies*

Many facilities outsource either all or part of the inspection, testing, and maintenance functions to fire protection service companies. These companies perform a variety of services, including inspection of portable and fixed systems, flow test of water systems, operational test of portable equipment, testing of fixed powder system, maintenance and repairs of portable and fixed systems, and design and installation of new systems.

It is important that the outside service company meet the facility pre-qualification criteria. As part of the review, the facility pre-qualification must determine if the service company is knowledgeable and experienced in the types of fire protection systems they will be servicing. Some states have licensing laws that require the service company designate a responsible managing employee who is licensed by the state. The service company should be able to provide evidence that the personnel performing inspections, testing, and maintenance are qualified to perform the work.

The qualification may include licenses and craft certification certificates issued by a recognized issuing authority.

#### *10.2.4.1. Prequalifying Service Companies*

Before any fire protection service company is selected to perform inspecting, testing, and maintenance on fire protection systems at the facility, they should be qualified. The service company qualifications may include:

- A written safety program that includes the following topics: new employee safety orientation, safety training, hazard communication, emergency action plan, fall protection, scaffolding, heat stress, personal protective equipment, electrical, first aid, and bloodborne pathogens.
- Acceptable injury and illness rates.
- Acceptable insurance modifier rate.
- Proof of insurance and workman's compensation coverage.

Refer to API RP 2220, *Improving Owner and Contractor Safety Performance* and API RP 2221, *A Manager's Guide to Implementing a Contractor Safety and Health Program* for information on the evaluation of all service contractors.

#### *10.2.4.2. Qualifications of Service Company Personnel*

The level of qualifications for personnel expected should be specified in the bid package or purchase order. The facility should require the service company to provide evidence that the personnel performing inspections, testing, and maintenance meet the qualifications. Safe work procedures should be extended to all service company personnel.

## 10.3. Inspection, Testing, and Maintenance Programs

Since the occurrence of an emergency cannot be predicted, it is essential that fire protection equipment be inspected, tested, and maintained to be certain they are operable at all times. To accomplish this, a rigorous fire protection equipment inspection program must be established.

Fire protection equipment, like all other equipment, deteriorates with time. It is also vulnerable to external influences such as corrosive environments, tampering, accidental damage, and careless use. Further, since fire protection equipment is used infrequently, it must be inspected and tested regularly to determine its condition, operability, and need for routine maintenance. Detecting an unsatisfactory condition prior to an emergency is far better than discovering it during the emergency.

Key components of a fire protection equipment inspection, testing, and maintenance program:

- Assign fire equipment inspection responsibility to a qualified, trained person.
- Document all fire protection equipment in the facility.
- Establish inspections/tests required and frequencies for each type of fire protection equipment.
- Select and train individuals to conduct the inspections and tests of fire protection equipment.
- Establish effective procedures for reporting and reviewing inspection results.
- Initiate prompt action to correct any noted deficiencies.
- Arrange qualified outside contractors for periodic testing and emergency maintenance of specialized fire and explosion protective systems.

The inspection, testing, and maintenance program details the level of inspection, testing, and maintenance for each type of fire protection system. The program should have written protocols for each type of equipment to be inspected and tested. The protocol should detail how to conduct the inspection and how to document the findings, including directions on reporting and correcting deficiencies found during the inspection and testing. Maintenance of the equipment should follow the manufacturers guidelines or established facility protocols.

### *10.3.1. Inspections*

Inspections consist of a visual check of the fire protection systems by qualified personnel to ensure the system is available for immediate use. Visual checks should be made using a checklist. The inspection should verify that the fixed and portable equipment is ready to be operated. The associated devices, fittings, piping, and valving are inspected to ensure they have not been tampered with and that there is no obvious deterioration, physical damage, or condition to prevent operation. Documentation of the visual inspections may vary from using a formal checklist to making a note in a log book. Formal documentation, in conjunction with a checklist, is the best method to ensure the equipment and the systems are being monitored and are in a "state-of-readiness."

### *10.3.2. Testing*

Testing involves the operation or activation of equipment or a device to verify it will perform as intended. Testing includes activation of alarms; test run of the fire pumps; and water flow and pressure testing of water systems, fire pumps, monitors, and hydrants. Testing also includes operation of fixed dry chemical, $CO_2$, and other suppression systems. Portable fire protection equipment, such as fire trucks and portable extinguishers should be included in the testing program. Testing procedures and protocols should meet accepted practices developed by recognized authorities, such as NFPA. Documentation of the tests should be written, reviewed by qualified personnel, and become part of the equipment history file.

### *10.3.3. Maintenance*

Maintenance includes a thorough check of equipment or a device to give maximum assurance that the equipment or device will operate effectively and safely. It may be necessary to disassemble and reassemble the equipment or device to ensure a through examination. Maintenance procedures should follow manufacture's recommended guidelines or accepted practices developed by recognized authorities, such as NFPA. Documentation of maintenance activities should be part of the facility's mechanical integrity program and all documentation should be part of the equipment history file.

### *10.3.4. Identification of Deficiencies*

Each facility needs a system to manage the deficiencies found during the inspections, testing, and maintenance of the systems. Managing deficiencies in fire protection systems is time-critical and an appropriate system must be developed to recognize, prioritize, and manage these deficiencies. It is important that the system addresses how to:

- Report the deficiency
- Determine if the deficiency affects the integrity of the system
- Communicate to the affected employees that the system is out-of-service
- Authorize the correction of the deficiency
- Track any corrective action to completion
- Communicate to affected personnel the deficiency has been corrected

### 10.3.5. Frequencies of Inspection, Testing, and Maintenance

Frequencies of inspection, testing, and maintenance should be based on codes and standards that apply to the equipment or system as well as operational experience. When developing inspection, testing, or maintenance frequencies, codes and standard should be consulted for the required frequency. Due to environmental conditions, physical conditions, or other factors, it may be necessary to conduct the testing more frequently than recommended by the manufacturer or applicable codes or standards. The integrity of fire protection systems is essential to ensuring they work during an emergency and provide the designed protection.

### 10.3.6. Documentation of Inspection and Testing

Records of inspection, testing, and maintenance of fire protection systems should be maintained. Records should indicate the procedures performed, the person performing the work, results of the inspection, testing, or maintenance, and the date the work was performed.

The original design information should be maintained by the owner for the life of the fire protection systems. Records of inspection, testing, and maintenance should be maintained for at least one year after the next scheduled inspection, testing, or maintenance.

Electronic management systems for maintaining records of inspection, testing, and maintenance are preferred. Bar coding, accompanied with the use of a personal data assistant (PDA) can significantly reduce the amount of time required for recordkeeping.

### 10.3.7. Impairment Handling

Fire protection system impairment occurs when a fire alarm or supervisory system is shut-off, damaged, fails, or is otherwise taken out-of-service completely or in part. These out-of-service conditions are called *impairments*. While process monitoring, control, safety, and security-entry systems also provide protective functions, this element of the fire prevention program is only concerned with impairments to fire protection systems and equipment. An essential element of the fire prevention program is a procedure for fire protection impairment handling.

While planned fire protection system impairments may be necessary during testing, maintenance, renovation, and new construction, it must be understood that when any fire protection impairment occurs, the facility is at increased risk of loss. Many large losses that have occurred may have remained small if the fire protection impairments were properly managed.

When a fire protection impairment is planned or occurs accidentally, precautionary measures are necessary to minimize the risk. These may include arranging temporary protection, reducing hazards, and ensuring continual and speedy progress on repair and restoration efforts.

The likelihood of a fire or explosion occurring while protection is impaired increases with the duration of the impairment. Thus, proper procedures and practices must be followed to minimize the duration and scope of planned and emergency impairments and to reduce the possibility of a hidden or unknown impairment.

## 10.4. Inspection and Testing Requirements

### *10.4.1. Fire Protection Systems and Equipment Covered*

Table 10-1 references the applicable NFPA code for the majority of the fire protection systems found in petrochemical and hydrocarbon processing facilities.

### *10.4.2. Water-Based*

Table 10-2 was developed using NFPA 25, *Inspection, Testing and Maintenance of Water-Based Fire Protection Systems*. Additional requirements may apply depending on the type of system and source of the water supply.

Dry-pipe, pre-action, and deluge systems are different types of sprinkler systems where the pipe is not filled with water. Generally, air fills the piping until a fire occurs. A dry-pipe, pre-action, or deluge valve then is opened and allows water to fill the piping. These types of systems are common in areas that are subject to freezing where it is not possible to heat the space. Since water left in the pipes will freeze, expand and cause damage, special procedures need to be implemented to ensure the integrity of the systems.

### *10.4.3. Fire Water Distribution System*

Fire water distribution systems should be inspected and tested regularly to ensure water can be delivered to all locations as required. Table 10-3 summarized the inspection and tests for components in a fire water distribution system.

### *10.4.4. Fire Pumps*

Fire pumps should have a weekly and annual inspection and test. Specific requirements for fire pump inspection, testing, and maintenance are contained in NFPA 25.

**Table 10-1**
*Applicable Fire Protection System NFPA References*

| System | NFPA Reference |
|---|---|
| Water Supply | NFPA 22—*Standard for Water Tanks For Private Fire Protection* |
| | NFPA 24—*Standard for the Installation of Private Fire Service Mains and Their Appurtenances* |
| Water Distribution | NFPA 13—*Standard for the Installation of Sprinkler Systems* |
| | NFPA 14—*Standard for the Installation of Standpipe and Hose Systems* |
| | NFPA 25—*Standard for the Inspection, Testing and Maintenance of Water-Based Fire Protection Systems* |
| | NFPA 291—*Recommended Practice for Fire Flow Testing and Marking of Hydrants* |
| Fire Pumps | NFPA 20—*Standard for the Installation of Centrifugal Fire Pumps* |
| | NFPA 24—*Standard for the Installation of Private Fire Service Mains and Their Appurtenances* |
| | NFPA 25—*Standard for the Inspection, Testing and Maintenance of Water-Based Fire Protection Systems* |
| | NFPA 750—*Standard on Water Mist Fire Protection System* |
| Detection and Alarm Systems | NFPA 25—*Standard for the Inspection, Testing and Maintenance of Water-Based Fire Protection Systems* |
| | NFPA 72—*National Fire Alarm Code®* |
| Fixed Systems | NFPA 11—*Standard for Low-Expansion Foam* |
| | NFPA 11A—*Standard for Medium- and High-Expansion Foam Systems* |
| | NFPA 12—*Standard on Carbon Dioxide Extinguishing Systems* |
| | NFPA 12A—*Standard on Halon 1301 Fire Extinguishing Systems* |
| | NFPA 13—*Standard for the Installation of Sprinkler Systems* |
| | NFPA 15—*Standard for Water Spray Fixed Systems for Fire Protection* |
| | NFPA 25—*Standard for the Inspection, Testing and Maintenance of Water-Based Fire Protection Systems* |
| | NFPA 750—*Standard on Water Mist Fire Protection System* |
| | NFPA 2001—*Standard on Clean Agent Fire Extinguishing Systems* |
| Portable Systems | NFPA 10—*Standard for Portable Fire Extinguishers* |
| | NFPA 1911—*Standard for Service Tests of Pumps on Fire Department Apparatus* |
| | NFPA 1914—*Standard for Testing Fire Department Aerial Devices* |
| | NFPA 1915—*Standard for Fire Apparatus Preventative Maintenance Program* |
| Miscellaneous | NFPA 80—*Standard for Fire Doors and Fire Windows* |
| | NFPA 221—*Standard for Fire Walls and Fire Barriers Walls* |

**Table 10.2**
*Water-Based Fire Protection Systems Inspection and Testing Requirements*

| Frequency | Item To Be Inspected or Tested |
|---|---|
| Daily | Enclosures around dry-pipe, pre-action, and deluge valves heated to at least 40°F (4.4°C) |
| Weekly | Valves free from damage, control valves (with seals) open, relief port on backflow device not discharging |
| Monthly | Gauges in good working condition and reading normal pressure, control valves (with locks or electronic supervision) open, fire department connections in good condition |
| Quarterly | Main drain test, master alarm test |
| Annually | Sprinklers free from corrosion, paint, and obstructions; pipe free from corrosion; control valves lubricated, closed and opened; trip test on dry-pipe, pre-action, and deluge systems |
| Annually (prior to freezing weather) | Antifreeze correct strength; adequate heat to areas where pipe prior to freezing is filled with water; low point drained in piping on dry-pipe, weather pre-action and deluge systems |

**Table 10-3**
*Inspection, Testing, and Maintenance for Fire Water Distribution Systems*

| Component | Description of Test | Frequency |
|---|---|---|
| Storage tank | ▪ In freezing weather, inspect tank temperature to ensure it is greater than 40°F (4°C). If the tank is equipped with a low-temperature alarm, then inspection can be weekly.<br>▪ Inspect water level and condition of water.<br>▪ Inspect exterior of tank, support structure, vents, piping for obvious damage.<br>▪ Test high and low level alarms. | Daily<br>Monthly<br>Quarterly<br>Yearly |
| Exposed piping | ▪ Inspect for leaks, damage, and restraints in place. | Yearly |
| Underground piping | ▪ Flush system following the guidance in Chapter 9. | Yearly |
| Hydrants | ▪ Inspect for proper operation, flush the hydrants.<br>▪ Inspect for leaks, worn threads, and accessibility.<br>▪ Lubricate valve and threads. | Yearly |
| Monitors | ▪ Inspect for proper operation, flush the monitors.<br>▪ Inspect for leaks and accessibility.<br>▪ Lubricate moving parts. | Yearly |
| Fire hose stations | ▪ Inspect hoses for leaks. | Yearly |
| Piping | ▪ Conduct flow test per NFPA 24. | 5 Years |
| Storage Tanks | ▪ Inspect interior of tank per NFPA 25. | 5 Years |

Weekly inspection and testing consists of visual inspection of oil and fuel levels and fire pump components, including the controller. A function test of the fire pump, driver and starting systems should be conducted. The fire pump should run for a minimum of 10 minutes for electric motor drivers and 30 minutes for other drivers.

Annual inspection, testing, and maintenance consist primarily of a full flow test of the pump. The flow test should be conducted at three points: no flow (churn), 100% of rated flow capacity and 150% of rated flow capacity. The results of the flow test should be compared to previous flow tests and the design curve. Increasing the engine speed beyond the rated speed of the pump at rated conditions is not an acceptable method for meeting the rated pump performance. If the pump curve is more than 10% below the last flow test, then this indicates a potential problem with the pump and additional maintenance may be required.

### 10.4.5. Foam Systems

All foam systems should be thoroughly inspected and checked for proper operation at least annually. The inspection should include performance evaluation of the foam concentrate or premix solution quantity or both. Test results that deviate more than 10% from those recorded in acceptance testing should be discussed immediately with the manufacturer. The inspection report should be documented with any deficiencies or recommendations included. Foam system inspections should include:

- Proportioning devices, their accessory equipment, and foam makers should be inspected.
- Fixed discharge outlets equipped with frangible seals should be provided with suitable inspection means to permit proper maintenance and for inspection and replacement of vapor seals.
- Above ground piping should be examined to determine its condition and to verify that proper drainage pitch is maintained. Underground piping should be spot-checked for deterioration at least every five years.
- Strainers should be inspected periodically and be cleaned after each use and flow test.
- Control valves, including all automatic and manual-actuating devices, should be tested at regular intervals.
- An annual inspection should be made of foam concentrates and their tanks or storage containers for evidence of excessive lumps or deterioration. Samples of concentrates should be sent to the manufacturer or qualified laboratory for quality condition testing.

Daily checks of foam systems should be made where freezing conditions exist to ensure the system is functional. Operating and maintenance instructions and layouts should be posted at control equipment with a second copy on file. All persons who are expected to inspect, test, maintain or operate foam-generating apparatus should be thoroughly trained. Training records should be kept up to date.

### 10.4.6. Portable Fire Extinguishers

A monthly visual inspection of fire extinguishers by a competent person should be made and documented. A trained person who has undergone the instructions necessary to reliably perform maintenance and has the manufacturer's service manual should service the fire extinguishers on at least an annual frequency.

Periodic inspection of fire extinguishers should include a check of at least the following items:

- Location in designated place
- No obstruction to access or visibility
- Operating instructions on nameplate legible and facing outward
- Safety seals and tamper indicators not broken or missing
- Fullness determined by weighing or lifting
- Examination for obvious physical damage, corrosion, leakage, or clogged nozzle
- Pressure gauge reading or indicator in the operable range or position
- Condition of tires, wheels, carriage, hose, and nozzle checked (for wheeled units)
- Appropriate label in place

The date the inspection was performed and the initials of the person performing the inspection should be recorded. Records should be kept on a tag or label attached to the fire extinguisher, on an inspection checklist maintained on file, or in an electronic system that provides a permanent record.

Stored-pressure types containing a loaded stream agent should be disassembled on an annual basis and subjected to complete maintenance. Prior to disassembly, the fire extinguisher should be fully discharged to check the operation of the discharge valve and pressure gauge. The loaded stream charge should be permitted to be recovered and reused, provided it is subjected to agent analysis in accordance with manufacturer's instructions.

A conductivity test should be conducted annually on all carbon dioxide hose assemblies. Hose assemblies found to be nonconductive should be replaced. Pressure regulators provided with wheel-typed fire extinguishers should be tested for outlet static pressure and flow rate in accordance with manufacturer's instructions.

Maintenance, servicing, and recharging should be performed by trained persons having available the appropriate servicing manuals, the proper types of tools, recharge materials, lubricants, and manufacturer's recommended replacement parts or parts specifically listed for use in the specific fire extinguisher.

Maintenance procedures should include a thorough examination of the three basic elements of a fire extinguisher:

- Mechanical parts
- Extinguishing agent
- Expelling means

Each fire extinguisher should have a tag or label securely attached that indicates the month and year the maintenance was performed and that identifies the person performing the service.

### 10.4.7. Dry Chemical Extinguishing Systems

On a monthly basis, inspection should be conducted in accordance with the manufacturer's maintenance manual. As a minimum, this visual inspection should include verification that the:

- Extinguishing system and nozzles are in the proper location for the hazard
- Manual actuators are unobstructed
- Tamper indicators and seals are intact
- Maintenance tag or certificate is in place
- System shows no physical damage or condition that might prevent operation
- Pressure gauge(s), if provided, is in operable range
- Nozzle blow-off caps, where provided, are intact and undamaged
- Protected equipment or the hazard has not been replaced, modified, or relocated

At least semiannually, maintenance should be conducted in accordance with the manufacturer's maintenance manual, including the following:

- A check to see that the hazard has not changed.
- An examination of all detectors, expellant gas container(s), releasing devices, piping, hose assemblies, nozzles, signals, and all auxiliary equipment.
- Verification that the agent distribution piping is not obstructed.

- Examination of the dry chemical (if there is evidence of caking, the dry chemical should be discarded and the system should be recharged in accordance with the manufacturer's instructions).
- All dry chemical systems should be tested, which should include the operation of the detection system, signals and releasing devices, including manual stations and other associated equipment. A discharge of the dry chemical normally is not part of this test.
- Each dry chemical system should have a tag or label indicating the month and year the maintenance is performed and identifying the person performing the service. Only the current tag or label should remain in place.

The inspection should be documented with any deficiencies or recommended included.

### *10.4.8. Carbon Dioxide Extinguishing Systems*

At least weekly, the liquid level gauges of low-pressure containers should be observed. If, at any time, a container shows a loss of more than 10%, it should be refilled, unless the minimum gas requirements are still provided.

A monthly inspection should be conducted to assess the system's operational condition.

At least semiannually, all high-pressure cylinders should be weighed and the date of the last hydrostatic test noted. If, at any time, a container shows a loss in net content of more than 10%, it should be refilled or replaced.

The following should be verified by competent personnel at least annually using available documentation:

- Check and test the carbon dioxide system for proper operation.
- Check that there have been no changes to the size, type, and configuration of the hazard and system.

All system hoses, including those used as flexible connectors, should be tested every five years (NFPA 12).

### *10.4.9. Clean Agent Systems*

Clean agent systems should be visually inspected semiannually. The inspection should consist of verifying the agent quantity and pressure are within allowable range.

The annual inspection consists of a thorough inspection and test for proper operation of the entire system, including the detection system. The enclosure should be inspected for new penetrations that could affect agent leakage and system performance. The enclosure inspection is not required if a documented

Management of Change program exists that includes managing changes in the enclosure.

Every five years, all hoses should be tested to 1½ times the container pressure at 130°F (54.4°C). Also, every five years the cylinders should be inspected per DOT requirements. For containers in continuous service, a visual inspection per CGA C-6, *Standard for Visual Inspection of Steel Compressed Gas Cylinders* may be conducted. The cylinder does not need to be emptied and cylinder should not be stamped.

For detailed information on inspection, testing, and maintenance refer to NFPA 2001.

### 10.4.10. Mobile Fire Equipment

Mobile fire equipment should be inspected, tested, and maintained in accordance with manufacturer's recommendations. Additional NFPA standards that may contain inspection, test and maintenance information are:

- NFPA 1901 *Standard for Automotive Fire Apparatus*
- NFPA 1911 *Standard for Service Tests of Fire Pumps Systems on Fire Apparatus*
- NFPA 1914 *Standard for Testing of Fire Department Aerial Devices*
- NFPA 1915 *Standard for Fire Apparatus Preventive Maintenance Program*
- NFPA 1961 *Standard for Fire Hoses*
- NFPA 1962 *Standard for the Care, Use, and Service Testing of Fire Hose*
- NFPA 1852 *Standard for Selection, Care, and Maintenance of Open-Circuit SCBA*

### 10.4.11. Fireproofing

Periodic inspection and testing maximizes the useful life of the fireproofing system. The manufacturer or applier may be invited to participate in the inspection. An inspection and testing program should include the following steps:

- Survey coatings for surface cracking, delamination, rust staining, or bulging.
- Survey coatings for signs of weathering (color change, powdering, thinning of coat).
- Selectively remove small sections of fireproofing to examine conditions at the face of the substrate and the surface of reinforcing wire. Repair the inspection area.

- Visually check for the loss of fireproofing materials as a result of mechanical abuse.
- When the fireproofing material is applied, coat and set aside several pieces of structural steel for periodic fire-testing over the expected life of the coating (this is not necessary with rigid box or flexible contaminant systems).
- Inspect to make sure that the fireproofing hasn't been removed for maintenance and not replaced.

Timely and consistent maintenance provides assurance that the system is physically in the condition intended. When more than hairline cracking appears, the openings should be cleaned out and filled with new material according to the manufacturer's instructions. Loss of bonding to the substrate may be detected by surface bulges or an abnormal sound when the surface is tapped with a light hammer.

In evident areas of bond failure, fireproofing should be removed and the substrate should be thoroughly cleaned and properly primed before new material is applied. If surface coating is required to prevent moisture from penetrating, it should be renewed at intervals recommended by the manufacturer. The previously listed inspections should be completed prior to renewal of coating so that defects are not hidden by the coating.

## 10.5. Inspection Checklist Examples

Checklists are a valuable tool for documenting the inspection, test, and maintenance of fire protection equipment and systems. The CD-ROM accompanying this book contains checklists developed by one company. Many small and medium size companies may not have the resources to develop these checklists.

Before using the checklists on the CD, your company should review the checklist and make sure that they are appropriate for your facility and fire protection equipment and systems.

# 11

# FIRE EMERGENCY RESPONSE

The objective of this chapter is to provide guidance on emergency response considerations, organization, plans, and training for response to fire.

Emergency response issues for fires should be addressed as part of a facility's fire protection strategy. *Emergency response* is defined as the efforts made by a coordinated group of personnel to assess, control and mitigate hazardous events. Emergency response issues can be complicated to manage and often require significant training and equipment. Emergency response organizations must also meet government and company requirements.

Emergency response capabilities have a significant impact on a fire protection and response strategy. One primary question that must be addressed is whether to *fight or not to fight fires* (other than incipient). Emergency response is one of the last layers of protection at a facility. If emergency response actions are necessary, then all the others layers of protection have not worked or have failed (CCPS, 2001a).

Emergency response can be either internal, external, or a combination of both. The degree of internal response can vary from the basic ability to properly respond with fire extinguishers and sound the alarm to a fully staffed and trained industrial fire brigade, with internal structure firefighting capability. External response may include local fire departments and mutual aid organizations. Before a facility relies on external response to perform firefighting actions within the facility, a relationship should be built with the responding organization. This relationship will allow the facility to better understand the capabilities of the external responders and the external responders to better understand the hazards they will face. It will also allow both parties to jointly agree on a practical protocol for command and decision making during a fire.

No matter how a company may provide emergency response, they must comply with federal, state, and local regulations. These regulations cover evacuation, fire brigades, hazardous materials response teams, confined space rescuers, and medical responders. These regulations generally apply to training, equipment, and other administrative matters affecting personnel and how they operate (CCPS, 1995a).

## 11.1. Considerations for Emergency Response Organizations

A decision analysis technique for assisting companies in deciding options for emergency response is contained in *Guidelines for Technical Planning for On-Site Emergencies*
(CCPS, 1995a). Factors that should be considered are divided into three broad categories:

- Response effectiveness
- Management issues
- Cost considerations

### *11.1.1. Response Effectiveness*

Table 11-1 provides response effectiveness factors that should be considered in determining the type of emergency response organization.

### *11.1.2. Management Issues*

Table 11-2 provides management considerations that should be included in the decision process for determining the type of emergency response organization.

### *11.1.3. Cost Evaluation Factors*

Table 11-3 identifies cost factors that should be considered for all emergency response organizations.

## 11.2. Develop Organization Plan

The fundamental requirement for effective emergency response is a plan that documents the elements of the emergency response process for a facility and the responsibilities of the personnel who will manage each element.

**Table 11-1**
*Response Effectiveness Factors*

| Factor | Considerations |
|---|---|
| Response Time | Rapid and safe response to an incident is essential for effective mitigation. The sooner forces mobilize and respond to the scene, the greater the chance of containing and limiting the effect of the incident. |
| Existing On-Duty Personnel | Determining and having enough trained people on duty at the facility that are available to respond effectively to an incident is essential. |
| Existing Qualifications/ Training | Consider the availability of qualified and trained personnel in existing organizations. Using an existing response team is sensible. |
| Site-Specific Knowledge | Effective responders need accurate knowledge of the materials involved and familiarity with the area in order to ensure effectiveness. |
| Existing Off-Duty Personnel | If a facility's on-duty staff is insufficient, then off-duty personnel must fill this gap or the facility can rely on mutual aid as backup. There will be a time delay associated with notification and travel of off duty personnel. |
| Availability of Response Team | There are times when a response team might be unable to respond because of other emergencies. A contingency plan must be developed for multiple emergencies. |
| Exposure to Fire Hazards | The use of personal protective equipment to prevent personnel exposure to hazards resulting from fires, such as toxic smoke. |
| Environmental Concerns | Contaminated water that runs offsite is a concern for lakes, rivers, drinking water, etc. |
| Existing Equipment | Examine suitability and availability of fixed and mobile fire suppression equipment. Suppression equipment requirements will vary with the types of hazards and configuration of the process units. Some organizations may already have suppression and special needs response equipment (HAZMAT response equipment, command vehicle, ambulance, etc.). Determine what type of suppression equipment will be needed. |

Emergency response planning should address reasonably anticipated emergency response events and the preparation for response to these events, including emergency response team staffing levels, necessary emergency response equipment, responder actions, training, and practice drills. Individual process unit emergency shutdown procedures should be coordinated with the site emergency response plan (ERP). The ERP should also include procedures for clean-up, decontamination, recovery, and the safe restarting of the process.

**Table 11-2**
*Management Considerations for Determining the Emergency Response Organization*

| Factor | Considerations |
|---|---|
| Control of Plant Emergencies | Some managers have a concern about losing control of their facility to public safety officials during emergencies. Many have built highly qualified teams to help avoid this potential. Public safety officials generally don't want to take control of an entire facility during an emergency. They usually won't do so as long as the facility has an effective emergency response program in place and a trustworthy relationship established. The question is: *Who is more qualified to mitigate the emergency, the owners of the process or the local fire department?* Ideally, this should be addressed prior to the emergency with preplanning and a joint philosophy developed between owners and external agencies. |
| Ongoing Commitment | Regardless of the option chosen, an emergency response program requires ongoing commitment and attention, including personnel training and equipment maintenance. If the community fire department is used, there must be ongoing dialog and training with the fire department on the chemicals and hazards associated with the facility. |
| Community Relations | There are several community relations considerations. For example, will a company gain credibility for helping improve a community response agency? Will a company gain even more credibility by taking care of its own problems without using community resources? Communication with the community about the hazards at the facility and what happens in the event of fire should be considered. |
| Ongoing Management/ Administration | Depending on the option chosen, a degree of administrative oversight is necessary. |

An effective ERP will:

- Identify emergency response authority and responsibilities
- Identify the Incident Command System (ICS)
- Describe the facility's emergency organization
- Identify emergency operation facilities and systems
- Address regulatory requirements
- Establish ongoing training and other preparedness programs
- Communicate specific response actions
- Identify company, community, and regulatory notification requirements
- Identify incident investigations and critique requirements
- Identify decontamination and safe start-up requirements

**Table 11.3**
*Cost Factors*

| Factor | Cost Considerations |
|---|---|
| Cost of equipment | What are the equipment costs? |
| Cost of training | What are the initial and ongoing training costs? |
| Ongoing medical surveillance | What are the costs and administrative requirements to provide annual physicals and maintain medical surveillance programs? |
| Refresher training costs | What are the costs and administrative requirements to provide annual refresher training programs? |
| Equipment maintenance costs | What are the costs and administrative burdens to test and maintain equipment? |
| Replacement cost | What are the costs of replacement of damaged processing equipment and loss of business with and without the level of emergency response being considered? |

- Address media relations and other support staff, e.g. engineering, logistics, etc.
- Identify actions to take in the event of natural hazards, such as windstorm, flooding, lightning, earthquakes, etc.
- Identify evacuation routes and procedures
- Identify shelter-in-place locations

The ERP should be written in a format that is easily understood by all personnel and be accessible to Incident Commanders (IC), responders, safety and health professionals, engineering staff, operations staff, and others involved in emergency response. Table 11-4 shows typical elements of an ERP.

The written ERP sets the stage for how well an incident is managed. All site personnel and community response groups must be trained on the ERP and on their prospective roles and responsibilities.

## 11.3. Outside Responders

Outside responders may be necessary to manage or assist during some emergencies. For some facilities, a community response organization is the primary emergency responder. Other facilities may utilize such community responders to augment their internal response organization or may call upon the emergency response organizations of nearby facilities to assist through a mutual aid agreement.

In considering the use of outside responders, the first issue is to decide what type of incidents they may be called to respond to and what role they will play when they respond. The second issue is to determine who will supply the Incident Command System (ICS) and how will facility emergency response, other employees, and all outside responding groups be integrated under the ICS.

For sites that exclusively use a community response organization, the threshold event for activating the community response organization should be defined for planning and decision purposes. Sites with in-plant response organizations my legally be required to notify the municipal fire department of events that they can control without outside assistance because of fire reporting and investigation responsibilities.

### 11.3.1. Integration of the Facility and Community Response Organization ICS

The facility must determine which agency will have jurisdiction. Some facilities may have internal emergency response organizations and an existing ICS. The facility may reside in a jurisdiction where the outside response agency is required or expected to be called on all events. In these situations, the community response organization is often required by law or regulation (or just expects) to use its ICS.

A plan to integrate the facility, community response organization, and/or mutual aid group ICS must be included in the ERP.

Where the community response organization supplies the incident command system, the role of the facility employees will need to be defined:

- *Who will represent the company and in what role in the ICS?*
- *Who will be the official spokesperson?*
- *Will employees simply evacuate after shutting down operating units to a preplanned state?*
- *Will some employees remain to provide advice to the incident commander or operations officer?*
- *Will trained and qualified employees assist the community fire brigade in suppression activities?*
- *Who will decide when it is safe to return control of the emergency site to the owner?*

Where facilities have internal emergency response organizations and an ICS, but reside in a jurisdiction where the outside response agency is required or expected to be called on all events or utilize community or mutual aid groups, the role of the internal emergency response organization must be defined:

**Table 11-4**

*Description of Key Sections of a Comprehensive ERP*

| Section | Title | Description |
|---|---|---|
| I | Basic Plan or Executive Summary | Provides an overview of the emergency management program including organization, responsibilities, emergency action levels, notification and facilities. Briefly describes potential emergencies, assumptions, credible incidents and other situations, ERP maintenance, and distribution. This might include organization statements required by OSHA, such as a fire brigade organizational statement. |
| II | Prevention Procedures | These procedures describe the actions to be taken to identify potential emergencies and to prevent or mitigate their effects. Some of these procedures may be covered by a safety plan or other document and they may be referenced in the ERP rather than reproduced. |
| III | Preparedness Procedures | All the activities taken to prepare the facility and personnel for an emergency, including training, drills and exercises, equipment acquisition, development of mutual aid agreements, community relations, annual reviews, and testing. |
| IV | General Response Procedures | The functional procedures that apply to any emergency, including alerting and warning, communications, evacuation, security, mutual aid, public information, reporting requirements, and procedures. These procedures provide details on the actions to be taken by the appropriate personnel and organizations, equipment to be used, and the operation of the facilities. Figure 11-1 describes a sample incident flow process. |
| V | Team Procedures | The emergency response team procedures that detail the responsibilities and duties of each ERT member. Responsibilities and duties are detailed by job description, Incident Commander, Fire Captain, fire brigade team member, fire equipment operator, hazardous material response specialist, and other associated emergency response members. |
| VI | Hazard-Specific Procedures | Procedures that explain the actions to be taken in response to a specific emergency, such as credible incidents (fire, chemical spill, etc.) and natural disasters. |
| VII | Recovery Procedures | Procedures that explain the necessary actions following the response to the emergency. These include incident investigation, employee support, community coordination, critiques, decontamination and recovery operations. |
| VIII | Appendices | Useful reference materials such as maps, fire water system diagrams, critical utility emergency shut off points, names and contacts for service providers should be included. |

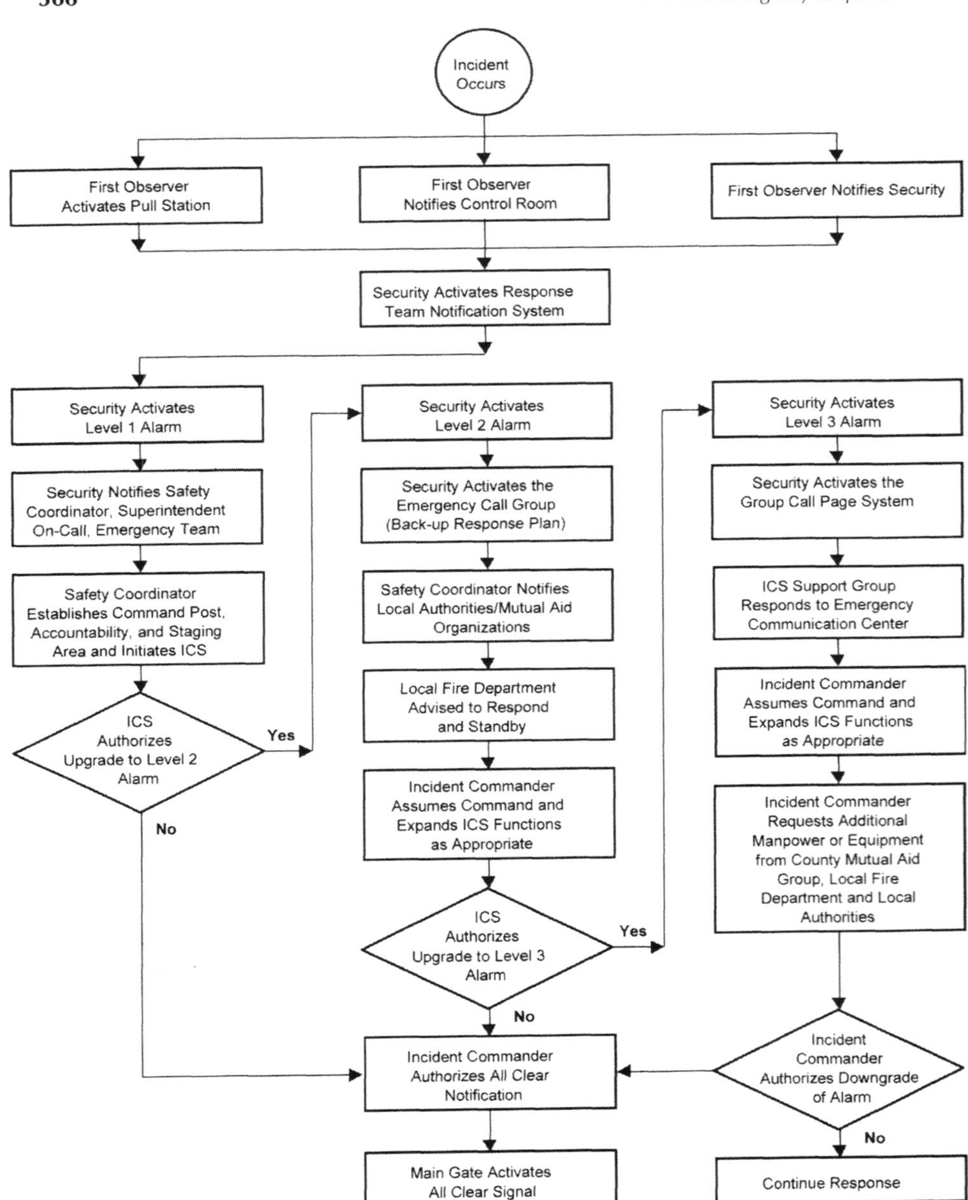


**Figure 11-1.** Example Response Flow Chart (modified from RRS, 2000)

- *Which agency has jurisdiction?*
- *What is the role of the community response organization?*
- *What will be the role of the site emergency response organization?*
- *How will the two groups be integrated?*
- *What is the function of the community incident commander?*
- *What is the function of the onsite incident commander?*
- *What is the role of the site incident commander and the community incident commander in the emergency mitigation decision process?*
- *Who communicates with the public and the press?*
- *Who will decide when it is safe to return control of the emergency site to the owner?*

> ***Whether the community response group has full jurisdiction or only aids in the mitigation of an incident, it is important that an understanding of roles and responsibilities is reached, agreed upon, and documented in the ERP.***

## 11.4. Training and Drills

Each employee must know what his/her responsibilities are and how to execute them during an emergency. This may include shutting down a process operation in a safe manner, evacuating a building or area, or coordinating the activities of a highly trained firefighting or hazardous materials response team. Training requirements can be found in government regulations (OSHA, EPA, and DOT) and in industry standards or good practices, such as NFPA 600 and API 2001 and 1200. *Technical Planning for On-Site Emergencies* (CCPS, 1995a) also provides recommended training requirements.

### *11.4.1. Training*

Emergency response training should be provided to all facility personnel. The level and detail of the training will depend on their role and responsibilities. For example, an office clerk may only need to be trained in how to report a fire and in the evacuation procedure. Ann operator who is a member of the emergency response team will require significantly more training.

In addition to recognizing and reporting emergencies, operating unit employees may be trained to take immediate steps to limit the scope of an emergency situation. This can be done when the emergency response team is notified and in route. Operators should be trained to safely shutdown their operations.

Employees working where the emergency is occurring should be trained to take steps to shutdown and isolate the emergency if possible. This should be done through such actions as closing valves that are at a safe distance or operating remotely activated controls. The appropriate personal protective equipment should be specified and worn for this action. The limits of what these employees may do in these situations should be clearly defined in the ERP and be included in their training. Considerations for emergency responder training include:

- The nature of the processes and the materials involved in an emergency situation
- The appropriate procedures for safely handling emergency scenarios
- The proper use and limitations of emergency response equipment
- The function of the ICS and how to work within it
- Appropriate rescue procedures
- Bloodborne pathogens procedures
- Respiratory protection procedures
- Decontamination procedures
- Personal Protective Equipment (PPE)

The following sections identify suggested training for the different levels of involvement in emergency response at the facility.

All training should be documented so that the facility can quickly provide a status of training to management or regulatory agencies. Adequate documentation will also assist in planning for training needs. The frequency of training for fire responders is shown in Table 11-5. A minimum number of hours or days of training should be specified.

#### *11.4.1.1. Facility and Operations Personnel*

Suggested training for facility personnel and operations personnel who are likely to witness or discover a fire, medical emergency, or hazardous substance release should include:

- How to report emergencies
- Evacuation procedures
- Shelter-in-place procedures
- How to use portable fire extinguishers
- Special hazards that may be encountered
- Hazardous/toxic materials
- Emergency shutdown and isolation

Unit operating procedures should include a section on the actions an operator should take if there is an emergency.

### 11.4.1.2. Incipient Fire Brigades

*Incipient fire brigades* are trained to deal with the initial or beginning stage of a fire and can control such a fire using portable fire extinguishers, Class II standpipe, or small hose systems without the need for protective clothing or breathing apparatus. Training should be conducted frequently enough to ensure that fire brigade members can perform their functions in a safe manner, but annually at a minimum.

Suggested training for incipient fire brigade and emergency response team members who respond to the emergency, perform initial fire suppression, and may be involved in prolonged exterior fire suppression activities includes:

- Fire behavior
- The phases of a fire
- Use of portable fire extinguishers
- Classes of fire (A, B, C and D)
- Fire control equipment
- Techniques used to control fires and emergencies (fires, spills, leaks)
- Alarm notification systems
- Special hazards that may be encountered
- Hazardous/toxic materials
- Radiation sources
- Electrical
- Storage and use of flammable liquids and gases

**Table 11-5**
*Frequency of Training*

| Individuals | Frequency |
|---|---|
| Facility and operations personnel | Prior to initial assignment and annually thereafter |
| Incipient | Per NFPA 600 |
| Interior structural | Per NFPA 600 |
| Hazardous Waste Operations and Emergency Response (HAZWOPER) | Prior to assignment and annual refresher thereafter, Reference OSHA 1910.120 (e) |
| Rescuers | Annually |
| Team officers | Annually |

- Water reactive substances
- Incipient fire brigade organizational structure
- Emergency pre-plans

### 11.4.1.3. Fire Brigade/Emergency Response Teams

*Fire Brigade/Emergency Response Teams* should be trained to:

- Implement the Incident Command System, including procedures for the notification and utilization of outside resources.
- Recognize the hazards and risks associated with working in specialized chemical and high-temperature protective clothing.
- Know the relationship between the ERP and local and state emergency response plans.
- Develop and implement a plan of action, including safety considerations, consistent with Standard Operating Procedures and within the capability of the available personnel, personal protective equipment, and control equipment.
- Evaluate the progress of the planned response to ensure that response objectives are being met safely, effectively, and efficiently and adjust the plan of action as necessary.

Various local, state, and government agencies have specified minimum training required. Additional training may be required depending upon the nature of the site and its exposures.

### 11.4.1.4. Interior Structural Fire Brigade

*Interior structural fire brigades* perform fire suppression activities or rescues in buildings involving fire beyond the incipient stage. For those members assigned to interior structural fire fighting, training should be conducted at least quarterly. The training requirements of NFPA 600 (Industrial Fire Brigades) provide the frequency and levels of training.

Interior structural fire brigade members perform fire suppression activities in buildings involving fire beyond the incipient stage. In addition to the training required for an incipient fire brigade, interior fire brigades require the following training:

- Fire water stream techniques
- Ventilation techniques
- Use of self-contained breathing apparatus
- Hazards of interior spaces
- Suggested training for hazardous material incident responders

#### *11.4.1.5. Hazardous Material Incident Responders*

All employees involved in hazardous material operations require HAZWOPER training. It is also required for employees who are involved in some level of response to plant emergencies involving hazardous substances. There are five levels of responders who must be trained:

*Level 1—First Responder Awareness Level:* individuals who are likely to witness or discover a hazardous substance release and who have been trained to initiate an emergency response sequence by notifying the proper authorities of the release. First responders take no further action beyond notifying the authorities of the release. All facility personnel should be trained in the safe evacuation and accounting of personnel in the event of a major fire, hazardous material, or gas release.

*Level 2—First Responder Operations Level:* individuals who respond to releases or potential releases of hazardous substances as part of the initial response for the purpose of protecting nearby persons, property, or the environment from the effects of the release. They are trained to respond in a defensive fashion without actually trying to stop the release. Their function is to contain the release from a safe distance, keep it from spreading, and prevent exposures.

*Level 3—Hazardous Materials (HAZMAT) Technician:* individuals who respond to releases or potential releases for the purpose of stopping the release. They assume a more aggressive role than a first responder at the operations level in that they will approach the point of release in order to plug, patch, or otherwise stop the release of a hazardous substance.

*Level 4—Hazardous Materials (HAZMAT) Specialist:* individuals who respond with and provide support to hazardous materials technicians. Their duties parallel those of the hazardous materials technician, however, their duties require a more directed or specific knowledge of the various substances they may be called upon to contain.

*Level 5—Incident Commander On-Scene Commander:* individuals who will assume control of the incident.

#### *11.4.1.6. Rescuers*

Suggested training for rescuers includes:

- Recognition of and rescue in a confined space
- Recognition of and rescue in atmospheric hazards
- Recognition of physical hazards
- Understanding of Confined Space Entry program
- Rescue techniques

- First aid and CPR
- Bloodborne pathogens

### 11.4.2. Drills and Exercises

Drills and exercises are used to verify that employees will respond as trained in the event of an emergency. *Drills and exercises are not a replacement for training*. The objectives of drills and exercises are to:

- Practice emergency procedures and skills
- Evaluate capabilities of response teams and/or individual knowledge of emergency responders
- Determine the overall effectiveness of both emergency response procedures and the ERP

Drills and exercises may range in scope from table-top simulations of emergency scenarios to a full simulation of an emergency with all potential response groups involved. Fire plans should also be included in these periodic drills.

The frequency of drills varies from facility to facility but, as a minimum, a facility should annually conduct at least one of the drills identified in Table 11-6. Many companies conduct a minimum of one drill per shift per year with the goal of achieving proficiency in the responders.

Additional information on developing and planning drills can be found in *Guidelines for Technical Planning for On-Site Emergencies* (CCPS, 1995a).

### 11.4.3. Critiques

A critique of response activities is a means to evaluate the effectiveness of the emergency response plans, procedures, and pre-plans. The critique should be conducted immediately after the completion of the emergency response activity, drill, or exercise. All emergency response members who participated in the emergency response activity should participate in the critique. Each response group, firefighters, HAZMAT, medical, support staff, and emergency control center should comment on how they were able to follow the emergency response plan, procedures, and pre-plans. Problems with any of the plans, procedures, or pre-plans should be discussed and recommendations developed to improve the response effectiveness.

The person responsible for conducting the critique (usually the Incident Commander or Emergency Control Center Director) will document the results of the critique and the recommendations generated. A designated management representative is responsible for addressing and tracking the recommendations to completion.

**Table 11-6**
*Drills*

| Type | Description | Recommende d Frequency |
|---|---|---|
| Table-top exercise | The purpose of a table-top exercise is to have participants practice problem-solving and resolve questions of coordination and assignment of responsibilities in a controlled environment. | 3–6 months |
| Walk-through drill | The purpose of a walk-through drill is to familiarize the fire brigade with the different units, physical hazards associated with the units, overview of fire pre-plan and location of fixed fire suppression equipment. | 6–12 months |
| Functional drill | The purpose of a functional drill is to test an individual component of the emergency response plan. The drill could test the communication system, emergency operations center, evacuation, or the shelter-in-place plan. | 12 months |
| Full-scale exercise | The purpose of a full-scale exercise is to test the effectiveness of the emergency response plan. A full-scale exercise should incorporate a high degree of realism, extensive involvement of resources and personnel, and an increased level of stress. The exercise should involve mutual aid groups, agencies, and outside responders. | 24–36 months |

## 11.5. Notification

A procedure must be established and included in the ERP for notification of an emergency situation to:

- Other personnel onsite
- On-site emergency response team
- The surrounding community
- Outside emergency response teams, community, or mutual aid
- Regulatory agencies

The notification may include the nature and location of the emergency.

While most emergency situations require notification of other site personnel, each emergency situation will typically not require notification to all personnel. An important part of emergency response planning should be the definition of types of situations that require notification and the extent of notification. In addition, the responsibility for that notification should be established in the emergency response plan and through training.

The means and the equipment for emergency notification should also be provided. The notification equipment could include a public address system, siren, phone system, plant radios, or pagers. The means of notification should be sufficient to alert potentially affected personnel at the facility. Since facilities often change, the capability of the systems to alert everyone must to be periodically evaluated.

Notification equipment should be periodically tested and may require scheduled preventive maintenance. This should be defined and integrated into the facility preventive maintenance programs.

## 11.6. Operating Procedures for Fire Emergency Response Equipment

Equipment used during emergency response should have operating procedures that are utilized for training emergency responders. Equipment manufacturers and sales companies are often excellent sources for these procedures. However, the nature of a particular facility, the processes, or the range of potential credible emergency situations expected often requires additional procedures that address the proper and safe use of the equipment in site-specific situations.

These procedures may also include pre-use inspection steps, clean-up after use, and periodic preventive maintenance that is appropriately performed by users of the equipment.

## 11.7. Fire Pre-Plans

Developing plans for response to credible fire events is an important emergency planning element. Different fire events require different responses. Complete extinguishment is the appropriate approach for some events. For others, the optimum approach may be to cool adjacent structures until the source of the fuel can be isolated or until the fuel is depleted. Still others may require that firefighters secure perimeter areas and allow the fire to burn itself out.

Factors that enter into the development of fire pre-plans include safety of firefighting personnel; environmental, regulatory and community consequences of allowing a fire to burn itself out; and costs and availability in having firefighting personnel standing by while a fire is burning itself out.

The term *fire pre-planning* describes the actual process of developing fire response tactics for emergency response personnel as well as the actions taken by operations. A fire pre-plan provides emergency fire responders an inventory of essential information necessary for developing tactical response at the onset of the emergency. Fire pre-plans should describe who is responsible for the actions to be taken. Fire pre-plans can take many shapes and should be part of the emergency response plan. A sample fire pre-plan is shown in Table 11-7 on the following two pages and is available on the accompanying CD.

Typical sections for the pre-plan include:

- General
- Structure
- Process equipment
- Suppression
- Utilities
- Protection systems
- Process shutdown/ instrumentation
- Response considerations

If the emergency response plan calls for use of outside response teams, community, or mutual aid, the fire pre-plans should be reviewed with them.

Once the fire pre-plans are finalized, they should become the subject of training and drills.

**Table 11-7**
*Example Fire Pre-Plan*

**ABC Chemical Corporation**
**Emergency Response Pre-Plans**
Somewhere, Texas

Review Date: 00-Jan-00

| **Equipment:** | **Unit:** | **Batch:** | **Emergency Contact:** |
|---|---|---|---|
| V-135 Reactor | S1B | 193 | Incident Commander |
| Location: | Muster Point: | | Emergency Contact Phone: |
| North and Northeast | ABC Control Room | | 000-000-000 |

| **Area Contents:** | **Special Hazards:** | **ERG Guide Number:** |
|---|---|---|
| Process Equipment | Sulfur Monochloride (S2CL) | 137 |

| **Storage Configuration:** | **Storage Quantity:** | **Storage Height:** | **Storage Width:** | **Storage Length:** |
|---|---|---|---|---|
| NA | NA | NA | 0 | 0 |

| **Type of Structure:** | **Dimensions (N/S):** | **Dimensions (E/W):** | **Structure Height:** | **Structure Stories:** |
|---|---|---|---|---|
| Interior | 100 | 250 | 30 | 2 |

| | | | |
|---|---|---|---|
| **Roof Construction:** | Fiberglass on steel | **Vertical Openings:** | None |
| **Wall Construction:** | Fiberglass on steel columns | **Venting:** | None |
| **Floor Construction:** | Metal grating on steel beams | **Vent Operation:** | NA |
| | | **Vent Operator Locations:** | NA |

| | | | |
|---|---|---|---|
| **North Egress Point:** | Door | **North Exposure:** | Propylene Compressors |
| **South Egress Point:** | Door | **South Exposure:** | PAO |
| **East Egress Point:** | Door, Bay Door | **East Exposure:** | IV Expansion |
| **West Egress Point:** | Door, Bay Door | **West Exposure:** | Rail Car Dock |

| **Raw Materials/Utilities:** | **Use:** | **Shutoff Location:** |
|---|---|---|
| S2C1 | V-135 Reactor | Near V-138 |
| Isobutylene (IC4) | V-135 Reactor | Near V-161 |
| Electricity | V-135 Reactor | In Switchgear #2 |
| Steam | V-135 Reactor | Near V-134 |

| **Fire Protection System:** | **System Design:** | **System Pressure:** | **System Control Valve Locations** |
|---|---|---|---|
| SIB Deluge | | | Valve house east side of building |

| **Monitor/Hydrant:** | **Flow:** | **Distance:** | **Drainage:** |
|---|---|---|---|
| M-134 | 700 | 100 | Lift Station #4 |
| M-233 | 700 | 150 | |
| | 0 | 0 | |

| **FW Pumps:** | **FW Pump Flow:** | **FW Pump Pressure:** | **FW Pump Drivers:** | **FW Pump Status:** |
|---|---|---|---|---|
| P101/102/201/202 | 2500 | 150 | All diesel driven | Standby |

| **Chemical:** | **Reacts:** | **Extinguishing Media:** |
|---|---|---|
| S2C1 | Reacts violently with water. | Dry Chem. Dry sand, CO2, H20. |

| | |
|---|---|
| **Special Precaution:** | Contact with water generates highly toxic fumes; HC1, chlorine gas, SO2 and H2S. |
| **PPE Requirements:** | Bunker gear, SCBA. |

**ABC Chemical Corporation**
**Emergency Response Pre-plans**
Somewhere, Texas

**Review Date:** 00-Jan-00

**Plot Plan:**

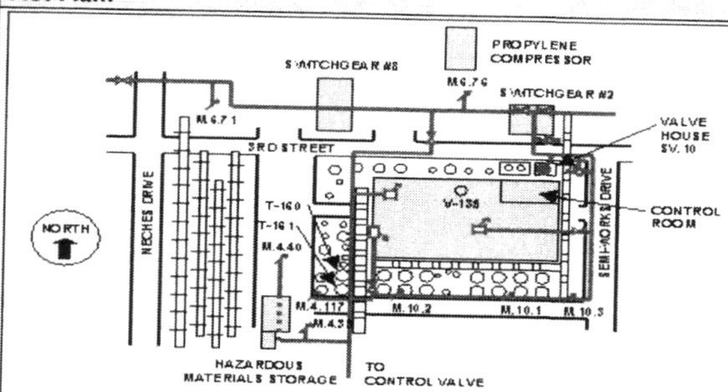


**Fire Procedures:**

- Isolate S2C1 to V135 with block valve near V138.
- Apply dry chemical, dry sand, or CO2 to fire.
- Do not use water on the S2C1 material itself.
- If large quantities of S2Dc1 combustible are involved, use water in flooding quantities as spray and fog.
- Apply water from as far a distance as possible.
- Use water to keep fire exposed equipment cool.
- Use water spray to knock down vapors.

**Isolation Procedures:**

- Isolate S2C1 to V135 with block valve near V138 at grade.
- Isolate IC4 to V135 with block valve in pipe rack near V161 with chain operator.
- Isolate electricity to V-135 in Switchgear #2.
- Isolate steam to V135 with block valve near V134.

**Environmental / Runoff Procedures:**

- Dig a pit, pond, lagoon, holding area to contain liquid or solid material
- Dike surface flow using soil, sand bags, foamed Polyurethane, or foamed concrete.
- Neutralize with agricultural lime (CaO), crushed Limestone (CaCO3), or sodium bicarbonate (NaHCO3).
- Apply water spray or mist to knock down vapors. Vapor knock down is corrosive or toxic and should be diked for containment.

**Operation of Mitigating Systems Procedures:**

- Manually operate SIB Deluge System with switches on east or west walls or from valve house.
- Use monitors and hand-held hose lines for exposure protection as necessary.

**Evacuation Distances**

All directions: 600 ft

Downwind: 2 miles

APPENDIX

# CASE HISTORIES

*What has happened before will happen again.*
*What has been done will be done again.*
*There is nothing new under the sun.*
—Ecclesiastes 1:9

## Introduction

The history of incidents within the chemical, petrochemical and hydrocarbon processing industries indicates the following general truths:

- Serious incidents are a major instigator of change.
- The greater the value of the loss, whether life or property, the greater the probability that change will occur.
- The effects of the change will not be sustained.
- Within any company or organization, the staff at the facilities generally undergoes a complete change. With their departure, the corporate memory of incidents and their causes also seems to disappear.

*Those who cannot remember the past are condemned to repeat it.*
—George Santayana

*If history repeats itself, and the unexpected always happens, how incapable must Man be of learning from experience.*
—George Bernard Shaw

## Case History 1 Large Vessel Explosion

In January of 1966, a large pressure vessel exploded at a French refinery, leading to fatalities and numerous serious injuries. At the beginning of the day shift, a team of three operators was sent to take a sample from a propane storage sphere. This involved draining the water that had collected at the bottom of the sphere through a draw-off line. After a short interval, a bang was heard and a powerful jet of propane gushed out. As the propane cloud began to escape, the water sprays on the sphere were activated. Operators could not close the draw-off valve because the autorefrigeration effect associated with the flashing propane had caused it to seize.

The operators were afraid to use the nearby telephone to raise the alarm or to start their truck because they could be ignition sources, so they walked to the pump house, which took 15 minutes. Fifteen minutes after that, the refinery firefighters arrived and again failed to close the valves. The refinery had a well-rehearsed evacuation plan, and effort to stop nearby traffic was initiated. A local road was not closed in time and, 35 minutes after the initial release, the propane cloud was ignited by a car on the road 4,120 ft (1,250 m) away.

The fire flashed back to the sphere surrounding it with flames. The water spray system for the sphere was designed to deliver only half the quantity of water normally recommended, and the supply was inadequate to meet the fire demand. The firefighters believed the relief valve would protect the vessel and so any available water was used to cool the neighboring spheres.

The ground under the sphere was level and any propane that did not immediately burn or disperse, collected under the sphere and burned. Ninety minutes after the fire started, the sphere BLEVEd. Ten out of the 12 firefighters within 160 ft (50 m) were killed. People 460 ft (140 m) away were badly burned, and some died later from burns. Altogether, 18 people died, about 80 were injured, and two thousand people were evacuated. The immediate area was abandoned, and during the next two days, five spheres and two other pressure vessels BLEVEd and three more were damaged.

Damage to the refinery was estimated at $4.6 million (1966 $), with outside damage estimated at another $2 million (1966 $).

***Lessons Learned:***

- The lack of drainage allowed escaping propane to accumulate under the spheres instead of draining away.
- Emergency response planning on cooling the sphere on fire and location of personnel was poor.
- Adequately designed water spray on the spheres was not provided.
- Two valve method for sampling was not used, thus closing a valve was not an option.

## Case History 2: Pipe Rupture Leads to an LPG Tank BLEVE

On November 19, 1984, a major fire and a series of catastrophic explosions occurred at the government owned and operated PEMEX LPG Terminal at San Juan Ixhuatepec, Mexico City.

Three refineries supplied the facility with LPG on a daily basis. The plant was being filled from a refinery 248 miles (400 km) away. Two large spheres and 48 cylindrical vessels were filled to 90% and 4 smaller spheres to 50%.

A drop in pressure was noticed in the control room and also at a pipeline pumping station. An 8-inch (20-cm) pipe between a sphere and a series of horizontal bullets had ruptured. Unfortunately, the operators could not identify the cause of the pressure drop. The release of LPG continued for ~5–10 minutes when the gas cloud, estimated at 656 ft × 492 ft × 6.6 ft (200 m × 150 m × 2 m) high, drifted to a flare stack. It ignited, causing violent ground shock.

Workers at the plant tried to deal with the escaping LPG and the emergency response, taking various actions. At a late stage, an emergency shutdown was initiated. About fifteen minutes after the initial release, the first BLEVE occurred. For the next hour and a half there followed a series of BLEVEs as the LPG vessels violently exploded. LPG was said to rain down and surfaces cov-

*Damage from catastrophic explosions and fires at a liquid petroleum gas (LPG) storage facility in Mexico City, 1984. (From the video "The Day the Sky Caught Fire," courtesy of Skandia Group Insurance Co. Ltd., Stockholm, Sweden.)*

ered in the liquid were set on fire. The explosions were recorded on a seismograph at the University of Mexico.

This incident resulted in 650 fatalities, more than 6,400 injuries, destruction of the terminal and many of the homes in the neighborhood located adjacent to the terminal.

***Lessons Learned:***

- The layout and spacing allowed for escalation.
- Fire protection was minimal.
- Understanding of the potential fire hazards were lacking.
- Emergency Isolation was not provided.

## Case History 3: Fire Turns into an Ecological Disaster

A large warehouse fire occurred on November 1, 1986 at a Swiss chemical facility. The warehouse, though originally built for storing machinery, was approved for agro-products and chemicals of flash points above 70°F (21°C).

The fire generated heavy smoke containing toxic materials, such as phosphoric esters and mercaptans, and the local population was warned to take refuge indoors and close their windows until the "all clear" was sounded seven hours later. Much more serious consequences resulted from the firefighters having to resort to water to prevent the fire from spreading because the use of foam had failed. In the absence of adequate provision for retention, over two million gallons (seven and a half million liters) of fire water drained into the River Rhine, carrying with it about 30 tons (27 tonnes) of chemicals from the warehouse, including about 3,330 pounds (150 kg) of highly toxic mercury compounds.

Although many people sustained minor, short-term effects such as headaches and nausea, no serious or long-term damage to the health of the local population was detected. Severe ecological damage was caused to the river over a distance of about 155 miles (250 km), including the death of large numbers of fish and eels.

The disaster caused much political disturbance throughout Switzerland and Germany, through which the Rhine flows for most of its length. Severe criticism was provoked by the company's long delay in warning the monitoring stations downstream.

***Lessons Learned:***

Measures taken subsequently to avoid recurrence included:

- Reducing the output of insecticides.
- Reducing the stocks of agro-chemicals in the warehouse.
- Eliminating all processes involving the use of phosgene.
- Discontinuing worldwide manufacture and sale of all products containing mercury.
- Reviewing the product range in respect of both economic and hazard criteria.
- Strengthening safety regulations for storing toxic and flammable substances.
- Installing two catch basins of capacities 5,000 and 2,500 $m^3$ for fire water retention.

## Case History 4: Exchanger Leaks, Burns Cooling Tower

In July of 1997, a cooling tower at an ammonia and urea plant, originally constructed in 1968, caught fire and was destroyed. The plant produced 1,450 tons/day (1,315 tonnes/day) of ammonia and 240 tons/day (218 tonnes/day) of urea. The cooling tower was a 5-cell, induced draft, cross flow unit. It was constructed of redwood with steel supports and fiberglass fill. The capacity of the cooling tower was 50,000 gallons (190,000 liters).

The plant was running smoothly when a common trouble alarm on the cooling tower fans sounded. The outdoor operators investigated, discovered a fire, called the fire department, and initiated a plant shutdown. The fire was extinguished in about 2 hours.

The fire was limited to the cooling tower. The pumps and metal piping were unaffected, but the fans, shrouds, top decking, and fiberglass piping were destroyed. The source of the fuel was found to be a leak in a heat exchanger due to fretting corrosion into the cooling water system. The exact cause of ignition could not be determined, but it was felt to be nonclassified electrical equipment on the tower. The damages (in 1997 dollars) totaled $2,052,000 and downtime was 55 days.

***Lessons Learned:***

- Cooling towers, even those in operation and full of water, can burn.
- Fire protection was not provided.
- Electrical area classification was inadequate.

## Case History 5: Insufficient Sprinkler Density

In August of 1994, a fire at a polyurethane foam plant resulted in extensive damage and one fatality. The site was composed of several interconnected buildings totaling about 120,000 ft$^2$ (11,000 m$^2$). Polyurethane foam is produced using "vertifoam" and "maxform" processes and cooled in an "envirocure" chamber. The main polymerization and gas-producing reactions are highly exothermic. The foam "buns" are produced and then stored in a solid piled configuration of about 12 ft (3.7 m) in height in an unheated area. Fire protection in the area was provided by a dry pipe sprinkler system.

An electrical surge resulted in off-ratio metering of the reactants or poor mixing. The foam was cooled in an "envirofoam" chamber. A defective section was noted and the operator was instructed to cut out the defect and place the bun outdoors. A short time later, it was discovered that the defective bun had not been moved outdoors, although it showed no signs of smoking or discoloration. An exothermic reaction continued until it autoignited.

The fire department used 11 firefighting vehicles and took about 9 hours to extinguish the fire. One employee died and four firefighters were injured. About 50% of the site suffered damage. The warehouse building, where the fire originated, was totally destroyed. An adjacent office building was also destroyed. Adjacent production and shipping areas were heavily damaged by smoke and heat. Damages from this incident totaled $8,295,000.

### *Lessons Learned:*

- The dry pipe system was probably too slow in operation and did not provide a density sufficient to extinguish the fire. The new system was designed for a sprinkler density of 0.6 gpm/ ft$^2$ (24 lpm/m$^2$) which was more than three times the original design density.
- Escalation occurred because of poor fire protection and poor spacing.
- Procedures for abnormal events need to be followed and reasons explained.

## Case History 6: Jet Fire

A jet fire occurred during a clean-out operation to remove residues from a batch still, known as "60 still base." This vessel had not been cleaned since it was installed in 1961. An operator dipped the sludge to examine it and reported to management that the sludge was gritty, with the consistency of soft butter. No sample was sent for analysis nor was the atmosphere inside the vessel checked for a flammable vapor. It was mistakenly thought that the material was a thermally stable tar.

In order to soften the sludge, which was estimated to have a depth of 14 in (34 cm), steam was applied to the bottom heating element. Advice was given not to exceed 194°F (90°C). Employees started the clean-out operation using a metal rake. The material was tar-like and had liquid entrained in it. Approximately one hour into the cleaning process, a longer rake was used to reach further into the still.

The vessel's temperature gauge in the control room was reported to be reading 118°F (48°C). Instructions were given to isolate the steam. A number of employees involved in the raking left the still base to get on with other tasks. One person left on the scaffold had stopped raking and noticed a blue light, which turned instantly to an orange flame. As he leapt from the scaffold, an incandescent conical jet erupted from the manhole. This projected horizontally toward the control building. A vertical jet of burning vapor shot out of the top rear vent to the height of the distillation column nearby.

The jet fire lasted for approximately one minute before subsiding to localized fires around the man-lid and buildings nearby. The force of the jet destroyed the scaffold, propelling the manhole cover into the center of the control building. The jet severely damaged this building and then impacted on the north face of the main office building causing a number of fires to start inside the building.

### *Lessons Learned:*

- Knowing the identity of the chemicals being handled and understanding their properties is imperative in the safe handling and siting of materials. Without knowing what the chemical is, the risks of fire, explosion, and toxic impact cannot be considered.
- Jet fires involve pressure and velocity, impacting far greater distances than pool fires. Jet fires typically last for a limited time.

## Case History 7: Internal Column Fire

On February 11, 2001 a fire occurred in a 250-foot-tall, 26–29-foot-diameter column. The fire happened during the replacement of the carbon steel structured packing, which involved cutting out the old column internals and removing each bed of packing. Despite persistent attempts to extinguish the fire, the incident progressed to the point where the column fell over early on February 12, 2001. Excellent procedures resulted in the 60+ people on the column being safely evacuated without injury

### *Lessons Learned:*

- This event was categorized as a "metal fire" with the structured carbon steel packing acting as the fuel. This is not a generally recognized and understood mechanism and was not identified in the PHA or FHA.
- Packing has a large surface area by design that is hard to clean and inspect. Even extraordinary cleaning processes (washout, steam-out, dry-out) such as were used in this incident may not completely clean combustible material off of the packing. Packing should always be considered to contain a combustible fuel source. In many cases, contaminants on the column internals may be pyrophoric.

- The metal itself should also be considered a combustible fuel source. As mentioned above, metal can ignite and burn. The amount of metal in packed beds is a *significant* fuel source.
- As a result of this investigation, all attempts should be made to remove the packing prior to performing hot work. If hot work is required, a thorough safety and fire protection review should be performed (including the prevention of and response to metal fires).

## Case History 8: Electrical and Instrumentation Room Explosion

On October 30, 1998, a flammable gas release led to gas migration to a non-electrically classified Electrical and Instrumentation (E&I) Room. The flammable gas was ingested into the exterior heating and ventilation inlet duct. When the 480 volt primary power switch, within the E&I room was remotely shut off, an explosion resulted. The E&I module explosion led to a subsequent fire in adjacent production buildings. There were no injuries to personnel. The root cause of the incident was determined to be the failure of management systems

to adequately understand and address the hazards. All of the fire and gas protection systems performed as designed.

At 1:14 p.m., the operator received a radio transmission from the PCC (Process Control Center) informing him that a "low" gas alarm had been received. Immediately, the operator was informed of two more gas alarms. Upon reaching the E&I module, the operator noted the sound of gas escaping somewhere inside the production module. At approximately 1:22 p.m. he climbed the stairs to the E&I control room and opened the door to check the fire and gas alarm panel which indicated a high gas atmosphere in the production modules. The operator then requested the Emergency Response Team be put on standby and went to the to inform three contract workers, in the area of the danger.

Once the contract personnel were safely off pad, the operator moved to a safe observation and contacted the PCC operator to order a process shutdown. The PCC operator acknowledged and replied that the shutdown signal had been sent. At approximately 1:28 p.m. an explosion occurred in the E&I module and soon after, fire was noted coming from the adjacent process modules.

### Lessons Learned:

- The root cause of the incident was determined to be the failure of management systems to adequately understand and address the hazards.

- The E&I module did not have gas detection which, if located in the air intake, would have significantly reduced the probability of an explosion. Gas detectors should be fitted in the HVAC supply to all nonhazardous areas where ingress of released hydrocarbon gas is credible.
- There was no automatic production shutdown from fire or gas (F&G) detection in production modules.
- There are inconsistencies around shutdown logic and controls which contributed to operator unfamiliarity. Additionally, alarm information was ambiguous and led to operator confusion.
- Personnel involved in operations and emergency response should be trained in the role of the various safety systems (e.g. F&G detection, emergency shutdown, HVAC). The practices for responding to F&G alarms should ensure that personnel do not enter known hazardous conditions.
- The management of projects to address safety related standardization across facilities needed to be reviewed for effectiveness.

APPENDIX

# UNDERSTANDING FIRES

## B1. Introduction

For something so familiar, fire is a surprisingly complex phenomenon. There are many excellent detailed references on the physics of fires, properties of burnable material, and the fundamentals of fire science. Rather than attempt to be a comprehensive guide, this Appendix will introduce some common fire terms and concepts and discuss in simple terms how they relate to fire protection.

Everyone knows what a fire is and can recognize a fire when they see one. However, the exact definition is often lost. So, what exactly is fire?

> ***Fire is a chemical reaction involving rapid oxidation (combustion) of a fuel.***

Oxidation can be defined in several different ways. The simplest definition is "the loss of electrons from an atom, compound or molecule." An oxidizer is a chemical compound (usually oxygen) that reacts with a reducing agent (the fuel) to create heat or power through chemical conversion in processes, such as burning or electrochemistry. By definition, during an oxidation-reduction reaction, the oxidizer is reduced and the reducing agent is oxidized.

Fire is a self-sustaining, exothermic oxidation–reduction reaction. The fire reaction usually involves oxygen which forms the oxides of the fuel. The most important examples in petrochemical and hydrocarbon processing facilities are combustion reactions of hydrocarbons with oxygen.

The products of complete combustion of hydrocarbons in air are water and carbon dioxide. However, combustion is rarely complete and byproducts are produced. These byproducts can be toxic and form dense black smoke. These byproducts give fire its characteristic yellow-orange flame. A flame is defined as the confines, not necessarily visible, of the combustion reaction.

## B2. Fire Triangle

Usually, fire takes place when heat comes in contact with a combustible material. If the combustible material is a solid or liquid, it must be heated to generate sufficient vapor to form a flammable mixture with the oxygen in air. If this flammable mixture is heated to its ignition point (ignited), combustion will occur. Three basic conditions are required for fire to take place. These are fuel, oxygen, and heat.

1. *Fuel*—the reducer; any combustible material, solid, liquid or gas. Most solids and liquids must vaporize before they will burn.
2. *Oxygen*—the oxidizer; sufficient oxygen must be present in the atmosphere surrounding the fuel for fire to burn.
3. *Heat*—sufficient energy must be applied to raise the fuel to its ignition temperature.

Fire can only occur when all three of the above elements are present and in the proper conditions and proportions. These three basic conditions are often represented as a fire triangle shown in Figure B-1. The combustion reaction itself is often included as a fourth central element of the fire triangle.

If one of the sides of the fire triangle is missing, the fire will not start. If one side is removed, the fire will be extinguished. The fire triangle forms the foundation for all methods of fire prevention and firefighting (NFPA, 1997).

Thus, the three sides of the triangle indicate how a fire may be fought. In general, for fires in the chemical process industry, the first method of firefighting is to eliminate the fuel. For leaks, this is done by closing valves. If this

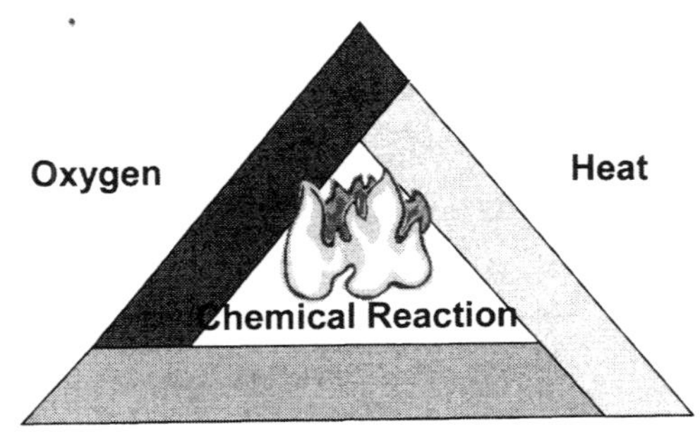


**Figure B-1.** Fire Triangle

can't be done or too much fuel has escaped, the second method is to remove heat. This is usually done by putting water on the fire. The third way to extinguish fire is to stop the supply of oxygen, typically achieved by the application of foam or inert gas.

### B2.1. Fuel

Fires are classified based on the type of fuel involved. Fuels are typically placed into three classes; ordinary combustibles (like wood and paper), flammable liquids, and combustible metals. For firefighting, a fourth fire class, electrical fires, is also considered. The four main classifications of fire are shown in Table B-1.

Fuel is typically removed from a fire by isolating liquid spills at source or, for ordinary combustible fires, by cutting a firebreak to prevent the spread of fire. Fuel available for a fire can also be reduced by cooling, which reduces the amount of vaporization of the fuel making less available for combustion.

### B2.2. Oxygen

For most fires of practical interest, oxygen is supplied by air, which contains around 21% oxygen by volume. It is not necessary to remove all the oxygen to extinguish a fire. Liquid fires can typically be extinguished by reducing the oxygen concentration to 12–16%. Solid fires require a greater reduction, below about 5% for surface smoldering and as low as 2% for deep-seated smoldering.

A common technique used to prevent ignition of flammable liquid in tanks is the provision of inerting systems, using a gas such as nitrogen. Once the fire

**Table B-1**
*Classification of Fires According to NFPA 10*

| Class | Description | Common Materials |
|---|---|---|
| Class A | Ordinary combustibles | Fires in ordinary combustible materials, such as wood, cloth, paper, rubber, and many plastics. |
| Class B | Flammable liquids | Fires in flammable liquids, combustible liquids, petroleum greases, tars, oils, oil-based paints, solvents, lacquers, alcohols, and flammable gases. |
| Class C | Electrical fires | Fires that involve energized electrical equipment where the electrical nonconductivity of the extinguishing media is of importance. (When electrical equipment is deenergized, fire extinguishers for Class A or Class B fires can be used safely.) |
| Class D | Fire in combustible metals | Fires in combustible metals, such as magnesium, titanium, zirconium, sodium, lithium, and potassium. |

has started, foams are typically used to form a blanket that separates the air from the fire.

### B2.3. *Heat*

Normally the heat to start a fire is supplied by an external source (ignition source). Once started, the heat from the fire reaction provides enough of its own heat to make the fire self-sustaining. A gas or vapor may be ignited by a small spark, while a solid may require a more intense source.

A fire needs to be supported by the ignition source until it is self-sustaining. For vapors, this time is short (<1 second), while for a solid combustible material it may be longer. Combustible fires often need application of heat for some time to be started, while a spark is sufficient to ignite vapors and start a fire.

Removing heat from a fire can also help in removing fuel, as cooling lowers the partial pressure of flammable vapor.

## B3. Common Terms for the Flammability of Materials

No single parameter defines flammability. Commonly used terms are flash point, flammability limits, autoignition temperature, minimum ignition energy and burning velocity.

The flash point of a substance is often treated as the principal index of flammability, especially for liquids. The lower the flash point, the more flammable the substance. The flammability hazard of a substance is also increased by:

- Wide flammability limits
- Low minimum autoignition temperature
- Low minimum ignition energy
- High maximum burning velocity
- Increasing the temperature of the fuel
- Oxygen enriched atmosphere

All substances in the form of liquids (and even many solids) possess a type of molecular motion that results in the escape of molecules from their surface in the form of vapor when they are not confined. When a liquid is left in an open container at room temperature, its molecules evaporate. When the liquid is confined in a partially full container that is *closed*, the molecules will continue to escape from the surface; however, because they cannot escape from the closed container, some of the molecules will return to liquid. Within a short time, an equilibrium will be achieved between the number of molecules escaping from the surface and those returning to the surface of the liquid.

When this equilibrium occurs, a certain pressure will be exerted in the empty space above the liquid in the closed container. This is called the vapor pressure of the liquid.

There is a continuous atmospheric pressure of 14.7 pounds per square inch absolute (1.01 newtons per square meter, or 1 bar) continuously being brought to bear on the surface of a liquid contained in an *open* container. As previously stated, a rise in temperature produces a rise in the vapor pressure above the liquid and, thus, a greater rate of escape of molecules (or vapor). If the temperature of a liquid (such as water in a open container) continues to rise due to the application of heat, large bubbles will begin to rise from the liquid and burst at its surface at a rapid rate. A thermometer immersed in the rapidly bubbling and vaporizing water would read 100°C (212°F); the thermometer would stay at this temperature as long as any liquid water remained in the container, and as long as heat continued to be applied to the container. Also, if the *upward* pressure of the vapor above the bubbling surface of the water was measured, it would equal the *downward* normal atmospheric pressure applied to the liquid in the open container. This is called the boiling point of the liquid water. At this point, the vapor pressure of the water equals the atmospheric pressure pressing upon it; as long as heat is supplied to it, the liquid boils in the attempt to release its molecules to the vapor state.

The boiling points of different types of liquids vary widely. They are an important physical characteristic both of liquids and of the many solids that melt to become liquids and then boil at a certain characteristic temperature.

Vapor density is a physical property of major importance to fire protection. Because the vapor density varies with the total weight of all the atoms in a molecule of the vapor of a substance, if the chemical composition of the substance comprising the vapor is known, then the weight or density of its vapor when compared to air can be determined as in the following relationship:

$$\text{Vapor density} = \frac{\text{Molecular weight of the substance being vaporized}}{\text{29 (the molecular weight of air)}}$$

From this formula, it can be seen that any vapor from a substance with a molecular weight of 29 will give a vapor density of 1.0. A substance with a higher molecular weight gives a vapor density over 1.0. Vapors tend to hug the ground for substances with a vapor density greater than 1.0. Substances with a vapor density less than 1.0 are lighter than air (NFPA, 1997).

### B3.1. Flash Point

Flash point is the minimum temperature at which the vapor above a liquid fuel will first support a combustion transient or "flash." Flash point is measured by a standardized test using a small quantity of liquid that is slowly heated until a flash is observed when an open flame is dipped down into a covered vapor space. Flash point temperatures are normally given as open cup or closed cup, based on the type of apparatus used to measure the flash point. The open cup flash point is normally a few degrees above the closed cup flash point.

As the temperature of a liquid increases, the partial pressure of the vapor above the liquid increases; hence, the percentage vapor composition above the pool increases. Liquid boils when its partial vapor pressure reaches the external pressure and the percentage of vapor reaches 100%. Flash point is when the vapor pressure of a substance is such that the concentration of vapor in air above the substance corresponds to the lower flammable limit. For flammable liquids, the term flammable is typically used for liquids with a flash point below 100°F (37.8°C), and the term combustible is used for liquids with a flash point above 100°F (37.8°C) (NFPA, 1997), but this in no way reflects the intensity of an ensuing fire.

### *B3.2. Fire Point*

Because it is an indicator of the hazard of a material, the flash point of a liquid is one of the most important fire characteristics of substances. At its flash point, a liquid continuously produces flammable vapors at the right rate and amount (volume) to give a flammable and even explosive atmosphere if a source of ignition should be brought into the mixture. Flammable liquids (like gasoline) with a flash point of –45°F (–42.8°C) continually give off vapors that can burn or explode (depending on the confinement of the mixture) at ordinary temperatures. However, fuel oil (such as that used in home-heating furnaces) with a flash point of 130°F (54.4°C) does *not* give off vapor that can burn until heated above its flash point (NFPA, 1997). However, when either material is ignited, an intense fire ensues.

A self-sustaining fire does not necessarily develop at the flash point. A closely related and less common term is fire point. Fire point is the temperature at which the initial flash becomes self-sustaining. For higher flash point materials, the fire point is usually a few degrees above the flash point.

As a rule of thumb, flash point can be thought of as the temperature above which a pool of liquid will ignite if a match or other small ignition source is dropped into it. If the temperature of the pool is below flash point, the pool will not ignite. From a safety perspective, a release of liquid below its flash point should not ignite even if it finds an ignition source.

The flash point and other important properties of some common materials are listed in Table B-2 (modified from NFPA 86, Table 7-5.2.2a and NFPA 325, 1994).

**Table B-2**
*Properties of Commonly Used Flammable Liquids in U.S. Customary Units*

| Solvent Name | Molecular Weight | Flash Point °F | autoigni tion °F | LEL% by Volume | UEL% by Volume | Boiling Point °F |
|---|---|---|---|---|---|---|
| Acetone | 58 | –4 | 869 | 2.5 | 12.8 | 133 |
| Ammonia | 17 | Gas | 1204 | 16 | 25 | –28 |
| Benzene | 78 | 12 | 928 | 1.2 | 7.8 | 176 |
| n-Butyl Alcohol | 74 | 98 | 650 | 1.4 | 11.2 | 243 |
| Carbon Disulfide | 76 | –22 | 194 | 1.3 | 50.0 | 115 |
| Cyclohexane | 84 | –4 | 473 | 1.3 | 8.0 | 179 |
| Ethane | 30 | –275 | 959 | 3.0 | 12.4 | –128 |
| Ethylene | 287 | –250 | 914 | 2.7 | 36 | –155 |
| Gasoline | Mix | –45 | 536 | 1.4 | 7.6 | Range |
| *n*-Heptane | 100 | 25 | 399 | 1.0 | 6.7 | 209 |
| *n*-Hexane | 86 | –7 | 437 | 1.1 | 7.5 | 156 |
| Kerosene (Fuel Oil #1) | Mix | 100–162 | 410 | 0.7 | 5.0 | Range |
| Methane | 16 | Gas | 1004 | 5 | 15 | –2,590 |
| Naptha (VM&P Regular) | Mix | 28 | 450 | 0.9 | 6.0 | 203–320 |
| Propane | 44 | –220 | 842 | 2.1 | 9.5 | –44 |
| *n*-Propyl Alcohol | 60 | 74 | 775 | 2.2 | 13.7 | 207 |
| Toluene | 92 | 40 | 896 | 1.1 | 7.1 | 231 |
| Turpentine | 136 | 95 | 488 | 0.8 | | 300 |
| Vinyl Acetate | 86 | 18 | 756 | 2.6 | 13.4 | 161 |
| o-Xylene | 106 | 88 | 867 | 0.9 | 6.7 | 292 |

### *B3.3. Flammability Limits*

Flammable vapor burns in air only over a limited range of fuel-to-air concentrations. The flammable range is defined by two parameters; the Lower flammable limit (LFL) and the upper flammable limit (UFL). These two terms are also called the lower explosive limit (LEL) and the upper explosive limit (UEL).

The *lower flammability limit* is the minimum proportion of fuel in air that will support combustion. The *upper flammable limit* is the maximum concentration of fuel in air that can support combustion. In popular terms, a mixture below the LFL/LEL is too "lean" to burn or explode and a mixture above the UFL/UEL is too "rich" to burn or explode.

For example, the lower flammability limit of methane in air at sea level is a concentration (by volume or partial pressure) of about 5%. The upper flammability limit is about 15% by volume or partial pressure. Heavier hydrocarbons tend to have lower LFLs. The LFL and UFL of some common hydrocarbons are given in Table B-2.

Flammability limits are not absolute, but are dependant on temperature, pressure, and other variables. Care must be exercised in using flammability limit data when conditions are different from ambient. For example, in reactors and thermal oxidizers.

An increase in temperature tends to widen the flammable range, reducing the LFL. For example, the LFL for methane in air is commonly quoted as 5%. As the temperature of methane increases to autoignition temperature, the LFL falls to around 3%. Stronger ignition sources can ignite leaner mixtures. Flammability limits also depend on the type of atmosphere. Flammability limits are much wider in oxygen, chlorine, and other oxidizers than in air (NFPA, 1997).

The flammability limit of a mixture of fuels may be calculated using the following equation:

$$\mathrm{LFL} = \left[ \sum_{i=1}^{n} \left( \frac{y_i}{\mathrm{LFL}_i} \right) \right]^{-1} \tag{B-1}$$

where

$y_i$ is the concentration of fuel component $i$ (mole fraction)

$\mathrm{LFL}_i$ is the lower flammable limit of fuel component $i$ (%vol)

Flammability limits can be narrowed by the addition of inert gases such as nitrogen or carbon dioxide.

### *B3.4. Autoignition*

The autoignition temperature of a substance is the lowest temperature at which a solid, liquid, or gas will spontaneously ignite resulting in self-sustained combustion without the need for an external ignition source. A material released from a process above its autoignition temperature will ignite. Autoignition temperatures of some common materials are shown in Table B-2.

Ignition of a mixture above its autoignition temperature is not instantaneous. The ignition time delay may be a fraction of a second for temperatures well in excess of the autoignition temperature; a few minutes when the material is just above its autoignition temperature.

Measurement of autoignition temperature can vary based on the size of the vessel (i.e., the test apparatus) holding the vapor and the strength of the ignition source. Because of these factors, there is often some variability in quoted autoignition temperatures. The convention is to use the lowest value found in literature.

Autoignition temperature is useful in determining the temperature of a hot surface required to ignite a mixture. The hot surface heats the gas and ignition occurs. The lowest temperature at which ignition can occur is the autoignition temperature. In most cases of interest, the hot surface ignition temperature is often significantly lower than the autoignition temperature due to surface effects.

Calculation of hot surface ignition temperature is complex and depends on many variables. From a safety perspective, it is often safest to assume that a surface with a normal temperature slightly below the autoignition temperature is a potential ignition source.

### *B3.5. Minimum Ignition Energy*

Minimum ignition energy is the lowest possible energy that will result in the ignition of a flammable mixture by an electrical discharge. The minimum ignition energy depends on the composition of the mixture and can be as low as 0.2 mJ for many common hydrocarbon fuels and even lower for reactive hydrocarbons like acetylene. This low energy threshold means even a small electrical spark or static discharge can ignite a hydrocarbon vapor cloud.

### *B3.6. Burning Velocity*

The rate of combustion of a flammable mixture is dependant on the burning velocity. Burning velocity is the speed with which a smooth, laminar flame advances into a stationary mixture of reactants. The burning velocity of most saturated hydrocarbons is 1.2–1.5 ft/sec (0.4–0.5 m/s). The burning velocity increases for olefins and acetylene (3.7 ft/sec (1.1 m/s) for acetylene) reaching 11 ft/sec (3.4 m/s) for hydrogen. The burning velocity is a function of the concentration of fuel, oxygen, temperature, and pressure of the mixture. The burning velocity of a hydrocarbon in pure oxygen can be an order of magnitude higher than in air.

A related term is flame speed. Flame speed is the speed with which a flame appears to move relative to a stationary observer. The flame speed can be much larger than the burning velocity due to expansion of the combustion products, instability and turbulent deformation of the flame. Flame speeds of 30–300 ft/sec (9–90 m/sec) are commonly observed for hydrocarbon–air mixtures. A gas phase detonation occurs when the flame speed exceeds the speed of sound in the burning vapor air mixture.

### *B3.7. Stoichiometric Ratio*

The stoichiometric ratio is the proportion of fuel and oxidizer that results in optimal combustion and maximum heat release. The optimal ratio is determined by finding the amount of air that will result in the products of the combustion reaction containing only water and carbon dioxide with no left over oxygen. Burning 100 standard cubic feet of methane requires 1,000 standard cubic feet of air for a stoichiometric mixture.

A mixture below it stoichiometric ratios is described as "lean." A mixture above its stoichiometric ratio is described as "rich." A "lean" mixture has extra oxygen along with the combustion products and a "rich" mixture has fuel remaining with the combustion products.

Fuel–air mixtures at or around stoichiometric concentration have the lowest autoignition temperature, lowest minimum ignition energies, and highest burning velocities.

## B4. Modes of Heat Transfer

Heat can be transferred by conduction, convection, and radiation. Conduction is heat transfer through a solid. Convection is heat transfer in a fluid. Radiation is heat transfer by electromagnetic radiation and does not require a transfer medium. Heat transfer is always from hot to cold.

The main mechanisms of heat transfer in a hydrocarbon are thermal radiation and direct flame contact. Heat transfer to personnel will cause burns. Heat transfer to equipment and structures can lead to failure of hydrocarbon containing equipment, which can further feed the fire.

### *B4.1. Conduction*

Conduction is the primary mode of heat transfer through solid material. Conduction occurs by two mechanisms:

- *Molecular motion*—molecules of higher energy (motion) impart that energy to adjacent molecules of lesser energy
- *Migration of free electrons*—this is primarily associated with heat transfer in metals

Mathematically, conduction is described by Fourier's law. In 1822, Fourier postulated that the rate of heat transfer is proportional to the temperature gradient present in a solid. Fourier's law can be expressed as:

$$Q = -kA\,DT \tag{B-2}$$

where

$A$ is the cross-sectional area of the solid that the heat is flowing through
$DT$ is the temperature gradient in the solid.
$K$ is the proportionality constant of the thermal conductivity ($k$) of the material

Thermal conductivity is a physical property of the solid through which the heat is being transferred. It is a measure of the material's ability to conduct heat. Insulators have a low thermal conductivity and conductors have a high thermal conductivity. The rate of heat transfer has magnitude and direction. This is represented mathematically by the negative sign that appears in Fourier's law of heat conduction.

Insulators or materials with low heat conduction are often used for fire protection coatings and thermal insulation. This thermal barrier prevents heat reaching and damaging structural steel and process vessels.

Conduction is also important as heat can be drawn away from the point of flame contact. This has an application in some large metal structures where the heat from the fire can be drawn away allowing the structure to survive the fire (NFPA, 1997).

## B4.2. Convection

Convection occurs when a gas or liquid exchanges energy with an adjacent solid. Convection is the term used to describe the motion or circulation that occurs in any gas or liquid as it is heated or cooled.

Convection is not a singular heat transport vehicle as are conduction and radiation. Instead, convection increases conduction by constantly circulating warmer material away from hot surfaces and replacing it with colder material. This increases the effective temperature difference, which increases the rate of heat transfer by conduction.

The fluid motion transfers energy away from the hot surface moving the energy to another cooler surface. If there is no fluid motion, then heat transfer occurs due to conduction. There are two types of convective heat transfer:

- *Forced convection*—fluid motion is induced by an external source such as a fan or pump
- *Natural convection*—heating a fluid results in natural convection heating

In natural convection, the temperature gradient in the fluid creates variations in density within the fluid. Gravity then causes the colder, denser fluid to sink, and the hotter less-dense fluid to rise. This movement results in the circular motion of warm fluid (i.e., this motion is convective heat transfer). In space where there is no gravity, natural convection does not occur.

The rate of convection heat transfer from a solid hot surface to a fluid is described by Newton's law of cooling. This relationship is valid for both forced and natural convection:

$$Q = hA\,(T_{surf} - T_{amb}) \qquad \text{(B-3)}$$

where

$H$ is defined as the convective heat transfer coefficient. This proportionality constant contains all the nonlinearities associated with convection
$A$ is the area of the surfaces in contact
$T_{surf}$ is the temperature of the hot surface
$T_{amb}$ is the ambient fluid temperature

For fires in open areas, heat is transferred by convection and rises away from the fire. However, for fires in building and other enclosed areas, the heat is transported through the building and collects on ceilings (NFPA, 1997).

### B4.3. Radiation

Radiation is fundamentally different from conduction as it describes the transfer of heat between two substances that are not in contact with each other. Like conduction, radiation is an independent form of heat transfer. Ignoring the conflicts of wave and quantum theory, radiation, refers to the transmission of electromagnetic energy through space.

All matter above absolute zero (–456.7°F) emits electromagnetic radiation. The exact process is a complex quantum physics phenomena. How much heat an object radiates is determined by the temperature of the object, the temperature of the surrounding environment, and the object's emissivity factor.

While the term radiation applies to the entire electromagnetic spectrum, heat transfer occurs only within a portion of the electromagnetic spectrum. The range of wavelength for thermal radiation is approximately from 0.3 to 50 microns and spans from infrared, visible light, to the lower end of the ultraviolet range. As the heat energy is transmitted via electromagnetic waves or photons, it travels at the speed of light. Radiant energy that strikes a surface can be:

- Reflected
- Absorbed
- Transmitted (for transparent material)

The rate of radiant thermal energy transfer between two bodies is described by the Stefan–Boltzman law. Originally proposed in 1879 by Joseph Stefan and verified in 1884 by Ludwig Boltzmann, the Stefan–Boltzmann law states that the emission of thermal radiative energy is proportional to the fourth power of the absolute temperature (Kelvin or Rankine):

$$E = \varepsilon \sigma T^4 \tag{B-4}$$

where
$T$ is the temperature of the emitter
$E$ is the emissive power flux of the surface (W/m$^2$)
$\sigma$ is the Stefan–Boltzmann constant (5.67E-10 W/m$^2$-k$^4$)

$\varepsilon$ is the emissivity of the surface

The rate of radiative heat transfer between two surfaces, $\alpha$ and $\beta$ is:

$$Q = \varepsilon\sigma\,(T\alpha^4 - T\beta^4) \tag{B-5}$$

Heat transfer does not occur until the electromagnetic waves or photons strike an object. This impact causes motion of the molecules on the surface. The heat generated is spread to the interior of the object through conduction (NFPA, 1997).

In many process fires, heat transfer by radiation is the dominant form of heat transfer. The heat radiated from a flame is emitted by gases, in particular the products of combustion and by soot. A flame in which the radiation is emitted solely from the gaseous products of combustion is termed nonluminous and a flame in which there is soot is termed luminous (i.e., yellow or visible).

Flames of some hydrocarbons, such as natural gas, contain relatively little soot, whereas heavier hydrocarbons, such as kerosene and crude oil, generate copious amounts of soot and smoke. The heat transferred to a target will depend on the:

- Size and shape of the flames
- Temperature of gases and soot in the flames
- Amount absorbed by the atmosphere as it travels to the target
- Size and orientation of the target

Radiant heat transfer can result in burns to personnel and can heat up unprotected process equipment and structural elements. If the heat is not dissipated by the application of cooling or conduction, the process equipment or structure may fail.

## B5. Effects of Fire Confinement

The knowledge of confined fires is incomplete, due to the complex physical and chemical processes that control confined fires. Expert advice should be sought on how to assess confined fires. The following provides an introduction to the subject and highlights the relevant factors that need to be considered. For the purposes of this section, a fire is considered to be confined if there is a roof above the release. There may or may not be additional side walls providing additional levels of confinement. In general, confining a fire will have two effects:

- Confining the flame and combustion products
- Restricting the ventilation supply of fresh air to the fire

### B5.1. Confinement of Combustion Products

In a confined fire, combustion products will be contained to a greater extent. For a pool fire or a vertical jet, the rising plume will impinge on the ceiling and spread radially in the form of a ceiling jet which may or may not extend to the edge of the module. The hot, less dense, combustion products will begin to accumulate at the top. The downward spread of the accumulating gases will be counteracted by the buoyancy forces of the hot gases and a well defined layer of hot gases (including flame) will develop and grow. This is known as the "hot gas layer." Below the hot gas layer, there will be a comparatively cool layer of almost clear air. If the ceiling is flat with no obstructions and clean openings extending up to the full height and there is good ventilation, then the hot gas layer may be relatively thin and it will spread easily across the ceiling and out through the openings.

If the ceiling is congested with intersecting beams, cable trays, and piping causing dead areas and turbulence, then the layer will be much thicker. If the openings do not extend to the full height, then the hot gas layer will descend below the top of the openings until the excess heat and smoke can escape. With poor ventilation, it descends even further.

If there are openings in the structure, as the hot gas layer builds up, gases will begin to escape through the upper part of the opening(s). Fresh air will be entrained through the lower part(s). Because a module ceiling will spread the flames across the ceiling, a significant part, if not all, of the ceiling beams and high level piping may be engulfed in flame. It is possible that this could lead to simultaneous failure of the ceiling beams. This could lead to the release of any inventories supported on the level above or of total structure collapse.

### B5.2. Restrictions to Ventilation

The amount of air that can be drawn into an enclosure to support combustion will depend on both the size and configuration of ventilation openings in the enclosure. A poorly ventilated fire will create more smoke and byproducts. Any unburnt fuel may also burn outside the enclosure, endangering adjacent areas. A fire in a structure may also cause other fuels to burn such as cables, paint, plastic equipment etc. These will add to the fuel load and to the smoke.

### B5.3. Heat Fluxes Within the Module

The first consideration is that the heat cannot escape as easily as in open air. Unless the fire is totally starved of oxygen, it would be at least as hot as the equivalent open air case; whether it is a pool, spray, or jet. The action of the roof and the plant will be to cause flame recirculation. This tends to increase the depth or thickness of the flames and, thus increases the radiation. In the jet and spray cases, the velocities close to the point of release may be slightly reduced, but the overall flame movement in the structures will give considerable heat transfer.

The hottest fires may be associated with those cases where the fire is big enough to give flames to fill at least half the structure volume, cases where it is stoichiometric or just under ventilated, and cases where the hot gas layer is 10 ft (3 m) or more deep. Heavier fuels would be less likely to give the hottest fires, as they may not receive enough heat feedback to vaporize the liquid and therefore they may be self limiting in terms of the burn rate. Where these conditions may be encountered, heat fluxes of 1320–1584 BTU/ft$^2$ (250 to 300 kW/m$^2$) may be experienced. In certain circumstances, (which are not yet fully understood) highly efficient combustion can occur with fluxes of 1848–2112 BTU/ft$^2$ (350–400 kW/m$^2$) and temperatures of 2,500°F (1,400°C).

Confined fires will spread the flames across the ceiling and, in general, this will be the hottest part of the fire. If the flame volume is more than 20% of the structure volume, then all of the ceiling beams and high level piping will be engulfed. It is possible that this could lead to simultaneous failure of all of the ceiling beams and upper module structure. This could lead to the release of any inventories supported on the level above or of total structure collapse.

## B6. Hazardous Chemicals and Processes

There are innumerable situations where gases, liquids, and hazardous chemicals are produced, stored, or used in a process that, if released, could potentially result in a hazardous fire condition. It is important to analyze all materials and reactions associated with a particular process, including production, manufacturing, storage, or treatment facilities. Each process requires analysis of the potential for fire.

Hydrocarbon fires are the principal concern in many processing facilities. There are many different types of hydrocarbon fires. The mode of burning depends on characteristics of the material released, temperature and pressure of the released material, ambient conditions, and time to ignition. Types of hydrocarbon fires include:

- Jet fire
- Fire or flash fire
- Pool fire
- Running liquid fire
- Boiling liquid expanding vapor explosion (BLEVE) or fire ball

Other fires that can occur in specific areas within a process plant include:

- Solid, for example cellulose fires involving material such as wood, paper, dust, etc.
- Warehouse fires
- Electrical equipment fires, e.g., transformer fires
- Fires involving oxygen, e.g., systems for oxygen addition to a fluid catalytic cracking (FCC) unit
- Fires involving combustible metals, e.g., sodium used as a coolant or catalyst
- Fires involving pyrophoric materials, e.g., aluminum alkyls used as catalysts

### B6.1. Gases

The release of a flammable gas or the vaporization of a liquefied flammable gas can lead to different types of fire scenarios dependent on the release mechanism and the point of ignition. Figure 5-2 on page 53 illustrates the different outcomes expected from a gas release. If ignition of a gas release does not occur immediately at the origin of the release, then a gas cloud can develop (the same situation can also occur above flammable liquid spills). A delayed ignition of the gas cloud can result in a flash fire in which the premixed (fuel and air) gas cloud burns rapidly, typically in a matter of seconds.

A flash fire can result in the flame burning back to the source of the release and continuing as a jet fire (i.e., a fire in which air is entrained into the issuing fuel and the flame burns along the interface of the mixing fuel and air streams). If the initial gas cloud is located among equipment or structures, then turbulent flow results in the acceleration of the flame and a deflagration may occur (CCPS, 1994). The confinement of the cloud can cause an explosion with large overpressure effects (blast waves). A flash fire will not have any blast effects.

### B6.2. Liquids

The release of a flammable liquid can lead to different fire outcomes as illustrated in Figure 5-3, page 55. Liquids may burn as a:

- Static (confined) pool
- Running pool (unconfined), i.e., gravity-fed or low pressure flow
- Spray/mist, jet, e.g., under high pressure

For a liquid to ignite, three conditions must be satisfied:

1. Liquid must vaporize
2. Vapor must mix with air

3. A piloted ignition source to obtain self-propagation of the flame must be available

Most liquids respond to a temperature rise through a thermodynamic phase change to gas. For ignition to occur, the fuel concentration in air must be in a range that defines a flammable mixture. These bounding limits are commonly referred to as the lower flammability limit (LFL) and upper flammable limit (UFL). These are the lowest and highest fuel concentrations in air (by volume) that will support flame propagation. Fuel concentrations below the LFL or above the UFL are too lean or rich, respectively, and will not support combustion.

## B6.3. Hazardous Chemicals

### B6.3.1. Unstable Chemicals

Unstable chemicals are subject to spontaneous reactions. Situations where unstable chemicals may be present include the catalytic effect of containers, materials stored in the same area with the chemical that could initiate a dangerous reaction, presence of inhibitors, and effects of sunlight or temperature change. Examples include acetaldehyde, ethylene oxide, hydrogen cyanide, nitromethane, organic peroxides, styrene, and vinyl chloride.

As an example, styrene polymerizes at ordinary temperatures and the rate of polymerization increases as temperature increases. The reaction is exothermic and becomes violent as it is accelerated by its own heat. Inhibitors are added to prevent the initiation of dangerous polymerization. When the styrene is used to fabricate materials, e.g., fiberglass resin, a catalyst may be added in the manufacturing process to initiate polymerization at a controlled rate. Any unbalance of these reactions in terms of quantities or temperatures could cause hazardous fire conditions.

### B6.3.2. Reactive Chemicals

Chemicals that are water or air reactive pose a significant fire hazard because they may generate large amounts of heat. These materials may be pyrophoric, that is, they ignite spontaneously on exposure to air. They may also react violently with water and certain other chemicals. Water-reactive chemicals include anhydrides, carbides, hydrides, and alkali metals (e.g., lithium, sodium, potassium).

Air-reactive chemicals include aluminum hydride, aluminum alkyls, and yellow phosphorous. Other reactive chemicals include alkalis, aluminum trialkyls, anhydrides, charcoal, coal, hydrides, certain oxides, phosphorous, and sodium hydrosulfate.

Charcoal, for example, will react with air under certain conditions at a sufficient rate to heat spontaneously and ignite. The smaller the dust particles, the greater the fire hazard.

#### *B6.3.3. Combustible Chemicals*

All organic chemicals are essentially combustible. Combustion of some chemicals, such as sulfur and sulfides of sodium, potassium, and phosphorous, result in the production of hazardous gases, in this case sulfur dioxide. Carbon black, lamp black, lead sulfocyanate, nitroaniline, nitrochlorobenze, and naphthalene are examples of combustible chemicals.

#### *B6.3.4. Oxidizers*

Oxidizers may not themselves be combustible, but they may provide reaction pathways to accelerate the oxidation of other combustible materials. Combustible solids and liquids should be segregated from oxidizers. Certain oxidizers undergo dangerous reactions with specific noncombustible materials. Some oxidizers, such as calcium hypochlorite, decompose upon heating or contamination and self-react with violent heat output. Oxidizers include nitrates, nitric acid, nitrites, inorganic peroxides, chlorates, chlorites, dichromates, hypochlorites, perchlorates, permanganates, persulfates and the halogens.

### *B6.4. Other Hazardous Effects*

Independent of combustibility or reactivity, chemicals may exhibit hazardous properties that affect fire protection, in particular firefighting. Chemicals may be inherently toxic or radioactive. In either situation, potential exposure of smoke/gases to personnel from the burning of materials exhibiting these characteristics needs to be addressed.

### *B6.5. Process Fires*

A thorough analysis must be performed to identify possible fire scenarios. Figure 5-3, page 55, depicts an example of an event tree for the release of a flammable material, showing the pathways that lead to the various types of process fires. It is assumed that the release occurs from a pressurized source such as a large vessel (for storage, reaction, batching etc.) or a pipeline (for transfer, fittings, instruments, etc.). The release may be due to a major failure (e.g., spontaneous tank failure) or a minor accident (e.g., breakage of a fitting). The type of failure will dramatically impact the release rate.

The release may be or may not be accompanied by immediate ignition. All ignition scenarios should be analyzed to uncover and address all relative hazards. With immediate ignition, a jet flame will result if the release is from a relatively small opening. Such a release could be either vapor, liquid or both and, if liquid, could also involve flashing of liquid into vapor and/or accumulation of liquid.

If the release is the result of a major spill and there is immediate ignition, the result is usually a pool fire. In an unignited liquid release, the spill will be accompanied by vaporization, liquid entrainment, and/or vapor dispersion. A delayed local ignition following accumulation may result in a pool fire whose characteristics are strongly influenced by the geometry of containment (or lack thereof).

APPENDIX

# COMPUTER TOOLS FOR DESIGN

## C1. Introduction

The use of computer fire models in the field of fire protection engineering has grown substantially through the 1990s as the results gain wider acceptance among designers and regulators. This phenomenon is enhanced by economical, yet powerful, computers, better and more sophisticated computer models, effective means of presentation, and an increasing base of modeling experts. Computer modeling is an integral part of the trend toward performance-based design.

A fire model is a physical or mathematical representation of burning or other processes associated with fires. Mathematical models range from relatively simple formula that can be solved analytically to extensive hybrid sets of differential and algebraic equations that must be solved numerically on a computer. Software to accomplish this is referred to as a computer fire model.

Reference is made to specific models throughout this appendix. These references are for example purposes only. There is no attempt to provide exhaustive lists of all types of models, nor are any particular models emphasized or specifically endorsed. The user must choose the appropriate model based on their experience, the need, and the availability of the model.

## C2. Evolution of Computer Fire Modeling

Computer fire models have been available in one form or another since the 1960s. They gained greater popularity in the 1970s and 1980s as more models became available and computer power increased; however, for the most part computer models were still limited to research work. By the 1990s, many types of fire models became available, often competing versions of the same type of model. Nelson provides an overview of the history of fire models (Nelson, 2002).

The most commonly used computer fire models simulate the consequences of a fire in an enclosure. Zone models, as well as computational fluid dynamics (CFD) models, are used for this purpose. While they are in wide use, enclosure models have limited application in assessing hazards in the petrochemical industry. They are briefly described in this Appendix for general reference purposes.

Other computer models and analytical tools are used to predict how materials, systems, or personnel respond when exposed to fire conditions. Hazard-specific calculations are more widely used in the petrochemical industry, particularly as they apply to structural analysis and exposures to personnel. Explosion and vapor cloud hazard modeling has been addressed in other CCPS Guidelines (CCPS, 1994). Again, levels of sophistication range from hand calculations using closed-form equations to numerical techniques.

Many times, hand calculations are performed with the aid of calculators, spreadsheet templates, or a basic computer program to make the process easier. For example, the hand calculations in Chapter 5 have been computerized for easier use (Hughes Associates, 2002).

## C3. Computer Model Applications

Computer fire models have found a wide range of uses in the modern fire protection field. There are two primary purposes of fire models: to reconstruct and analyze a fire, or to evaluate the fire safety design of a structure or process. The uses include research and development, code and regulatory compliance, fire hazard analyses, litigation and forensic studies, structural design, smoke control design, and egress design.

There are no widely adopted techniques for assessing and validating computer fire models. ASTM has established a subcommittee on fire modeling, which developed guides that cover specific issues pertinent to computer fire modeling:

- ASTM E1355 addresses evaluating the predictive capability of fire models
- ASTM E1472 provides guidelines for documenting fire models
- ASTM E1591 describes procedures to obtain input data for fire models
- ASTM E1895 addresses uses and limitations of computer fire models

Of particular interest are the guidelines in ASTM E1355, which identify the following steps for evaluating models:

1. Define the scenarios for which the evaluation is to be conducted
2. Validate the theoretical basis and assumptions used in the model
3. Verify the mathematical and numerical robustness of the model
4. Evaluate the model, i.e., quantify its uncertainty and accuracy

A simplistic model, the **DETACT-QS** model used to predict the thermal response of detectors and sprinklers, has been subjected to the ASTM E1355 guidelines as a test case (Janssens, 2002). Over five years of effort have been dedicated to this evaluation, showing the difficulty in performing a detailed, recognized validation process.

Most of the models identified in the following sections have referenceable documentation and generally have been developed in the public domain. As such, they have relatively wide spread recognition. The final section of this Appendix includes privately developed models specific to the petrochemical industry. They are provided for informational purposes only, and no claim is made related to their applicability and performance since public domain documentation generally is not available.

## C4. Compartment Fire Simulations

Generally, compartment fire simulation models predict the fire development in a compartment under varying conditions. These types of simulations are useful for estimating tenability criteria, thermal insult to the compartment, and the likelihood of fire spread from one compartment to another. These types of models can be further subdivided into three categories based on their approach to simulating the fire environment: the zone model, the field model, and the post-flashover model.

### *C4.1. The Zone Model*

A zone model calculates the fire environment by dividing each compartment in the model into two homogeneous zones. One zone is an upper hot smoke zone that contains the fire products. The other zone is a lower, relatively smoke-free zone that is cooler than the hot zone. The vertical relationship between the zones changes as the fire develops, usually via expansion of the upper zone. The zone approach evolved from observations of such layers in full-scale fire experiments. While these experiments show some variation in conditions within the zones, the variations are most often small compared to the difference between the zones themselves.

Zone models may estimate the upper and lower layer temperature, the interface location between zones, the oxygen concentration, the carbon monoxide concentration, the visibility, and flows in and out of openings in the compartment as a function of time. This information may be useful for evaluating the tenability of a compartment or determining when flashover may occur in a space. A zone model may look at one room with a single opening or multiple rooms with many openings.

The input requirements for zone models vary dramatically depending on the model and the desired information. Generally, input for the zone model is relatively modest compared to the requirements for the field-type models. The compartment geometry and opening dimensions are needed to define the space and the surroundings. The thermal properties of the compartment boundaries are needed to estimate the heat loss through the walls, ceiling, and floor. The fire size must be entered, though the models may modify the heat release rate as the oxygen concentration in the compartment is reduced by the fire. Some zone models account for effects of mechanical ventilation, which means that the fan flow rate and the location of the vent inlets and outlets are required. Some examples of zone models are **CFAST** (CFAST, 1993), **FASTLite** (FASTLite, 1996), and **BRI-2** (BRI2, 2000).

### *C4.2. The Field Model*

Field models estimate the fire environment in a space by numerically solving the conservation equations (i.e., momentum, mass, energy, diffusion, species, etc.) as a result of a fire. This is usually accomplished by using a finite difference, finite element, or boundary element method. Such methods are not unique to fire protection; they are used in aeronautics, mechanical engineering, structural mechanics, and environmental engineering. Field models divide a space into a large number of elements and solve the conservation equations within each element. The greater the number of elements, the more detailed the solution. The results are three-dimensional in nature and are very refined when compared to a zone-type model.

Although field models generate detailed estimates of the effects of compartment fire environments, they are time consuming because of the enormous number of computations that are made. Some effects, such as radiation, can only be treated on a simplified level. In many instances, such a detailed solution is not needed, and simpler approaches are appropriate. Field models are useful to resolve problems that yield excessively conservative results when calculated using other model types.

Like zone models, field models require a description of the compartment geometry and the openings within the compartment. Field models are not limited to compartments, however. They may be used to simulate such phenomena as open plumes and unique configurations such as tunnels and shafts. The fire's heat release rate must also be specified. Field models do not always modify the heat release rate as the oxygen level decreases. Hence, care should be taken that the heat release rate can be supported. Heat loss to the compartment boundaries is also calculated using the thermal properties of the bounding materials. Some examples of field models are **JASMINE** (Cox and Kumar, 1987), **FLOW3D** (Portier et al, 1996), **PHOENICS** (Spalding) **SOFIE** (Rubini, 2002), and the **Fire Dynamics Simulator** (**LES3D**) (Baum, 2000).

### C4.3. *The Post-Flashover Model*

Post-flashover fire models calculate the time-temperature history in a compartment by solving simplified forms of the energy, mass, and species equations. The concentration of various gaseous constituents can be monitored as well as vent flows. Some post-flashover fire models allow mechanical ventilation to be factored in the calculation. These types of models are most useful for determining the time-temperature exposure to a structure for a specific compartment and fuel load. Such time-temperature histories can be used for assessing the possibility of structural failure or fire spread to adjacent compartments.

The input requirements for post-flashover types of models can be quite broad. Besides the compartment and vent dimensions, detailed fuel combustion characteristics are often needed. The fuel characteristics include the fraction of carbon, hydrogen, nitrogen, and oxygen that make up the fuel, the burning efficiency, and the quantity of fuel available for burning. Mechanical ventilation flow rates and the material properties of the compartment boundaries may be necessary. Some models can account for the heat transfer through the boundaries in detail, and may even allow the user to supply time-dependent material properties. An example of a post-flashover fire model is **COMPF** (Babrauskas, 1979).

## C5. Egress/Evacuation Models

The egress and evacuation simulations are not truly fire models. They were developed in response to the need to evaluate impact of fires on the occupants of a building. Most egress models describe a structure as a network of paths along which the occupants travel. The occupant travel rates are derived from people movement studies and are often stochastic. Factors that are included in the travel rates are the age and ability of the occupant, crowding, and the type of travel path.

Egress models generally require a detailed plan of the layout or area considered. Additionally, the occupant load, the type of occupants, and the smoke conditions are input requirements. The models can be used to predict the egress time from the floor or deck level, stair egress time, congestion locations, areas that need more exits, and the impact of exit locations in a space. Some examples of egress models are **EVACNET** (Kisko and Francis, 1985), **EXITT** (Levin, 1988), and **EXIT89** (Fahy, 1995).

## C6. Smoke Movement Models

Smoke movement models are used to estimate smoke spread through a building. Some models are suited for modeling smoke control systems under varying exterior and interior conditions. These models are similar to hydraulic sprinkler models to the extent that they model rooms and spaces as nodes and the openings as links. Links, which are analogous to a pipe, have a pressure drop that depends on the type of opening and the flow rate through the opening. A model estimate is obtained by balancing the pressure and flow rates.

Some smoke movement models can track the dispersion of gaseous species such as toxic gases. Smoke movement models may be an economical alternative to full-scale smoke control acceptance testing, and can be used to assess the impact of a wide range of parameters, such as different wind directions, exterior temperatures, and window openings. These models are usually used in larger, multilevel structures. Unless there is a graphical interface, the input process can be tedious because all of the compartment interconnections must be entered. Normally, the volume of a space, the size and flow coefficients for the openings, and the exterior/boundary conditions are needed for a smoke movement analysis. Some examples of these models are **CONTAM96** (Walton, 1997), **Airnet** (Walton, 1989), and **MFIRE** (Chang et al., 1990).

A relatively new simulation technique, **ALOFT** (McGrattan, 1987) has been developed for radiation and smoke movement from large pool fires.

## C7. Thermal/Structural Response Models

Thermal/structural response models are related to field models in that they numerically solve the conservation of energy equation, though only in solid elements. Finite difference and finite element schemes are most often employed. A solid region is divided into elements in much the same way that the field models divide a compartment into regions. Several types of surface boundary conditions are available: adiabatic, convection/radiation, constant flux, or constant temperature. Many of these models allow for temperature and spatially dependent material properties.

The structural fire endurance of a structural system is a measure of its ability to resist collapse during exposure to a fire. The thermal/structural response models evaluate the time-temperature history within a solid exposed to a fire environment. The time-temperature history, or design fire exposure, can be a standardized fire curve, such as the ASTM E119 and the UL1709 curves, or it can be estimated using a compartment fire model. Structural steel and concrete members are most commonly analyzed using these types of models, though floor and roof assemblies can also modeled. These models are exceptionally well suited for determining the required thickness of fire insulation needed to protect against a design fire. They can also be used with structural models to calculate deflections in a floor/deck or in the framing.

The thermal properties that are necessary to perform a structural analysis are the thermal conductivity and the specific heat. The density is required, and phase-change data may be needed depending on the type of problem considered. The exposure time-temperature history is input at each time step or is interpolated by the program. Some examples include **FIRES-T3** (Iding et al, 1977) and **TASEF** (Wickstrom, 1999).

## C8. Conglomerate/Miscellaneous Fire Models

Various calculation tools and other types of fire models are available that do not fit neatly into the above categories. Among these are multicalculation packages, flame spread models, and glass breaking simulations.

Two of the more popular multicalculation software packages are **FPETOOL** (Nelson, 1990) and **FASTLite** (Portier, 1996). These two software packages contain more than a dozen fire protection calculation tools. They each also include a compartment fire model. The calculation tools are useful for obtaining first order approximations to fire-related aspects. The routines include vent flows, radiant heating of objects, sprinkler actuation, timed egress, and plume dimensions. Most of the tools are computerized versions of hand calculations and are useful in that capacity.

Another type of fire model that has yet to gain popularity and widespread use is the flame spread model. Flame spread models use material properties from small-scale fire tests to estimate fire spread over different types of surfaces. One of the drawbacks of these models is that the surfaces are limited to a few specific geometries, such as a vertical wall and a ceiling, or corner and a ceiling configuration. They are still primarily used as research tools, but may eventually be used to generate more realistic heat release rates in compartment room fire models. Typically, these models usually require the thermal properties of the surface material, the ignition temperature, and a lateral flame spread parameter. Some models divide a surface into smaller areas whereas others use moving surface areas to simulate the burning process. An example of a flame spread model is **CFHAT** (Lattimer, 2000).

Glass breaking models are sometimes used to predict when a window will fail in a compartment fire. This is important in some types of fire simulations because when a window breaks, the ventilation conditions in a compartment can change drastically. In some cases, the added oxygen supply can worsen a fire to the point that flashover will occur. Glass breaking models use a combination of heat transfer and thermal stress equations.

## C9. Fire Models and Analytical Tools Specific to the Petrochemical Industry

The hazards common to the petrochemical industry involve flammable liquids and gases. Many of the fire models that can be used to assess these hazards fall under the categories mentioned above. The models discussed below are fire models or conglomerate models that have been designed specifically to address fire hazards presented in the petrochemical industry.

A comprehensive listing of public domain and private sector models available for petrochemical hazard analysis is provided by The Center of Marine and Petroleum Technology (Spouge, 1999). This listing includes both public domain and private sector models/programs having varying degrees of access or restriction to users (e.g., some require licenses or are available "on request"). No public domain references are specifically cited.

### *C9.1. Public Domain/Unrestricted*

**FIREX**—The FIREX model is a program designed for most fire situations involving the accidental release of a gas or liquid resulting in an explosion. Capabilities of FIREX include smoke production, visibility, and temperature response of steel. SINTEF NBL-Norwegian Fire Research Laboratory, and INTELLEX GmbH were the original developers of FIREX.

**FRED**—Shell Research has developed FRED (Fire Release Explosion Dispersion) as a PC based model for accidental release of gas or liquid. Areas of application include fire, dispersion, and explosion. Shell also has produced **SEAFIRE**, a consequence model for fires from subsea pipelines.

### *C9.2. Restricted*

**EFFECTSGIS**—This is a conglomerate package for modeling chemical releases. These releases are linked to appropriate physical phenomena models, such as pool fire models to predict the consequences on humans. The program is equipped with an internal mini-GIS system to enable calculation results, such as heat radiation, to be overlayed onto maps.

**KAMELEON FIRE E-3D**—This model is a program specifically designed to deal with hydrocarbon fires in the form of both liquid pool fires and gas jet fires.

KAMELEON was validated in the SINTEF enclosed pool fire experiments. Availability of this model is restricted.

**PHAST** (Process Hazard Analysis Software Tool)—This is a conglomerate package for gas dispersion and fire modeling. PHAST is capable of calculating the formation of a cloud or pool to final dispersion calculating concentrations, fire radiation, toxicity, and explosion overpressure endpoints.

**REDIFEM**—This fire model has applications including steady state releases of compressible gas/vapor, incompressible liquid and transient release from a gas vessel, Gaussian Plume models, continuous free momentum, BLEVE, and confined and unconfined vapor cloud explosions. REDIFEM is reported to have internal validation with ISO 9001 and checked against PHAST and FRED.

Additional models and software are identified in *A Guide to Quantitative Risk Assessment for Offshore Installations* (Spouge, 1999) which address offshore risk analysis, explosion modeling, evacuation and rescue analysis, reliability analysis, accident databases, event tree analysis, and safety management.

APPENDIX

# SAMPLE FIRE PRE-PLAN

## Storage Tank Fire

| Task | Predetermined Assignments |
|---|---|
| ■ Activate Emergency Alarm Notification System and take defensive action until relieved by a more qualified employee. (Use Site notification system) | First Responder |
| ■ If trained to do so and confident that you are capable, activate any fire monitors and/or fixed fire water systems in the immediate area and attempt to cool the tank and surrounding equipment.<br>■ If the tank is receiving material or product, close a valve on the inlet line at a safe location.<br>■ Stand by to direct the Emergency Response Team (ERT) to scene<br>■ Use appropriate Personal Protective Equipment (PPE) | First Responder |
| ■ Activate Incident Command system(ICS) | Incident Commander |
| ■ Upon activation of ICS:<br>■ Gain status of situation<br>■ Determine if evacuation or shelter-in-place is required<br>■ Determine emergency level (0, 1, or 2)<br>■ Notify Communications Center of emergency response level (0, 1, or 2)<br>■ If evacuation/shelter-in-place is required, notify Communications Center<br>■ Determine what resources need to be mobilized<br>■ Determine if Mutual Aid is required | Incident Commander |

| Task | Predetermined Assignments |
|---|---|
| ▪ Activate appropriate plant emergency alarm | Communication Center |
| ▪ Activate emergency response notification system based on emergency response level | Communications Center |
| ▪ If requested by ICS, place Mutual Aid group(s) on standby and/or request response to staging area | Communications Center |
| ▪ Nonessential personnel report to safe assembly areas. Stop all hot work and confined space entry | Nonessential Personnel |
| ▪ Begin hazard assessment and PPE requirements | Incident Commander |
| ▪ Develop tactical plan | Incident Commander |
| ▪ Communicate plan to ERT leaders | Incident Commander |
| ▪ Assign Safety Officer and Sector Officers | Incident Commander |
| ▪ Open staging area and coordinate with responding mutual aid groups | Staging Officer |
| ▪ Communicate tactical plan to ERT members | ERT Officers |
| ▪ Start fire suppression | ERT Officers |
| ▪ Monitor progress of suppression efforts | Incident Commander |
| ▪ Communicate with Unit/Facility Management or Emergency Operations Center (EOC) if activated | Incident Commander |
| ▪ Monitor control of emergency<br>▪ Gain status of situation<br>▪ Determine what resources have been mobilized<br>▪ Determine if additional resources are required<br>▪ Complete necessary agency notification<br>▪ Prepare to handle the media<br>▪ If personnel are taken to hospital, send management representative to hospital | Unit/Facility Management<br>or<br>EOC if activated |
| ▪ Once fire has been extinguished, monitor the area to ensure no flash backs | ERT Officers |
| ▪ Determine if "ALL CLEAR" can be issued | Incident Commander |
| ▪ Issue "ALL CLEAR" | Communications Center |
| ▪ Decontaminate personnel, equipment, and fire equipment | Incident Commander |

| **Task** | **Predetermined Assignments** |
|---|---|
| ▪ Restore emergency equipment to "State-of-Readiness" | Incident Commander |
| ▪ Complete required documentation of event | Incident Commander |
| ▪ Schedule critique of event with ERT members | Incident Commander |
| ▪ Form accident investigation team | Operations Manager |

# REFERENCES

## American Petroleum Institute (API) References


API Publication 2026. 1998. *Safe Access/Egress Involving Floating Roofs of Storage Tanks in Petroleum Service*. 2nd Edition. American Petroleum Institute, Washington D.C.

API Publication 2030. 1998. *Application of Water Spray Systems for Fire Protection in the Petroleum Industry*. 2nd Edition. American Petroleum Institute, Washington D.C.

API Publication 2207. 1998. *Preparing Tank Bottoms for Hot Work*. 5th Edition. American Petroleum Institute, Washington D.C.

API Publication 2218. 1999. *Fireproofing Practices in Petroleum and Petrochemical Processing Plants*. 2nd Edition. American Petroleum Institute, Washington D.C.

API Publication 2510A. 1996. *Fire Protection Considerations for the Design and Operation of Liquefied Petroleum Gas (LPG) Storage Facilities*. 2nd Edition. American Petroleum Institute, Washington D.C.

API RP 500. 1997. *Recommended Practice for Classification of Locations for Electrical Installations at Petroleum Facilities Classified as Class 1, Division 1 and Division 2*. 2nd Edition. American Petroleum Institute, Washington D.C.

API RP 520. 2000. *Sizing, Selection, and Installation of Pressure-Relieving Systems, Part 1-Sizing and Selection*. 7th Edition. American Petroleum Institute, Washington D.C.

API RP 521. 1997. *Guide for Pressure-relieving and Depressuring Systems*. 4th Edition. American Petroleum Institute, Washington D.C.

API RP 554. 1995. *Process Instrumentation and Control*. 1st Edition. American Petroleum Institute, Washington D.C.

API RP 556. 1997. *Fired Heaters and Steam Generators*. 1st Edition. American Petroleum Institute, Washington D.C.

API RP 752. 1995. *Management of Hazards Associated with Location of Process Plant Buildings, CMA Manager's Guide*. 1st Edition. American Petroleum Institute, Washington D.C.

API RP 2001. 1998. *Fire Protection in Refineries*. 7th Edition. American Petroleum Institute, Washington D.C.

API RP 2003. 1998. *Protection Against Ignitions Arising Out of Static, Lightning and Stray Currents*. 6th Edition. American Petroleum Institute, Washington D.C.

API RP 2016. 2001. *Guidelines and Procedures for Entering and Cleaning Petroleum Storage Tanks*. 1st Edition. American Petroleum Institute, Washington D.C.

API RP 2021. 2001. *Management of Atmospheric Storage Tank Fires*. 4th Edition. American Petroleum Institute, Washington D.C.

API RP 2220. 1998. *Improving Owner and Contractor Safety* Performance. American Petroleum Institute, Washington D.C.

API RP 2221. 1996. *A Manager's Guide to Implementing a Contractor Safety and Health* Program. American Petroleum Institute, Washington D.C.

API Specification 12B. 1995. *Bolted Tanks for Storage of Production Liquids*. 14th Edition. American Petroleum Institute, Washington D.C.

API Specification 12D. 1994. *Field Welded Tanks for Storage of Production Liquids*. 10th Edition. American Petroleum Institute, Washington D.C.

API Specification 12F. 1994. *Shop Welded Tanks for Storage of Production Liquids*. 11th Edition. American Petroleum Institute, Washington D.C.

API Standard 620. 2001. *Design and Construction of Large, Welded, Low-pressure Storage Tanks*. 10th Ed. American Petroleum Institute, Washington, D.C.

API Standard 650. 1998. *Welded Steel Tanks for Oil Storage*. 10th Ed. American Petroleum Institute, Washington, D.C.

API Standard 2015. 2001. *Requirements for Safe Entry and Cleaning of Petroleum Storage Tanks*. 6th Edition. American Petroleum Institute, Washington D.C.

API Standard 2510. 2001. *Design and Construction of Liquefied Petroleum Gas Installations (LPG)*. 8th Edition. American Petroleum Institute, Washington D.C.

## Center for Chemical Process Safety (CCPS) References

CCPS. 1992. *Guidelines for Hazard Evaluation Procedures*. 2nd Edition. American Institute of Chemical Engineers, New York.

CCPS. 1993a. *Guidelines for Auditing Process Safety Management Systems*. American Institute of Chemical Engineers, New York.

CCPS. 1993b. *Guidelines for Engineering Design for Process Safety*. American Institute of Chemical Engineers, New York.

CCPS. 1994. *Guidelines for Evaluating the Characteristics of Vapor Cloud Explosions, Flash Fires, and BLEVEs*. American Institute of Chemical Engineers, New York.

CCPS. 1995. *Guidelines for Process Safety Documentation*. American Institute of Chemical Engineers, New York.

CCPS. 1995a. *Guidelines for Technical Planning for Onsite Emergencies*. American Institute of Chemical Engineers, New York.

CCPS. 1995b. *Guidelines for Technical Management of Chemical Process Safety*. American Institute of Chemical Engineers, New York.

CCPS. 1995c. *Guidelines for Safe Storage and Handling of Reactive Materials*. American Institute of Chemical Engineers, New York.

CCPS. 1996a. *Guidelines for Integrating Process Safety Management, Environment, Safety, Health and Quality*. American Institute of Chemical Engineers, New York.

CCPS. 1996b. *Inherently Safer Chemical Processes: A Life Cycle Approach*. American Institute of Chemical Engineers, New York.

CCPS. 1996c. *Guidelines for Evaluating Process Plant Buildings for External Explosions and Fires*. American Institute of Chemical Engineers, New York.

CCPS. 1998. *Guidelines for Safe Warehousing of Chemicals*. American Institute of Chemical Engineers, New York.

CCPS. 1999. *Guidelines for Consequence Analysis of Chemical Releases*. American Institute of Chemical Engineers, New York.

CCPS. 1999a. *Avoiding Static Ignition Hazards in Chemical Operations*. American Institute of Chemical Engineers, New York.

CCPS. 2000. *Guidelines for Chemical Process Quantitative Risk Assessment*. American Institute of Chemical Engineers, New York.

CCPS. 2001a. *Guidelines for Layers of Protection Analysis*. American Institute of Chemical Engineers, New York.

CCPS. 2001b. *Guidelines for Making EHS an Integral Part of the Process Design*. American Institute of Chemical Engineers, New York.

CCPS. 2002. *Guidelines for Analyzing and Managing Security Vulnerabilities of Fixed Chemical Sites*. American Institute of Chemical Engineers, New York.

CCPS. 2003a. *Understanding Explosions*. American Institute of Chemical Engineers, New York.

CCPS. 2003b. *Guidelines for Facility Siting and Layout*. American Institute of Chemical Engineers, New York.

CCPS. 2003c. *Guidelines for Investigating Chemical Process Incidents*. American Institute of Chemical Engineers, New York.

CCPS. 2003d. *Essential Practices for Managing Chemical Reactivity Hazards*. American Institute of Chemical Engineers, New York.

## National Fire Protection Association References

NFPA 10. 1998. *Standard for Portable Fire Extinguishers*. National Fire Protection Association, Quincy, MA.

NFPA 11. 1998. *Standard for Low-Expansion Foam*. National Fire Protection Association, Quincy, MA.

NFPA 11A. *Medium-and High-Expansion Foam*. National Fire Protection Association, Quincy, MA.

NFPA 12. *Standard on Carbon Dioxide Extinguishing Systems*. National Fire Protection Association, Quincy, MA.

NFPA 12A. *Standard on Halon 1301 Fire Extinguishing Systems*. National Fire Protection Association, Quincy, MA.

NFPA 13. 1999. *Standard for the Installation of Sprinkler Systems*. National Fire Protection Association, Quincy, MA.

NFPA 14. *Standard for the Installation of Standpipe, Private Hydrant, and Hose Systems*. National Fire Protection Association, Quincy, MA.

NFPA 15. 2001. *Standard for Water Spray Fixed Systems for Fire Protection*. National Fire Protection Association, Quincy, MA.

NFPA 16. *Standard for the Installation of Deluge Foam-Water Sprinkler and Foam-Water Spray Systems*. National Fire Protection Association, Quincy, MA.

NFPA 17. *Standard for Dry Chemical Extinguishing Systems*. National Fire Protection Association, Quincy, MA.

NFPA 20. *Standard for the Installation of Centrifugal Fire Pumps*. National Fire Protection Association, Quincy, MA.

NFPA 22. *Standard for Water Tanks for Private Fire Protection*. National Fire Protection Association, Quincy, MA.

NFPA 24. *Standard for the Installation of Private Fire Service Mains and their Appurtenances*. National Fire Protection Association, Quincy, MA.

NFPA 25. *Standard for the Inspection, Testing, and Maintenance of Water-Based Fire Protection Systems*. National Fire Protection Association, Quincy, MA.

NFPA 30. 2000. *Flammable and Combustible Liquids Code*. National Fire Protection Association, Quincy, MA.

NFPA 45. *Standard on Fire Protection for Laboratories Using Chemicals*. National Fire Protection Association, Quincy, MA.

NFPA 49. 1994. *Hazardous Chemicals Data*. National Fire Protection Association, Quincy, MA.

NFPA 50A. 1999. *Standard for Gaseous Hydrogen Systems at Consumer Sites*. National Fire Protection Association, Quincy, MA.

NFPA 50B. 1999. *Standard for Liquefied Hydrogen Systems at Consumer Sites*. National Fire Protection Association, Quincy, MA.

NFPA 58. *Standard for the Storage and Handling of Liquefied Petroleum Gases*. National Fire Protection Association, Quincy, MA.

NFPA 59. *Standard for the Storage and Handling of Liquefied Petroleum Gases at Utility Gas Plants*. National Fire Protection Association, Quincy, MA.

NFPA 68. 2002. *Guide for Venting of Deflagrations*. National Fire Protection Association, Quincy, MA.

NFPA 69. 2002. *Standard on Explosion Prevention Systems*. National Fire Protection Association, Quincy, MA.

NFPA 70. *National Electrical Code®*. National Fire Protection Association, Quincy, MA.

NFPA 72. *National Fire Alarm Code®*. National Fire Protection Association, Quincy, MA.

NFPA 75. 1999. *Standard for the Protection of Electronic Computer/Data Processing Equipment*. National Fire Protection Association, Quincy, MA.

NFPA 77. *Recommended Practice on Static Electricity*. National Fire Protection Association, Quincy, MA.

NFPA 80. *Standard for Fire Doors and Fire Windows*. National Fire Protection Association, Quincy, MA.

NFPA 80A. *Recommended Practice for Protection of Buildings from Exterior Fire Exposures*. National Fire Protection Association, Quincy, MA.

NFPA 85. National Fire Protection Association, Quincy, MA.

NFPA 86. *Standard for Ovens and Furnaces*. National Fire Protection Association, Quincy, MA.

NFPA 101®. *Code for Safety to Life from Fire in Buildings and Structures*. National Fire Protection Association, Quincy, MA.

NFPA 214. 2000. *Standard on Water-Cooling Towers*. National Fire Protection Association, Quincy, MA.

NFPA 221. *Standard for Fire Walls and Fire Barrier Walls*. National Fire Protection Association, Quincy, MA.

NFPA 241. *Standard for Safeguarding Construction, Alteration, and Demolition Operations*. National Fire Protection Association, Quincy, MA.

NFPA 251. 1999. *Standard Methods of Tests of Fire Endurance of Building Construction and Materials*. National Fire Protection Association, Quincy, MA.

NFPA 291. *Recommended Practice for Fire Flow Testing and Marking of Hydrants*. National Fire Protection Association, Quincy, MA.

NFPA 325. 1994. *Guide to Fire Hazard Properties of Flammable Liquids, Gases, and Volatile Solids*. National Fire Protection Association, Quincy, MA.

NFPA 326. 1999. *Standard Procedures for the Safe Entry of Underground Storage Tanks*. National Fire Protection Association, Quincy, MA.

NFPA 329. 1999. *Recommended Practice for Handling Underground Releases of Flammable and Combustible Liquids*. National Fire Protection Association, Quincy, MA.

NFPA 484. National Fire Protection Association, Quincy, MA.

NFPA 496. 1999. *Standard for Purged and Pressurized Enclosures for Electrical Equipment*. National Fire Protection Association, Quincy, MA.

NFPA 550. 1995. *Guide to the Fire Safety Concept Tree*. National Fire Protection Association, Quincy, MA.

NFPA 600. *Standard for Industrial Fire Brigades*. National Fire Protection Association, Quincy, MA.

NFPA 750. *Standard on Water Mist Fire Protection Systems*. National Fire Protection Association, Quincy, MA.

NFPA 780. 2000. *Standard for the Installation of Lightning Protection Systems*. National Fire Protection Association, Quincy, MA.

NFPA 820. 1999. *Standard for Fire Protection in Wastewater Treatment and Collection Facilities*. National Fire Protection Association, Quincy, MA.

NFPA 850. 2000. *Recommended Practice for Fire Protection for Electric Generating Plants and High Voltage Direct Current Converter Stations*. National Fire Protection Association, Quincy, MA.

NFPA 906. *Guide for Fire Incident Field Notes*. National Fire Protection Association, Quincy, MA.

NFPA 921. *Guide for Fire and Explosion Investigations*. National Fire Protection Association, Quincy, MA.

NFPA 1561. *Standard on Fire Department Incident Management*. National Fire Protection Association, Quincy, MA.

NFPA 1852. *Standard for Selection, Care, and Maintenance of Open-Circuit SCBA*.

NFPA 1901. Standard for Automotive Fire Apparatus. National Fire Protection Association, Quincy, MA.

NFPA 1911. *Standard for Service Tests of Pumps on Fire Department Apparatus*. National Fire Protection Association, Quincy, MA.

NFPA 1914. *Standard for Testing of Fire Department Aerial Devices*. National Fire Protection Association, Quincy, MA.

NFPA 1915. *Standard for Fire Apparatus Preventative Maintenance Program*. National Fire Protection Association, Quincy, MA.

NFPA 1961. *Standard for Fire Hose*. National Fire Protection Association, Quincy, MA.

NFPA 1962. *Standard for the Care, Use, and Service Testing of Fire Hose, Including Couplings and Nozzles*. National Fire Protection Association, Quincy, MA.

NFPA 2001. *Standard on Clean Agent Fire Extinguishing Systems*. National Fire Protection Association, Quincy, MA.

NFPA. 1997. *Automatic Sprinkler Systems Handbook*, 7th Ed. National Fire Protection Association, Quincy, MA.

NFPA. 1997. *Fire Protection Handbook*. 18th Edition. National Fire Protection Association, Quincy, MA.

NFPA. *Industrial Fire Protection Handbook*. National Fire Protection Association, Quincy, MA.

## General References

AISC. 1978. *Specification for the Design, Fabrication, and Erection of Structural Steel for Buildings*. American Institute of Steel Contractors, Chicago, IL.

ANSI/IEEE. Publication 979. *Guide for Substation Fire Protection*. American National Standards Institute, Washington, DC.

ASTM E. *Boiler and Pressure Vessel Code*. American Society for Testing and Materials, West Conshohoken, PA.

ASTM E. 814. *Standard Method of Fire Tests of Through-Penetration Fire Stops*. American Society for Testing and Materials, West Conshohoken, PA.

ASTM. E1529. 2000. *Standard Test Methods for Determining Effects of Hydrocarbon Pool Fires on Structural Members and Assemblies*. American Society for Testing and Materials, West Conshohoken, PA.

ASTM. E119. *Methods for Fire Test of Building Construction and Materials*. American Society for Testing and Materials, West Conshohoken, PA.

Avallone, E.A., Baumeister, T. 1996. *Marks' Standard Handbook for Mechanical Engineers*. The McGraw Hill Companies. New York.

Babrauskas, V. 1979. *COMPF2 - A Program for Calculating Post-Flashover Fire Temperatures. NBS TN 991*. National Bureau of Standards, Gaithersburg, MD.

Babrauskas, V. 1983. *Estimating Large Pool Fire Burning Rates*. Fire Technology, (19).

Babrauskas, V. 2002. *Burning Rates. The SFPE Handbook of Fire Protection Engineering*. 3rd Edition. National Fire Protection Association, Quincy, MA.

Barry, T.F. 1995. *An Introduction to Quantitative Risk Assessment in Chemical Process Industries*. The SFPE Handbook of Fire Protection Engineering. 3rd Edition. National Fire Protection Association, Quincy, MA.

Baum, H.R. 2000. *Large Eddy Simulation of Fire*. Fire Protection Engineering, (6).

Beyler, C.L. 2002. *Fire Hazard Calculations for Large Open Hydrocarbon Fuel Fires*. The SFPE Handbook of Fire Protection Engineering. 3rd Edition. National Fire Protection Association, Quincy, MA.

Bonn, J.C. 1991. *Goal-Setting Design for Fire and Gas Detection Systems*. SPE 23305. First International Conference on Health, Safety and the Environment. Society of Petroleum Engineers, Inc.

BP America. 2002. *Fire and Gas Detection*. Document No. SPC-FP-00001. BP America, Inc.

Bryan, J.L. 1986. *Defining Damageability – The Examination, Review, and Analysis of the Variables and Limits of Damageability for Buildings, Contents, and Personnel from Exposure in Fire Incidents*. Symposium on Quantitative Fire Hazards Analysis. Society of Fire Protection Engineers, Boston, MA.

Brzustowski, T.A. 1971. *Predicting Radiant Heating from Flares*. EE 15ER.71. Esso Engineering Research and Development Report.

Brzustowski, T.A. 1973. *A New Criterion for the Length of a Gaseous Turbulent Diffusion Flame*. Combustion Science and Technology, (6).

Brzustowski, T.A. 1976. *Flaring in the Energy Industry*. Progress Energy Combustion Science.

Brzustowski, T.A. 1977. *Flaring: State of the Art*. Loss Prevention, (11).

Brzustowski, T.A. and Sommer, E.C. 1973. *Predicting Radiant Heating from Flares*. Proceedings of the API Division of Refining, (53).

Brzustowski, T.A., Gollahalli, S.R., Gupta, M.P., Kaptein, M. and Sullivan, H.F. 1975. *Radiant Heating from Flares. ASME Paper 75-HT-4*. American Society of Mechanical Engineers (ASME), New York, NY.

Brzustowski, T.A., Gollahalli, S.R., Gupta, M.P., Kaptein, M. and Sullivan, H.F. 1975a. *The Turbulent Hydrogen Diffusion Flame in a Cross Wind*. Combustion Science and Technology, (11).

CGA G-5.4. 1992. *Standard for Hydrogen Piping Systems at Consumer Locations*. Compressed Gas Association, Chantilly, VA.

CGA C-6.19. *Standard for Visual Inspection of Steel Compressed Gas Cylinders*. Compressed Gas Association, Chantilly, VA.

Chambers, G.D. 1977. *Flight Line Extinguisher Evaluation*. DOD-AGFSRS-76-9. US Air Force Report.

Chang, X., Laage, L.W., and Greuer, R.E. 1990. *A User's Manual for MFIRE: A Computer Simulation Program for Mine Ventilation and Fire Modeling*. The Peoples Republic of China Bureau of Mines. Xian Mining Institute. Mining Technical University, Houghton, MI.

Coffee, Robert D. 1979. *Cool Flames and Autoignitions*. 13th Annual AIChE Loss Prevention Symposium, April 2–5, 1979.

Cowley, L. and Prichard, M. 1990. *Large-Scale Natural Gas and LPG Jet Fires and Thermal Impact on Structures. GASTECH 90 Conference*. Amsterdam.

Cox, G., and Kumar, S. 1987. *Field Modeling of Fire in Forced-Ventilation Enclosures*. Combustion Science and Technology, Volume 52.

Donkelaar, P. Van. *Protection Waterways from Run-off Pollution*. Number 149. Fire International.

Drysdale, D. 1998. *An Introduction to Fire Dynamics*. McGraw-Hill, New York, NY.

EPRI Project 1843-2. 1985. *Turbine Generator Fire Protection by Sprinkler System*.

FM Data Sheet 1-1. *Firesafe Building Construction and Materials*. Factory Mutual Insurance Company, Norwood, MA.

FM Data Sheet 1-6. *Cooling Towers*. Factory Mutual Engineering Corporation, Norwood, MA.

FM Data Sheet 1-10. *Smoke/Heat Venting in Sprinklered Buildings*. Factory Mutual Engineering Corporation, Norwood, MA.

FM Data Sheet 1-19. *Fire Walls, Subdivisions, and Draft Curtains*. Factory Mutual Engineering Corporation, Norwood, MA.

FM Data Sheet 1-20. *Exposure Protection*. Factory Mutual Engineering Corporation, Norwood, MA.

FM Data Sheet 1-20. *Protection against Exterior Fire Exposure*. Factory Mutual Engineering Corporation, Norwood, MA.

FM Data Sheet 1-21. *Fire Resistance of Building Assemblies*. Factory Mutual Engineering Corporation, Norwood, MA.

FM Data Sheet 1-22. *Criteria for Maximum Foreseeable Loss Fire Walls and Space Separation*. Factory Mutual Engineering Corporation, Norwood, MA.

FM Data Sheet 1-23. *Protection of Openings in Fire Subdivisions*. Factory Mutual Engineering Corporation, Norwood, MA.

FM Data Sheet 1-51. *Fire-Retardant Coatings and Paints for Interior Finish Materials*. Factory Mutual Engineering Corporation, Norwood, MA.

FM Data Sheet 5-28. *Spacing of Facilities in Outdoor Chemical Plant*. Factory Mutual Engineering Corporation, Norwood, MA.

FM Data Sheet 5-32. *Electronic Data Processing Systems*. Factory Mutual Engineering Corporation, Norwood, MA.

FM Data Sheet 5-49. *Gas & Vapor Detectors*. Factory Mutual Engineering Corporation, Norwood, MA.

FM Data Sheet 7-14. *Fire & Explosion Protection for Flammable Liquid, Gas & Liquefied Gas Processing Equipment & Supporting Structures*. Factory Mutual Engineering Corporation, Norwood, MA.

FM Data Sheet 7-32. *Flammable Liquid Operations*. Factory Mutual Engineering Corporation, Norwood, MA.

FM Data Sheet 7-43. *Loss Prevention in Chemical Plants*. Factory Mutual Engineering Corporation, Norwood, MA.

FM Data Sheet 7-44. *Spacing of Facilities in Outdoor Chemical Plants. Smoke Detectors*. Factory Mutual Engineering Corporation, Norwood, MA.

FM Data Sheet 7-45. *Chemical Process Control and Control Rooms*. Factory Mutual Engineering Corporation, Norwood, MA.

FM Data Sheet 7-46. 1999. *Chemical Reactors and Reactions*. Factory Mutual Engineering Corporation, Norwood, MA.

FM Data Sheet 7-83. *Drainage Systems for Flammable Liquids*. Factory Mutual Engineering Corporation, Norwood, MA.

FM Data Sheet 8-24. *Idle Pallet Storage*. Factory Mutual Engineering Corporation, Norwood, MA.

FM Data Sheet 17-0. 2002. *Maintenance*. Factory Mutual Engineering Corporation, Norwood, MA.

FM Data Sheet 10-3. 2002. *Hot Work Management*.

FM Data Sheet. *Construction*. Factory Mutual Engineering Corporation, Norwood, MA.

FM Data Sheet 5-20. 2001. *Electrical Testing*. Factory Mutual Engineering Corporation, Norwood, MA.

FM Data Sheet 4-0. 2002. *Special Protection Systems*. Factory Mutual Engineering Corporation, Norwood, MA.

Fahy, R.F. 1995. *EXIT89: An Evacuation Model for High-Rise Buildings – Recent Enhancements and Example Applications* International Conference of Fire Research and Engineering. Society of Fire Protection Engineers, Boston, MA.

Gollahalli, S.R, Brzustowski, T.A. and Sullivan, H.F. 1975. *Characteristics of a Turbulent Propane Diffusion Flame in a Cross-Wind*. Transactions of CSMC, (3).

Harada, K., Nii, D., Tanaka, T., and Yamada, S. 2000. *Revision of Zone Fire Model BRI2 for New Evaluation System*. NISTIR 6588. US/Japan Government Cooperative Program on Natural Resources (UJNR). 15th Joint Panel Meeting Proceedings. San Antonio, TX. Fire Research and Safety. National Institute of Standards and Technology, Gaithersburg, MD.

Harwell Laboratory. *General-purpose Computational Fluid Dynamics (CFD) Code*. United Kingdom, Harwell Laboratory.

HSE. 2001. *Reducing Risk, Protecting People, HSE's Decision-Making Process*. Health and Safety Executive, UK.

Heskestad, G. 1981. *Peak Gas Velocities, and Flame Heights of Buoyancy-Controlled Turbulent Diffusion Flames*. Eighteenth Symposium on Combustion. The Combustion Institute, Pittsburgh, PA.

Heskestad, G. 1983. *Luminous Height of Turbulent Diffusion Flames*. Fire Safety Journal, Volume 5, No. 2.

Heskestad, G. 2002. *Fire Plumes*, Flame Height and Air Entrainment. The SFPE Handbook of Fire Protection Engineering. 2nd Edition. National Fire Protection Association, Quincy, MA.

IEEE Standard 634. *Testing of Fire Rated Penetration Seals*. Institute of Electrical and Electronic Engineers.

IEEE Standard 500. *Guide to Collection and Presentation of Electrical Electronic, Sensing Component and Mechanical Equipment*. Institute of Electrical and Electronic Engineers.

IEEE. 1985. *Reliability for Nuclear-Power Generator Station*. Institute of Electrical and Electronic Engineers.

IChemE. 1989. *Calculation of the Intensity of Thermal Radiation from Large Fires*.

Iding, R.H., Nizamuddin, Z., and Bresler, B. 1977. *A Computer Program for the Fire Response of Structures – Thermal Three-Dimensional Version*. Department of Civil Engineering. University of California. Berkeley, CA.

Idling, R.H., Nixamuddin, S., and Bresler, B. 1977. *UCB FRB 77-15*. University of California, Berkeley, CA.

Incropera, F.P., and DeWitt, D.P. 1995. *Fundamentals of Heat and Mass Transfer*. 2nd Edition. John Wiley & Sons, New York, NY.

IRInformation IM.1.0.1 to IM.1.14.0. 1999. *A Management Program for the Protection of Property, Production and Profits*. Industrial Risk Insurers.

IRInformation IM.1.1.0. 1998. *Impairment to Fire Protection Systems*. Industrial Risk Insurers.

IRInformation IM.1.11.0. 1998. *Fire Protection and Security Surveillance*. Industrial Risk Insurers.

IR Information IM.1.12.0. 1998. *Fire Protection Equipment Inspection*. Industrial Risk Insurers.

IR Information IM.1.14.0. 1998. *Proper Housekeeping*. Industrial Risk Insurers.

IRInformation IM.2.5.1. 2000. *Fireproofing for Hydrocarbon Fire Exposures*. Industrial Risk Insurers.

IRInformation IM.2.5.3. 2000. *Drainage for Outdoor Oil and Chemical*. Industrial Risk Insurers.

ISGOTT. 1996. *International Safety Guide for Oil Tankers and Terminals*. ISGOTT. 4th Edition. Witherby and Company, London, England.

Janssens, M.L. 2002. *Evaluating Computer Fire Models*. Fire Protection Engineering, (13).

Jones, W.W., Forney, G.P., Peacock, R.D., and Reneke, P.A. 2000. *A Technical Reference for CFAST: An Engineering Tool for Estimating Fire and Smoke Transport*. Technical Note 1431. National Institute of Standards and Technology, Gaithersburg, MD.

Kalghatki, G.T. 1983. *The Visible Shape and Size of a Turbulent Jet Diffusion Flame in a Crosswind. Combustion and Flame*, (52) pp. 19–106.

Kisko, T.M., and Francis, R.L. 1985. *EVACNET+: A Computer Program to Determine* Optimal *Building Evacuation Plans. Fire Safety Journal*, pp. 9, 211.

Lattimer, B.Y., Wright, M., and Cutonilli, J. 2000. *Composite Fire Hazard Analysis Tool for Topside Structures (CFHAT) – Version 3." NSWCCD-TR-65-2000/32*. U.S. Navy Surface Warfare Center Carderock Division, Washington, DC.

Lawson, J.R., and Quintiere, J.G. 1985. *Slide Rule Estimates of Fire Growth*. National Bureau of Standards, NBSIR 85-3196, Washington (June 1985).

Lees, F.P. 1996. *Loss Prevention in the Process Industries*. ISBN 0-408-10697-2. Butterworths & Co. Ltd., United Kingdom.

Levin, B. M. 1988. *EXITT – A Simulation Model of Occupant Decisions and Actions in Residential Fires. NBSIR 88-3753*. National Bureau of Standards, Gaithersburg, MD.

Little, Arthur D. 1989. *The Environmental, Health and Safety Auditor's Handbook*. Arthur D. Little, Inc.

Magnussen, B.F. et al. *Kameleon II: A Transient, 3-Dimensional Computer Program for Fluid Flow, Heat- and Mass Transfer*. NTH/SINTEF. *Norwegian Institute of Technology (NTH)*. Trondheim, Norway: Division of Thermodynamics.

Marsh Risk Consulting. 2001. *Large Property Damage Losses in the Hydrocarbon-Chemical Industries – A Thirty Year Review*. 19th Edition. Marsh Risk Consulting.

McCaffrey, B. 1995. *Flame Height*. The SFPE Handbook of Fire Protection Engineering. 2nd Edition. National Fire Protection Association, Quincy, MA.

McGrattan, K.B., Baum, H.R., Walton, W.D., and Trelles, J. 1997. *Smoke Plume Trajectory* from In Situ Burning of Crude in Alaska – Field Experiments and Modeling of Complex Terrain." *NISTIR 5958*. National Institute of Standards and Technology, Gaithersburg, MD.

Munson, R.E. *Safety Considerations for Layout and Design of Processes Housed Indoors*. E.I. Du Pont de Nemours and Co., Wilmington, Delaware.

Nelson, H.E. 1990. *FPETOOL: Fire Protection Engineering Tools for Hazard Estimations. NISTIR 4380*. National Institute of Standards and Technology, Gaithersburg, MD.

Nelson, H.E. 2002. *From Phlogiston to Computational Fluid Dynamics*. Fire Protection Engineering.

Norstrom, G.P. 1996. *Property Insurance Considerations in Loss Prevention Expenditures*. AIChE Loss Prevention Symposium, Volume 20, New Orleans, LA.

OSHA 29CFR 1910.106. *Control of Ignition Sources*. Occupational Safety Hazard Association.

OSHA 29CFR 1910.165. *Employee Alarm Systems*. Occupational Safety Hazard Association.

Paulsson, M. 1983. *TASEF-2*. Lund Institute of Technology.

Peacock, R.D., Forney, G. P., Reneke, P., Portier, R., and Jones, W.M. 1993. *CFAST, the Consolidated Model of Fire Growth and Smoke Transport*. NIST Technical Note 1299. National Institute of Standards and Technology, Gaithersburg, MD.

Perry, R.H., Green, D.W. 1997. *Perry's Chemical Engineers' Handbook*. Seventh Edition. McGraw-Hill. New York.

Physics, Chemistry, and Technology. 1927. *International Critical Tables of Numerical Data*. Mc-Graw Hill, New York, NY.

Pintea, D., and Janssen, J.M. *Evaluation of the Thermal Part of the Code SAFIR by Comparison with the Code, TASEF.*

Portier, W.R., Peacock D.R., and Reneke A.P. 1996. *"FASTLite: Engineering Tools for Estimating Fire Growth and Smoke Transport."* NIST SP 899. National Institute of Standards and Technology.

Pratt, Tom. 1997. *Electrostatic Ignitions of Fires and Explosion*, AIChE, New York, 1997.

Putorti, A. D., McElroy, L.J., and Madrzykowski, D. 2001. *Flammable and Combustible Liquid Spill/Burn Patterns*. NIJ Report 604-00. National Institute of Justice, Washington, DC.

Rasbash, D.J. 1986. *The Extinction of Fire with Plain Water: A Review*. Proceedings from the First International Symposium on Fire Safety Science, C.E. Grant and P.J. Pagni, eds. National Institute of Standards and Technology, Gaithersburg, MD.

Roberts. 1982. *Thermal Radiation Hazards from Releases of LPG from Pressurized Storage*, Fire Safety Journal, Volume 4, pages 197–212.

RRS. 2002. *Process Hazards Analysis Leaders Training Course*. Risk, Reliability and Safety Engineering. Houston, Texas.

RRS. 2000. *Process Safety Management Training Course*. Risk, Reliability, and Safety Engineering. Houston, Texas.

Rubini, Dr. P. 2002. E-mail communication: p.a.rubini@cranfield.ac.uk. School of Mechanical Engineering. Cranfield University, Cranfield, Bedfordshire, England.

SANDOZ, United States Fire Administration, Sherwin-Williams Paint Warehouse Fire, 1987.

Scheffey J.L., Jonas, L.A., Toomey, T.A., Byrd, R., and Williams, F.W. 1990. *Analysis of Quick Response Fire Fighting Equipment on Submarines–Phase II, Full Scale Doctrine and Tactics Tests. NRL Memorandum Report 6632*. Naval Research Laboratory, Washington, DC.

Scheffey, J.L. 2002. *Foam Agents and AFFF Design Considerations*. The SFPE Handbook of Fire Protection Engineering. 3rd Edition. National Fire Protection Association, Quincy, MA.

Schwab, R.F. 1997. *Dusts*. Fire Protection Handbook. 18th Edition. National Fire Protection Association, Quincy, MA.

Shokri, M. and Beyler, C. 1989. *Radiation from Large Pool Fires*. Journal of Fire Protection Engineering, Volume 1, No. 4.

SINTEF. 1997. *Handbook for Fire Calculations and Fire Risk Assessment in the Process Industry*. 3rd Edition. SINTEF – NBL. Scandpower AS, Norway.

Society of Fire Protection Engineers. 2002. *The SFPE Handbook of Fire Protection Engineering*. 2nd Edition. Society of Fire Protection Engineers, Bethesda, MD.

Society of Fire Protection Engineers. 1999. *SFPE Engineering Guide, Assessing Flame Radiation to External Targets from Pool Fires*. Society of Fire Protection Engineers, Bethesda, MD.

Society of Fire Protection Engineers. 2000. *SFPE Engineering Guide, Predicting 1st and 2nd Degree burns from Thermal Radiation*. Bethesda, MD: Society of Fire Protection Engineers.

Spalding D.B. *A General Purpose 3-D Transient Fluid Dynamics Code*. CHAM, Ltd., United Kingdom.

Spouge, J. 1999. *A Guide to Quantitative Risk Assessment for Offshore Installations*. Publication 99/100. Aberdeen, UK: The Center for Marine and Petroleum Technology.

UL 58. *Standard for Steel Underground Tanks for Flammable and Combustible Liquids*. Underwriters Laboratories Inc., Northbrook, IL.

UL 80. *Standard for Steel Inside Tanks for Oil Burner Fuel*. Underwriters Laboratories Inc., Northbrook, IL.

UL 142. *Standard for Aboveground Atmospheric Storage Tanks for Flammable and Combustible Liquids*. Underwriters Laboratories Inc., Northbrook, IL.

UL 263. *Fire Tests of Building Construction and Materials*. Underwriters Laboratories Inc., Northbrook, IL.

UL 1709. *Standard for Rapid Rise Fire Tests of Protection Materials for Structural Steel*. Underwriters Laboratories Inc., Northbrook, IL.

UL 2080. *Standard for Fire Resistant Tanks for Flammable and Combustible Liquids*. Underwriters Laboratories Inc., Northbrook, IL.

UL 2085. *Standard for Protected Aboveground Tanks for Flammable and Combustible Liquids*. Underwriters Laboratories Inc., Northbrook, IL.

United Kingdom Chemical Industries Association. 1993. *An Approach to Categorization of Process Plant Hazards and Control Building Design*.

Walton, G.N. 1989. *AIRNET – A Computer Program for Building Airflow Network Modeling*. NISTIR 4072. National Institute of Standards and Technology, Gaithersburg, MD.

Walton, G.N. 1997. *CONTAM96 User's Manual*. NISIR 6056. National Institute of Standards and Technology, Gaithersburg, MD.

Wickstrom, U. *An Evaluation Scheme of Computer Codes for Calculating Temperature in Fire Exposed Structures*. Interflam '99. International Interflam Conference, 8th Proceedings. Interscience Communications Ltd., London, England.

# GLOSSARY

***Active fire protection***—A fire protection system or device that requires moving parts, detectors, instruments, electrical or other power or utilities.

***Approved***—Acceptable to the authority having jurisdiction.

***Atmospheric dispersion***—The low momentum mixing of a gas or vapor with air. The mixing is the result of turbulent energy exchange, which is a function of wind and atmospheric temperature profile.

***Autoignition temperature***—The minimum temperature at which combustion can be initiated without an external ignition source.

***Battery limit***—The perimeter of the process area as defined by the edges of equipment and the additional area required for equipment maintenance (i.e., bundle pulling area for exchangers).

***Boiling Liquid Expanding Vapor Explosion (BLEVE)***—A blast resulting from the sudden release and nearly instantaneous vaporization of a liquid under greater-than-atmospheric pressure at a temperature above its atmospheric boiling point. The material may be flammable or nonflammable. A BLEVE is often accompanied by a fireball if the contained liquid is flammable and its release results from vessel failure.

***Boil over***—An event in the burning of certain oils in open top tank when, after a long period of quiescent burning, there is a sudden increase in fire intensity associate with expulsion of burning oil from the tank.

***Business interruption***—An unplanned break in production.

***Business risks***—The risk that production or other business objective will be impaired because of unplanned incidents.

***Cementitious mixtures***—As defined by UL in "Spray Applied Fire Resistive Materials (SFRM)", cementitious mixtures are binders, aggregates and fibers mixed with water to form a slurry conveyed through a hose to a nozzle where compressed air sprays a coating; the term is sometimes used for materials (such as sand and cement) applied by either spray or trowel.

***Codes***—A code of practice is a set of standards agreed on by a group of people who do a particular job. These set of rules are accepted as general principles.

***Combustion***—Exothermic chemical reaction with oxygen as a primary reagent.

***Combustible liquids***—A liquid having a flash point at or above 100°F (37°C). Combustible liquids shall be subdivided as follows—

- Class II liquids shall include those having flash point at or above 100°F (37°C) and below 140°F (60°C).
- Class IIIA liquids shall include those having flash points at or above 140°F (60°C) and below 200°F (93°C).
- Class IIIB liquids shall include those having flash point at or above 200°F (93°C).

***Conceptual design***—The initial design of a project when basic parameters are known but final details have yet to be developed.

***Consequence***—The direct, undesirable result of an accident sequence usually involving a fire, explosion, or release of toxic material. Consequence descriptions may be qualitative or quantitative estimates of the effects of an accident in terms of factors such as health impacts, economic loss, and environmental damage.

***Consequence analysis***—The analysis of the expected effects of incident outcome cases independent of frequency or probability.

***Consequences***—A measure of the expected effects of incident outcome case.

***Continuous release***—Emissions that are long in duration compared with the travel time (time for the cloud to reach location of interest) or averaging or sampling time.

***Control of burning***—The application of water spray to equipment or areas to reduce the rate of burning and limit the heat release from a fire until the source of fuel can be shut-off or the fire can be extinguished.

***Cryogenic liquid***—A refrigerated liquid gas having a boiling point below –130°F (–90°C) at atmospheric pressure.

***Detection systems***—A mechanical, electrical, or chemical device that automatically identifies the presence of a material or a change in environmental conditions such as pressure or temperature.

***Dike***—An embankment of earth or rock built to act as a barrier blocking passage of material to surrounding areas.

***Directional incident outcome***—An incident outcome whose consequences produce an effect zone determined by a given wind direction.

***Domino effects***—The triggering of secondary events, such as toxic releases, by a primary event, such as an explosion, such that the result is an increase in consequences or area of an effect zone. Generally only considered when a significant escalation of the original incident results sometimes known as a knock on event.

***Emergency response***—The personnel and equipment response to an incident with the objective of controlling the incident and minimizing adverse impacts.

***Emergency Shutdown System (ESD)***—The safety control system which overrides the action of the basic control system when predetermined conditions are violated.

***Envelope***—Three-dimensional space enclosing the fire area.

***Event***—An occurrence involving equipment performance or human action, or an occurrence external to the system that causes system upset. In this book, an event is associated with an incident either as the cause or a contributing cause of the incident or as a response to the initiating event.

***Explosion***—A release of energy that causes a pressure discontinuity or blast wave.

***Exposed equipment***—Equipment subject to fire damage, usually from a source other than the equipment being protected.

***Exposure protection***—Application of water to a surface to minimize damage and prevent failure from a fire in an adjacent area.

***Extinguishment***—Cessation of the combustion reaction.

***Failure frequency***—The number of failure events that occur divided by the total elapsed calendar time during which those events occur or by the total number of demands, as applicable.

***Failure probability***—The probability, a value from 0 to 1, that a piece of equipment will fail in a given time interval or on demand.

***Failure rate***—The number of failure events that occur divided by the total elapsed operating time during which these events occur or by the total number of demands, as applicable.

***Fault tree***—A method for representing the logical combinations of various system states which lead to a particular outcome (top event).

***Fault Tree Analysis (FTA)***—Estimation of the hazardous incident (top event) frequency from a logic model of the failure mechanisms of a system.

***Fire***—A combustion reaction accompanied by the evolution of heat, light, and flame.

***Fireball***—The atmospheric burning of a fuel-air cloud in which the energy is mostly emitted in the form of radiant heat. The inner core of the fuel release consists of almost pure fuel whereas the outer layer in which ignition first occurs is a flammable fuel-air mixture. As buoyancy forces of the hot gases begin to dominate, the burning cloud rises and becomes more spherical in shape.

***Fireproofing***—Materials or application of materials to provide a degree of fire resistance to protected substrates.

***Fire resistance rating***—The time period that a specific fireproofing design will protect structural supports for equipment, piping and so forth from collapse, when exposed to a fire of specified intensity. The fire intensity is usually represented by a time–temperature curve.

***Flammable liquids***—A liquid having a flash point below 100°F (37°C) and having a vapor pressure not exceeding 40 psia at 100°F (37°C) (Class I liquids).

***Flammable limits***—The minimum (lower flammable limit, LFL) and maximum (upper flammable limit, UFL) concentration of combustible vapor in air that will propagate a flame.

***Flammable mass***—The mass of flammable vapor within a vapor cloud that will burn on ignition.

***Flash fire***—The combustion of a flammable vapor and air mixture in which flame passes through that mixture at less than sonic velocity such that negligible damaging overpressure is generated.

***Flash point***—The temperature at which a vapor-air mixture above a liquid is capable of undergoing combustion after ignition from an external energy source. (Fire point is the temperature at which the reaction will be sustained.)

***Foam***—A stable aggression of small bubbles of lower density than oil or water, which shows tenacious qualities in covering and clinging to vertical or horizontal surfaces. Foam flows freely over a burning liquid surface, forming a tough, air-excluding continuous blanket to seal volatile combustible vapors from access to air.

***Foam concentrate***—A foaming agent in concentrated liquid form, as received from the foam manufacturer.

***Frequency***—The number of occurrences of an event per unit of time.

***F-N curve***—A plot of cumulative frequency verses consequences (often expressed as number of fatalities). A societal risk measure.

***Generic data***—Data that are built using inputs from all of the plants within a company or from various plants within the CPI, from literature sources, past CPQRA reports and commercial data bases.

***Hazard***—A chemical or physical condition that has the potential for causing damage to people, property, or the environment.

***Hazard evaluation***—The analysis of hazardous situations associated with a process or activity, using qualitative techniques to identify weaknesses in design and operation.

***Hazard and Operability Study (HAZOP)***—A technique to identify process hazards and potential operating problems using a series of guide words to study process deviations.

***High fire potential equipment***—Equipment that is considered to have a high fire potential. Some examples are as follows—

- Fired heaters that process liquid or mixed-phase hydrocarbons, under the following conditions:
  —Operation at temperatures and flow rates that are capable of causing coking within the tubes.
  —Operation at pressures and flow rates that are high enough to cause large spills before the heater can be isolated.
  —Charging of potentially corrosive fluids.
  —Pumps with rated capacity over 200 US gpm that handle liquids or combustible liquids above or within 15°F (8°C) of their flash point temperatures.
- Pumps with a history of bearing failure or seal leakage (where engineering revisions have been unsuccessful at eliminating these as significant potential fuel sources).
- Pumps with small piping subject to fatigue failure.

- Reactors that operate at high pressure or might produce runaway exothermic reactions.
- Compressors, together with related lube-oil systems.
- Specific segments of process piping handling flammable liquids or gases in mixtures known to promote pipe failures through erosion, corrosion, or embrittlement. These include hydrocarbons, streams that may contain entrained catalyst, caustics, acids, hydrogen, or similar materials where development of an appropriate scenario envelope is feasible.
- Vessels, heat exchangers (including air cooled exchangers), and other equipment containing flammable or combustible liquids over 600°F (315°C) or their autoignition temperature, whichever is less.
- Complex process units such as catalytic crackers, hydrocrackers, ethylene units, hydrotreaters, or large crude distilling units typically containing high fire potential equipment.

***Historical incident data***—Data collected and recorded from past incidents.

***Incident***—An unplanned event with the potential for undesirable consequences.

***Individual risk***—The risk to a person in the vicinity of a hazard. This includes the nature of the injury to the individual, the likelihood of the injury occurring and the time period over which the injury might occur.

***Initiating event***—The first event in an event sequence.

***Inherently safer***—A term applied to a component, system or facility in which potential dangers have been removed rather than designed for. Inherent safety is incorporated during development, design, or engineering.

***Intumescent fireproofing***—A passive material that undergoes a chemical reaction when exposed to high heat or direct flame impingement; that protects against the heat from a fire without additional intervention.

***Isopleth***—A plot of specific locations (in the three spatial coordinates—*x, y, z*) downwind from the release source that is corresponding to a concentration of interest (e.g., fixed by toxic load or flammable concentration).

***Jet discharge***—The release of a vapor and/or liquid at sufficient pressure such that significant air entrainment results.

***Jet fire***—A fire type resulting from fires from pressurized release of gas and/or liquid.

***Layout***—The relative location of equipment and buildings within a given site.

***Likelihood***—A measure of the expected probability or frequency of occurrence of an event.

***Listed***—Equipment, materials, or services included in a list published by an organization that is acceptable to the authority having jurisdiction and concerned with evaluation of products or services, that maintains periodic inspection of production of listed equipment or materials or periodic evaluation of services, and whose listings states that either the equipment, material, or services meets identified standards or has been tested and found suitable for a specified purpose.

***Low fire potential equipment***—The following are examples of equipment considered to have a low fire potential:

- Pumps that handle Class IIIB liquids below their flash points.
- Piping within battery limits which has a concentration of valves, fittings, and flanges.
- Heat exchangers that may develop flange leaks.

***Mastic***—A pasty material used as a protective coating or cement.

***Maximum individual risk***—The individual risk to the person(s) exposed to the highest risk in an exposed population.

***Mitigation***— Lessening the risk of an accident event sequence by acting on the source in a preventative way be reducing the likelihood of occurrence of the event, or in a protective way be reducing the magnitude of the event and/or the exposure of local persons or property.

***Mitigation factors***—Systems or procedures, such as water sprays, foam systems, and sheltering and evacuation, which tend to reduce the magnitude of potential effects due to a release.

***Off-site exposure***—People, property, or the environment located outside of the site property line that may be impacted by an on-site incident.

***Passive Fire Protection (PFP)***—A barrier, coating, physical condition (i.e., spacing), or other safeguard that provides protection against the heat from a fire without the need for action to be taken.

***Point source model***—A thermal energy model based on representing the total heat release as a point source.

***Pool fire***—The combustion of vapors evaporating from the open surface of a contained or uncontained volume of liquid.

***Prevention***—The process of eliminating or reducing the probability of the hazards or risks associated with a particular activity. Sometimes used to denote actions taken in advance to reduce the likelihood of an undesired event.

***Probability***—The expression for the likelihood of occurrence of an event or an event sequence during an interval of time, or the likelihood of the success or failure of an event on test or on demand.

***Property line***—The boundary line between two pieces of property.

***Process structure***—A type of industrial building or framework designed to accommodate a specific processing operation and its equipment, vessels and machinery on multiple levels within it.

***Protective system***—Systems such as pressure relief valves that function to prevent or mitigate the occurrence of an incident.

***Radiant heat***—Infrared radiation emitted from a source.

***Regulation***—A principle, rule, or law designed to control or govern conduct.

***Risk***—A measure of economic loss or human injury in terms of both likelihood and the magnitude of the loss or injury.

***Risk analysis***—The development of a quantitative estimate of risk based on engineering evaluation and mathematical techniques for combining estimates of incident consequences and frequencies.

**Risk assessment**—The process by which the results of a risk analysis are used to make decisions either through a relative ranking of risk reduction strategies or through comparison with risk targets. Risk assessment is often defined as the qualitative estimation of probability and consequence of an incident or incidents.

**Risk contour**—Lines that connect points of equal individual risk around the facility.

**Risk management**—The systematic application of management policies, procedures, and practices to the tasks of analyzing, assessing, and controlling risk in order to protect employees, the general public, and the environment as well as company assets while avoiding business interruptions.

**Sensitivity**—The sensitivity of a measure to a parameter is defined as the change in the measure per unit change in that parameter.

**Societal risk**—A measure of risk to a group of people. It is most often expressed in terms of the frequency distribution of multiple casualty events.

**Standard**—Something, such as a practice or a product that is widely recognized or employed, especially because of its excellence.

**Storage tanks**—Vessels or containers used to hold process materials. The following are several types of storage tanks—

- *Atmospheric tank*—a storage tank designed to operate at any pressure from atmospheric up to 3.45kPa gage (0.5 psig).
- *Cone roof tank*—an atmospheric storage tank with a fixed cone-shaped roof.
- *Cone roof tank with internal floating cover*—an atmospheric storage tank with a fixed cone-shaped roof and an internal floating cover of one of the following designs:
  1. Metal (steel or aluminum) pan-type cover.
  2. Thin-skin metal deck on floats.
  3. Aluminum honeycomb sandwich panel cover.
- *Covered floating roof tank*—an atmospheric storage tank with a fixed roof and steel internal floating roof of the single-deck pontoon or double-deck design.
- *Open-top floating roof tank*—an atmospheric storage tank with a steel floating roof of the pontoon, buoy, or double-deck design and without a fixed roof.
- *High-pressure tanks*—a storage vessel designed to contain material at pressures above 15 psig.
- *Low-pressure tanks*—a storage tank designed to contain material at pressures above 3.45 kPa gage (0.5 psig) but not to exceed 103.4 kPa gage (15 psig).
- *Pressure vessel*—a vessel designed to operate at pressures above 103.4 kPa gage (15 psig) in accordance with Section VIII of the ASME Boiler and Pressure Vessel Code.

**Substrate**—The underlying layer being protected by a fireproofing barrier layer.

**Top event**—The unwanted event or incident at the "top" of a fault tree that is traced downward to more basic failures using logic gates to determine its causes and likelihood.

**Transmissivity**—The fraction of radiant energy that is transmitted from the radiating object through the atmosphere to a target. The transmissivity is reduced due to absorption and scattering of energy by the atmosphere itself.

***Turnaround***—A time during which a unit is shut down for repair and maintenance after a normal run, before it is returned to operation.

***Unavailability***—The probability the fault event exists at a specified time. The percentage of time that one or more fault events prevent effective operation of a device or system.

***Uncertainty***—A measure, often quantitative, of the degree of doubt or lack of certainty associated with an estimate of the true value of a parameter.

***Wind rose***—A plan view diagram that shows the percentage of time the wind is blowing in a particular direction.

***Watch Dog Timer (WDT)***—A program control monitor that indicates when a program execution has exceeded a prescribed time period.

***Water spray***—Water discharged from nozzles, specially designed to give a certain pattern, particle size, velocity, and density at a given application rate.

***Worst credible incident***—The most severe incident, considering only incident outcomes and their consequences, of all identified incidents and their outcomes, that is considered plausible or reasonably believable.

# INDEX

Printed in the USA/Agawam, MA
July 17, 2015

619326.001

SUPPLEMENT TO

# Guidelines for Fire Protection in Chemical, Petrochemical, and Hydrocarbon Processing Facilities

Center for Chemical Process Safety
American Institute of Chemical Engineers
3 Park Avenue
New York, NY 10016-5991



**Getting started:**
Insert CD into CD-ROM drive
Open Windows Explorer and select the file you wish to open.

**Assistance:**
If you have difficulties contact CCPS at (212) 591-7319 from 9:00 AM to 4:30 PM Monday through Friday